## DATE DUE

# *CRITICAL SYSTEMIC PRAXIS FOR SOCIAL AND ENVIRONMENTAL JUSTICE*

## Contemporary Systems Thinking

**Series Editor:** Robert L. Flood
Monash University
Australia

---

COMMUNITY OPERATIONAL RESEARCH
OR and Systems Thinking for Community Development
Gerald Midgley and Alejandro Ochoa-Arias

CRITICAL SYSTEMIC PRAXIS FOR SOCIAL AND ENVIRONMENTAL JUSTICE
Participatory Policy Design and Governance for a Global Age
Janet McIntyre-Mills

DESIGNING SOCIAL SYSTEMS IN A CHANGING WORLD
Bela H. Banathy

GUIDED EVOLUTION OF SOCIETY
A Systems View
Bela H. Banathy

METADECISIONS
Rehabilitating Epistemology
John P. van Gigch

PROCESSES AND BOUNDARIES OF THE MIND
Extending the Limit Line
Yair Neuman

SELF-PRODUCING SYSTEMS
Implications and Applications of Autopoiesis
John Mingers

SOCIOPOLITICAL ECOLOGY
Human Systems and Ecological Fields
Frederick L. Bates

SYSTEMIC INTERVENTION
Philosophy, Methodology, and Practice
Gerald Midgley

---

A Continuation Order Plan is available for this series. A continuation order will bring delivery of each new volume immediately upon publication. Volumes are billed only upon actual shipment. For further information please contact the publisher.

# CRITICAL SYSTEMIC PRAXIS FOR SOCIAL AND ENVIRONMENTAL JUSTICE

## *Participatory Policy Design and Governance for a Global Age*

*Janet McIntyre-Mills*
*Flinders Institute of Public Policy and Management*
*Flinders University*
*Adelaide, Australia*

Kluwer Academic / Plenum Publishers
New York, Boston, Dordrecht, London, Moscow

Library of Congress Cataloging-in-Publication Data

McIntyre-Mills, Janet J. (Janet Judy), 1959–
Critical systemic praxis for social and environmental justice: participatory policy design and governance for a global age/Janet McIntyre-Mills.
p. cm. — (Contemporary systems thinking)
Includes bibliographical references and index.
ISBN 0-306-48074-3
1. Social problems—Research. 2. Systems analysis. 3. Social planning—Citizen participation. 4. Economic policy—Citizen participation. 5. Political planning—Citizen participation. 6. Social problems—Australia—Case studies. I. Title: Policy design and governance for a global age. II. Title. III. Series.



HN29.M3685 2004
361.1′072—dc22

2003061970

ISBN: 0-306-48074-3



233 Spring Street, New York, New York 10013

http://www.wkap.nl

10 9 8 7 6 5 4 3 2 1

A C.I.P. record for this book is available from the Library of Congress



Printed in the United States of America

This book is dedicated to Olive Veverbrants Peltharre,
an Arrernte woman who has crossed many boundaries
and the young people and their families who took part
in the dialogue to help co-create this monograph.
'May the road rise to meet you'.

My thanks to my husband Michael, who unselfishly
travelled many kilometers to make this book possible and
to all things that connect me with the spirit of life.

## My Philosophical Debt to My Teachers

*To the Arrernte grandmothers who*
*shared their wisdom, my thanks and*
*to the Indigenous and non Indigenous people who shared their stories about the way things are and appear to be to them.*
*My thanks for the writings of C. West Churchman that helped me to shrug off categorical thinking and*
*Finally, my thanks to the young people who spoke their mind.*

*We know about systems*
*Indigenous thinking is systemic and has been for many thousand years*
*We don't like 'social capital'*
*Wellbeing and spirituality matter.*
*We can ask what, why and how questions about our own way of doing things. This is useful, but we must also ask who has the right to ask.*
*Culture, gender, age, class need to be swept into all considerations of what constitutes knowledge.*

*If we avoid hard questions, by not allowing them to be asked, we miss opportunities.*
*Hard questions are the basis of knowledge (in our opinion and in a loose Popperian sense).*
*Avoidance may however serve a purpose at a particular time and place.*
*Exploring the similarities and the differences is exciting*
*My Indigenous teachers hinted that holons, the shape of perceptions can be written conceptually*
*Or the stories can be memoed in the landscape, which is where oral culture stores its heritage.*

*You can only heal your own culture*
*Your own divided thinking*
*Indigenous people must heal themselves. Space for difference and space for sharing is equally important.*

*So what is my role (if any) in this postcolonial landscape of transcultural, transdisciplinary thinking?*

*It is perhaps to heal my own thinking*
*through recollections and reflections as well as through dialogue.*
*It is about praxis in and on the past, present and future.*

*To find the parallels between the concepts of self and other and to locate the areas of difference and to explore why these differences exist.*

*What can we learn from one another?*
*Is there a role for facilitating thinking in marginal spaces?*
*If I see myself as the ' bloody academic' then I could be tempted to usurp the role of ex duco (meaning to lead out, the Latin root of educator).*
*The challenge is to listen and to lead one another to a thinking space that is both spiritual and practical.*
*If spirituality is understanding that human beings are part of the land and that we need to tread carefully, this is indeed very practical and a useful basis for defining problems. This is the simple and complex response to challenges.*
*Writing this book resonates with my fears when rock climbing or sailing, it is pushing the boundaries and requires a leap of faith.*
*It also requires an acceptance that not all things can or should be taken for granted.*

## Some Thoughts on Categories, Continuities and Mandalas: Light,/ Radiance, Darkness/Shadow[1] are Defined in Terms of Each Other

*Our little systems have their day*
*They have their day and cease to be*
*They are but broken lights of thee. (Tennyson)*

*Interpret Leibniz's 'apperception' to mean the ability of the system designer to design the system from many points of view- to design science as a management system or design physics as a psychology. To the extent that a system fails in its apperception, it is less than God, i.e., it is an imperfect monad. (Churchman 1971, 75)*

*[I]n everything there is nothing*
*nothing belongs to everything*
*everything belongs to everything*
*everything contains nothingness,...*

*in every system are all systems.[2] (Churchman 1979: 186)[3]*

*'radiance is meaning' that spans difference and radiance is the energy that communicates (derived from Churchman 1982: 55)*

[1] All light casts a shadow and we need to unfold the implications for the way we understand ourselves and others. In Jungian terms the shadow self should be examined and the shadow of theory and practice should be similarly examined at a societal level.
[2] Churchman cites the work of Anaxagoras as a basis for this idea and relates how biology students discover the analogy of nucleus and electrons and energy within a cell and reflect that this is the systemic pattern of the universe.
[3] Hence since they share nothing, then the 'null class' is included in every class. The same logical process occurs at the other end of the spectrum of classes, the universal class of all classes. To keep the rules exceptionless, every class must be included in the universal class. But if we allow a slight shift in wording, the language becomes mystical and perhaps Zen-like.

# *Preface "Looking at my town"*
## *A poem by Olive Veverbrants (2001)*

*Watching for snakes, watching for ants*
*Watching storm clouds gather*
*This is Darwin weather*
*Local newspaper*
Racist, ignorant, illiterate
*'s 's 's 's in all the wrong places*
*au fait printed ofay by news professionals.*

*Positions Vacant advertising camel Handler,*
*Eagerly filled by pukka sahib*
*(how times have changed)*
*Camel cup photos, jockeys with tea towels head–dresses*
*Thundering down the strait;*
*Henley-on Todd, annual race in dry river.*
*Hairy legs instead of oars*
*Alice crew in Sydney- to -Hobart*
*The Frigate Arunta*
*Proudly acknowledges its home port Alice Springs*
*Arrernte elders welcome the New Millenium in.*

*Country town, barefoot Indigenous people*
*Busily food shopping*
*Or just sitting around*
*Queuing in banks*
*Pushing babes in stroller.*

*Tourist bums in all sized shorts*
*All kinds of hats*
*Stockman, Akubra, straw,*
*String under chin;*
*Baseball caps,*
*Teenagers imaging Yanks*
*Nike boots, over-sized tees,*
*Floppy shorts in black and grey.*

*Fast food on every corner.*
*Kentucky Chicken, Hungry Jacks*
*Pizza Haven, Get your Big Mac!*
*Shopping trolleys*

*Lumpy with plastic bags*
*Leaving the checkout*
*With cross eyed wheels,*
*Uneven cobblestones of CBD*
*I watch my step.*

*Into Toyota 4WD*
*Holden Commodore-*
*Hermannsberg Mercedes*
*Honda: all guzzling fuel*
*As fast as the tankers deliver.*

*Spruce cops in starched Khaki*
*Police cages cruising around*

*Town of new-born babes*
*On fathers' shoulders*
*Like sleeping koalas glued to a gum*
*Babes in strollers everywhere*
*Busy post office*
*Locals collecting*
*Tourists mailing didges*[1] *home.*
*Tourists at deli*
*Discussing in own language*
*Kangah-ruew or cam-emelle steak.*

*Out of work teenagers*
*Black Madonnas, straight hair scraped in severe bun*
*Keeping busy in the Plaza.*
*Security 'heavies' patrol up and down.*

[1] Colloquialism for didgeridoo, an Indigenous instrument that is sold for commercial purposes in Alice Springs, despite not being a traditional instrument in this area.

*Shade, shade*
*We all look for shade*
*To park under*
*So we don't return to an oven.*
*The sun beats down*
*Hot bitumen*
*Catching and swallowing dust and debris*
*In its melting face.*
*The entire continent spoiled.*
*Yet the original inhabitants*
*kept the equilibrium*
*for sixty thousand years.*
*The land was their Mother.*
*Early morning and every evening*
*pristine galahs in pink and gray*
*gather on telegraph wires and*
*parkland lawns chattering*
*till they swoop the next day.*
*Tiny orange-breasted finches settle on garden wire supports*
*changing notes of music written by an unseen hand;*
*and on the nature strip fast food containers and plastic bags, contents intact*
*(didn't make it to the tip)*
*discarded VE green–cans amongst dust and burrs.*
*The once proud, lean, healthy desert people*
*outcasts now, lookers on, passing time in alcoholic haze or rage.*
*One year follows another. Death intervenes.*
*Multiple grief is the name.*
*Funerals are the game that breaks the monotony.*
*Oh Kwmentyeye*
*Women keening,*
*backdrop to this spectacle.*
*Nothing draws a crowd like a funeral.*
*Three the week before last,*
*church packed to overflowing,*
*Always; as many outside as in.*
*Solemn elders, chief mourners*
*Sisters, auntie's, grannies,*
*quiet children in funereal best.*
*The congregation sings a hymn in Arrernte.*
*Tears flow. I am one with my people*
*The Lutheran pastor cradles the grievers.*
*The church bell tolls.*

# *Acknowledgments*

My thanks to those who work with the boundaries (rather than within them) to promote systemic thinking and paradigm dialogue, particularly Norma Romm. Thanks to all those with whom I have discussed the boundaries of what constitutes research and knowledge, in particular, my thanks to those who encouraged and bolstered my flagging spirit in dusty places.

My thanks to Adelaide Dlamini, an Indigenous healer who introduced me to the African notion of ubuntu and the holism concept twenty-two years ago.

My thanks to the Arrernte grandmothers, particularly Olive Veverbrants and the Healthy City Co-ordinators, Michael Sparks and Dr Pat Mowbray for their insights and to all those in Alice Springs who so kindly helped me to understand their points of view by spending time telling me their stories. My thanks to Adam Jamrozik, whose focus on the nature of welfare and democracy has stimulated my thinking and to John Janzen, a social anthropologist, whose teaching and ideas have had a lasting influence since he supervised my early work.

My thanks to Susanne Bagnato for her collaboration on the diagrams and book cover and to Susan Goff for her support and commentary on sections of this work and to the members of the post colonial study group at Flinders, who helped me locate my space. My thanks to my editor, Henry Gomm who made this work possible.

# Contents

# *CRITICAL SYSTEMIC PRAXIS FOR SOCIAL AND ENVIRONMENTAL JUSTICE*

# 1

# *Introduction: Axial Themes United in Space and Time*

> But the planner's laboratory is not just concerned with testing one hypothesis; it is also going about the business of transportation, education, nutrition, and protection...
>
> *(Churchman 1979: 58)*
>
> ... One option is to maintain the spirit of the classical laboratory by collecting just those data that appear relevant and can be obtained objectively; The other option, the harder one, is to recognise that the unpredictable human is an essential aspect, and to begin to invent a methodology in which human bias is a central aspect...
>
> *(Churchman 1979: 62)*

The global age is systemic and digital. It is undeniable that it has implications for the way we educate future practitioners in the human services. We need a different kind of education, because although as caterpillars we have potential, it needs to be expanded through the journey of learning from one another. It is not about *'merely making caterpillars go faster, when what we need are butterflies (Banathy 1991 in Norum 2001: 330). We get butterflies by examining and challenging' how people think, what they believe, and how they see the world'* (Senge, 1999 in Norum 331). Designing for the future requires an ability to think creatively and reflexively and to operate in terms of systems, not compartments. This has implications for organisational management, governance, as well as social and environmental policy. Communication styles need to be open and flexible so as to represent and take into account multiple meanings. This requires going beyond an interest in socio-technical systems (Jackson 1991). It requires asking not only about what tasks we choose to undertake or how we undertake the process, but also our rationale for an approach (Flood and Romm 1996). Critical questioning enables us 'to unfold' the meanings of multiple stakeholders and to 'sweep in' a range of issues and implications that could be relevant for problem solving and without this process could remain either invisible or devalued (Goff 2002, personal communication).

'The world is an amazing place' is the leitmotif for a television station in Australia that strives to give a more complex overview of diverse social and cultural viewpoints. More issues than we can easily comprehend are not usually addressed in the short news items and documentaries. Life is complex and in a global, digital age where the different images of life are juxtaposed, there is an increasing need to address the complexity that is everyday life. We, as human beings who actively engage in daily life, must become adept at crossing boundaries to make sense of the personal and public montage of images and reality. The challenge for education and governance is to communicate and co-create shared meaning that addresses complexity.

Internationally, globalisation has been paradoxically translated into colonization, economic rationalist development, nationalism and closure born of a fear of the implications of globalisation and global markets for the least powerful countries, regions and interest groups. Social problems in terms of the economic rationalist approach are increasingly individualised and citizenship models[4] emphasise the responsibility of individuals and families.[5] Many current human service and social policy models in Australia (like elsewhere) are non-systemic.[6] They continue to be based on compartmentalised approaches that are short-term and discipline based. These include psychological, medical, education, crime prevention and economic approaches.[7]

The case study could not have been undertaken in a more appropriate place because Alice Springs is a borderland of cultures, isolated but

[4] According to Thomson and McMahon (1996) who quote Marshall (1963: 78) these include 'liberty of the person, freedom of speech, thought and the right to own property and to conclude valid contracts and the right to justice'. Political citizenship is 'the right to participate in the exercise of political power, as a member of a body invested with political authority or as an elector of the members of such a body' (Marshall 1963: 78). They quote social citizenship as defined by Marshall (1963: 78) as 'the right to a modicum of economic welfare security regardless of the position on the labour market and the right to share to the full the social heritage and to live the life of a civilised being, according to the standard prevailing in society'.

[5] Wilson, J. Thomson, J. and McMahon (1996) *The Australian Welfare State* key documents and themes. Macmillan, Melbourne.

[6] See Kettner et al (1985).

[7] See *Systemic Practice and Action Research Journal*, Plenum Press, New York and London.

linked with the outside world historically by camels, later by the telegraph and currently by the Internet. Access to the means of communication in the past and currently is the challenge, as is access to a host of other resources. The study explores the articulation of town/desert, developed/undeveloped and modern/traditional contexts. The changing cultural role and attitude to commodity and technology (as it pertains to life chances) are central to the analysis. Cars, computers, the television, telephone, radio and medical technology are central for communication and access to services. Finally, 'life on the machine' is the last option for Indigenous people who rely on dialysis for survival; as such it is symbolic of the technocratic approach to problem solving Devitt and McMasters (1998).

This study attempts to show the impacts of development and globalisation at the local level in one of the most isolated, land locked locations in the world. It is about a 'research for planning' experience in a small town in Central Australia. To set the scene, the Alice Springs municipal area is located within a wider region that it serves. The perception of the quality of life, the services and infrastructure reveal a range of different points of view associated with different interest groups with widely different socio-demographic characteristics. These interest groups have very different life chances and ideas about the world and their rights and responsibilities. It is for this reason that an analysis of perceptions of citizenship is introduced because it pertains to the way people perceive governance and their community. The assessment is thus about both the users and providers of services. The research addresses (i) social policy, service delivery, organisational management and the need for integration across sectors and disciplines to address the development challenges as linked systems and (ii) the need to establish dialogue between citizens and government[8] and to build in the transcultural perceptions of citizens to inform policy and planning at the level.

Readers may choose to read chapter 1 and the conclusion first as an overview. Chapters 2–4 are for those interested in critical systems theory and methodology. Chapter 4 gives a sense of the links between the researcher and the research context and would be of interest to students and practitioners undertaking transcultural practice and participatory action research. Chapters 5–6 give details of the case study and provide the background for the systemic analysis and policy suggestions in chapters 7 and 8. The chapters 5 and 6 would be useful for social policy planners in particular.

[8] According to ATSIC (1995) cited in Fuller, D. (1996) *Aspects of Indigenous Economic Development*, Paper presented to Regional Development Program Forum.

## 1.1 *The Relevance of an Ideographic Case Study and Narrative Approach as a Vehicle for Critical Systemic Praxis (CSP)*

Alice Springs has been seen as one of the most isolated parts of Australia. Although transport and communications have opened up 'The Alice', the non-Indigenous 'frontier mentality' celebrated in the historical literature on the inland of Australia (Idriess 1933) comprising toughness and resilience in an unforgiving environment remain a badge of honour for the pioneer residents who have lived here for generations (or at least for some years). This frontier mentality contributes to shaping governance. As does the Indigenous culture of resistance and persistence (Keefe 1988).[9] An intolerant environment has created a strong cultural response, based on living at the boundaries of what is achievable. The tendency for people to pass through Alice Springs or to stay for short periods of time and to 'become experts overnight' has led to local people being suspicious about ideas from outside, based on other sets of circumstance. In some instances this suspicion has been justified and the lessons and implications of trusting outsiders, also known as 'blow ins' are circulated and become part of the Indigenous and non-Indigenous long-term resident's sense of place and history. Policy suggestions have been co-created through dialogue, but policy ideas have been drawn (in part) from the Healthy Settings Movement (HSM) using a version of Wenger's (1998) community of practice approach that has enjoyed widespread support in other parts of Australia and internationally. This integrated approach[10, 11] drawing reflection on and with a range of approches such as Cox, Winter, King et al, is ideally suited to addressing the socio-economic and environmental issues identified by means of qualitative and quantitative indicators of local need. The ideographic case study attempts to demonstrate the value of critical systemic thinking and practice. Suggestions are made through this

[9] This 'exemplifies the notion of separateness through culture as a form of both resistance and persistence' to use Keefe's (1988) terminology. The history of Aboriginality needs to be interpreted in terms of invasion, slaughter and inability to control labour, exclusion from citizenship, segregation and exclusion from property rights (Hollinsworth 1996).

[10] Indigenous social, cultural, political and economic concerns are expressed through:
Native Title Claims, The Land Act 1992, Amendments to the Pastoral Land Act 1992, Aboriginal Land Rights Act 1976 (NT), Sacred Sites Act 1989 and Aboriginal Heritage Act 1984 which is being amended.

[11] The Meek shall inherit the earth. Weekend Australian Review April 3–4 1999.

reflexive narrative that resonate with the idea that we need to be mindful of the whole and not parts of the system.[12] Respect for the whole system is based on spirituality[13] and respect for the sacred, in Indigenous terms. Although the details refer to a particular place and time, this suggested holistic policy approach could be a source of ideas for systemic designers.[14]

This is a story about living and doing systemic thinking and practice in Alice Springs (Mparntwe, *or caterpillar dreaming*) as a participant (social policy observer and researcher). Systemic thinking, practice and policy challenges boundaries through attempting to co-create shared meanings. The boundaries that I address in this research are: (i) between the narratives of the researcher and the researched, (ii) research integrity and bureaucratic management (in this respect I wrestled with the internal and external enemies of politics, religion, morality and aesthetics or perceptions of quality, to use Churchman's 1979 categories), (iii) conceptual boundaries within and across knowledge narratives, practice and policy areas and (iv) social, political and spatial boundaries. Alice Springs is a place where working the boundaries is an essential first step for achieving social rights and responsibility for all citizens. The mental and physical geography is mapped out by boundaries. This story is an attempt to present narratives as multilayered, multitextured perceptions of reality. In some instances the stories

[12] 'Struggles for social justice and cultural autonomy by Indigenous Australians have constituted some of the most far-reaching challenges to the Australian State. In the last twenty years Aborigines have gained official recognition as a people and support for self-management and self-determination policies. These apparent successes have resulted in an incorporation of Indigenous communities and their politics into mainstream institutions in ways that can actually increase state supervision and threaten cultural independence. Partly this contradiction arises from the need to create peak bodies able to represent Aboriginal issues at the highest levels of government which run counter to the localised and land-based social networks which have enabled Indigenous values to be maintained under welfare colonialism ....' This quotation exemplifies the notion of separateness through culture as a form of both resistance and persistence to use Keefe's (1988) terminology. The history of Aboriginality needs to be interpreted in terms of invasion, slaughter and inability to control labour, exclusion from citizenship, segregation and exclusion from property rights (Hollinsworth 1996).

[13] Cox, E. (1995) *A truly civil society* NSW ABC books, Sydney. Boyer lectures. Winter, I. (2000) *Social capital and public policy in Australia*, Australian Institute of Family Studies.

[14] King, R., Bently, M., Baum, F. and Murray, C. '*Community Groups, Health Development and Social Capital* 1999 Poster presentation at 31st Annual PHAA Conference: 'Our place, our health: Local values and global directions'.

overlap, in others they diverge. In all instances they can be seen to complement one another and to present a more complex representation of reality. Where possible different narratives are explored through dialogical vignettes that attempt to co-create areas of shared meaning. These common denominators form the basis for participatory policy design.

The axial issues are public and private barriers to conceptual space, geographical space and cyberspace that result in dramatically different life chances. The scope of the ideographic case study in Central Australia is socio-cultural, economic, political and environmental health and thus the quality of life and governance of citizens. If it is assumed that problems are complex, multifaceted and systemically linked, then critical policy research needs to address issues contextually in conceptual terms and in terms of a specific space and time. The discourses of all the participants (including the researcher make a co-created reality that needs to be understood through dialogue in which some shared ideas are constructed. Understanding concepts, emotions, assumptions and values are vital for building trust, in order to establish dialogue for problem solving. Thus this systemic approach strives to unpack the complexity layer by narrative layer using interactive dialogue for design and planning. The complexity of stories is maintained. Rather than simplifying or aggregating them in data chunks, the meanings are explored to reveal underlying patterns of meaning. Social planning is based on personal knowledge bases (Polanyi 1962) that span life experience and formal learning. To add to the complexity, the researcher is recursively linked (Giddens 1991) to the research context. As researchers we have the potential to shape our worlds, just as our worlds have the potential to shape us. This is discussed in more detail in chapter 4.

Alice Springs is selected as ideal for studying the articulation of the personal and the public at the local, national and international policy level and the impact of marginalisation in social, conceptual, geographical and cyberspace. Marginalisation has the potential to be translated into class/cultural withdrawal and nationalism. In Alice Springs class/culture are proxies written in the socio-demographic patterns of disadvantage. The politicisation of culture (as a response to being excluded from co-creation of designs for the future) needs to be addressed through policies that take cognizance of power and empowerment, in order to ensure that 'self–other' (Fine in Denzin and Lincoln 1994) distinctions do not become power plays between haves and have-nots (merely expressed in terms of reparation).

The narratives explore the dynamics of a town underpinned by shared stories, but paradoxically riven by divisions: social, political and economic. These divisions are spelled out in terms of (i) the social indicators of quality of life and (ii) the social health outcomes of citizens and (iii) by the social context that shaped the entire research process. These dynamics are closely linked with the paradoxical changes on the world stage. Many of which are not acknowledged or taken into account by the decision-makers that perhaps unknowingly design systems that are based on binary oppositional data and thus develop partial knowledge narratives. The challenge is to understand the potential of continuity and common denominators that underpin knowledge narratives based on wider, co-created discourses that support a governmentality able to manage diversity in an accountable manner.

This case study attempts to develop the middle ground needed between macro-level evolutionary systems designs and micro-level organisational systems designs through a medium range community and regional study. The local and the global are articulated through a modified WHO healthy city/settings approach based on the Ottawa Health Charter of 1986 and Agenda 21, employing participatory action research (PAR), self reflexive management tools for thinking and caring (that place the researcher within the research frame) and fieldwork building on, but diverging from, the sociological tradition of the Chicago School[15] and its social anthropological roots. This critical systems thinking and practice or *critical systems praxis* (CSP) *stresses the links between transcultural thinking and practice*. It is applied to a particular case in order to develop grounded theory and practice to address social and environmental justice pertaining to sustainable health, education and employment, irrespective of age, gender or culture.

Throughout, the case issues relate to wider ones, whilst the uniqueness of the situation is explicated. Case studies per se are useful for a number of reasons. They give interpretive depth, indicate new perspectives and areas that have previously been unanticipated. Case studies have been used increasingly as a means to develop critical insights into the links between the insider and the outsider view points. In this study the hinges across self and other are extended to include the environment. Further, I am part of the subject matter in this study and part of the communication dynamics. Paradoxes of meaning are unfolded in the process of working the boundaries.

The following chapters develop a systemic approach to public policy issues that have continued to be intractable because of a lack of emphasis on systemic and transcultural understanding. By systemic I mean an approach that 'unfolds and sweeps in' variables (as per Ulrich 2001) to address the

[15] See the work of Whyte, W.F. 1955. Street Corner Society 2nd Edition. University of Chicago Press, Chicago.

tensions between the notion of structures and constructs. Unfolding through dialogue attempts to ascertain contextual axial themes through analysis that employs retroductive logic but takes into account the narrative layers that give different constructs of the way society is understood and structured. The recursive (as per Giddens), dynamic nature of the research process is acknowledged. The role of the researcher as actor/change agent is acted upon/changed by the context. This is central to this story.

A sustained linked example helps to increase the accessibility of theory and methodology. It addresses:

(1) The interdisciplinary and trans sectoral approach[16] to problem solving to address poverty through health, education and employment creation.
(2) Theory, method and strategies to address governance and enhance participatory democracy that is as mindful of the needs of people as it is of the environment.

To sum up, the concepts of citizenship and life chances are explored (by means of a multi-method, multi-site approach that is underpinned by the principles of participatory action research) within the context of a diverse population (characterised by a transient but regionally stable Indigenous population and an extremely diverse non-Indigenous population). Culture and language, level of education, age and gender underpin different life chances in terms of social well being (within both Indigenous and non-Indigenous sectors of the population). Life chances are examined in terms of both qualitative and quantitative indicators of

[16] Trans sectoral means working across sectors and across organisations using matrix teams. These approaches are referred to as an integrated or systemic approach that follows the World Health Organisation's Ottawa Health Charter of 1986. It spelt out the links between health and development and is the basis for a new approach to development. The links between health, socio-economic and political development have been further underlined by means of the Healthy City Approach (see Davies and Kelly 1993). The World Health Organisation (WHO) has now extended this to the Healthy Environment approach. The integration across disciplines and sectors is recognised as the basis for bringing about change. Throughout Australia and the world, towns and cities have joined this *global* development movement for benchmarking their levels of development. A wide range of socio, cultural, political, economic and environmental indicators of development and empowerment are developed locally with reference to a wider national and global context. This integrated approach is ideally suited to addressing the socio-economic and environmental issues identified by means of qualitative and quantitative indicators of local need.

local need derived and adapted from WHO healthy city/environment literature.

The case study explores the social dynamics contextually. Alice Springs has several forms of governance. Besides the Commonwealth Government, territory government, there are multiple local organisations competing for resources and political support, for example Arrernte Council (for Arrernte residents), Alice Springs Town Council (perceived as serving the interests of rate payers), Tangentyere Council (for town camp residents), Pitantjara Council, Central Land Council (serves interests of all Indigenous Groups). It can be argued that diversity is best served through diverse local governance opportunities and that open communication across areas of governance can help to co-create shared interests. The challenge is to avoid lapsing into hide-bound bureaucracies and to remain open to forming matrix responses to issues. Class, citizenship rights and responsibilities and the way in which these concepts are linked with experience and life chances are explored. As such it goes beyond the current debates concerning so-called 'victim mentality' that tend to limit social analysis to an individual or at the micro-level. This approach considers broader issues systemically and critically and argues that through appropriate communication, designs can be steered to address issues.

This monograph strives to give a sense not merely of the marginalised who are striving for self-determination, but an overview of the dynamics of power, location and dislocation and the way people make sense of their lived experiences.

The relationship between human beings and commodities has been limited in Alice Springs by concentrating mostly on the technological slice of knowledge and not enough on the 'strategic' and 'communicative' aspects of knowledge (as per Habermas 1984) that are necessary to complete understanding that can lead to meaningful policies that empower all the participants. Some Indigenous communities decided to become alcohol-free or 'dry communities', whilst others decided to run canteens once liquor prohibition by government had ended. (D'Abbs 1998).[17] The blame for alcohol misuse is often shifted on the basis that people have the freedom to decide, without sweeping in the context of the decision-making. The nexus of an imposed welfare mentality (see Rowse 1998) and the tendency for desert cultures the world over to consume available resources (CARPA Newsletter No 27 1998) and to expect that societal members share whatever is available, means that accessing social security and consuming

[17] Peter d'Abbs (1998) Paper presented to Alcohol Availability Workshop, 2nd November.

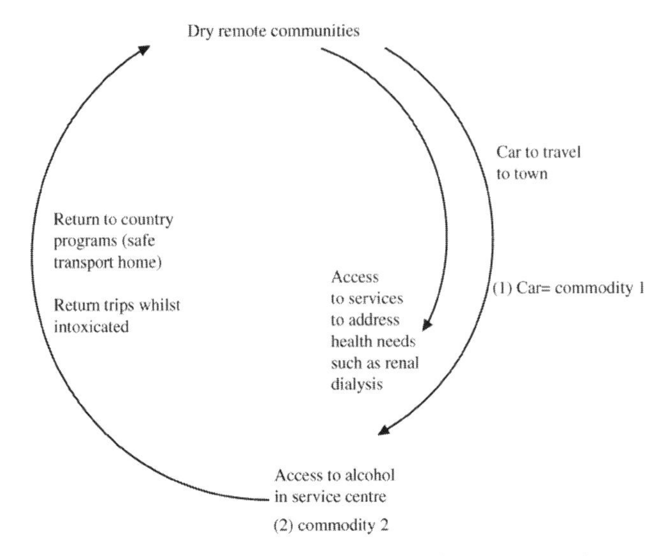


*DIAGRAM 1.1* The Service Centre to a Region with Dry, Remote Communities.

resources (with an emphasis on an immediate not a future time-frame) combine to create a receptive environment for alcohol misuse (Diagram 1.1).

The most extreme examples of 'technology gone wrong' or the non-systemic application of technology is the use of the car, and the dialysis machine. The car is used by the residents of dry remote communities to buy 'grog' (*commodity one* in the system) in town where it is consumed through binge drinking. The petrol (*commodity two* in the system) is inhaled as another means to alter consciousness. Excessive misuse of petrol and alcohol lead to physical and mental disability. Diabetes and kidney damage are a systemic consequence that leads to 'life on the machine' or kidney dialysis as a tri-weekly routine, becomes a way of life. To be human at this stage requires access to the machine and a life of quasi cyborgism. If development knowledge were expanded to include 'technical', 'strategic' and 'communicative' knowledge then it would lead to better problem solving. The heart of the issue is that a sense of rights and responsibilities at a personal and public level need to be cultivated. Systemically all the aspects of health and development need to be swept in at an early stage.

In Alice Springs we have an example of the intersection of a nexus of commodities. It is represented diagrammatically in Diagram 1.2.

'Doing the wrong thing right' (Ackoff and Pourdehnad 2001) has been 'perfected' to create an economy that supports alcohol sales, car sales and the helping professionals. An increase in the sales of cars offsets the

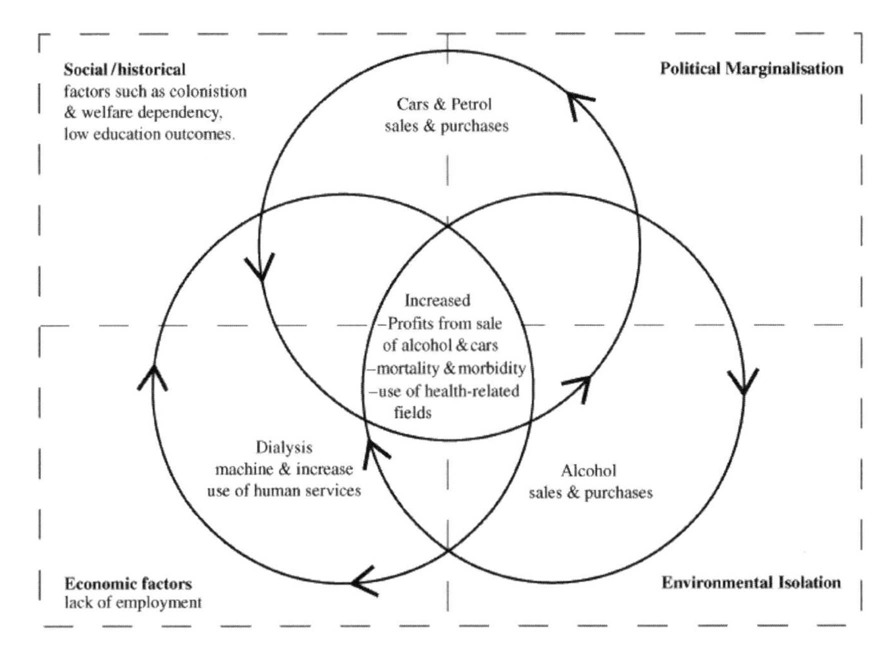


*DIAGRAM 1.2* A Nexus of Commodification as a Result of Historical and Social Shapers.

loss of cars through accidents. Development expenditure is focused on a system that has 'fine-tuned' the commodification of grog, petrol, cars, biomedical and human services supporting dependency and life on a machine. This is perceived by Indigenous informants as a deadly but profitable triumvirate, perfected to create an economy that supports alcohol sales, car sales and petrol sales, whilst the employment opportunities for the helping professionals (both Indigenous and non-Indigenous) is based on misery and addiction.

This book is about 'unfolding' and 'sweeping in' (as per Singer and Churchman 1979, 1981 and Ulrich 2001) the issues that can be explained retroductively as historical, economic, intergenerational violence associated with marginalisation, alcohol and poverty. An economy that supports the class/culture system is written in the socio-demographic patterns of disadvantage (educational outcomes, unemployment and incarceration), morbidity and mortality and life chances. That is why the intervention to break the interlinked cycles has to be at the level of regional governance so that the notion of rights and responsibilities can be understood. The shadow of a life of intoxication and then a life on a machine hangs over an Indigenous community that seems to be constantly grieving. This community comprises widely linked extended families within language groups, but has deep divisions across other language groups. Mental and physical well-being is affected by ongoing violence associated with the effects of colonisation and ongoing racial tension in the community, intergenerational violence associated with alcohol and debt within families and competition across Indigenous interest groups for resources. Self-medication (using alcohol and other drugs) is seen as an excuse to make the unbearable bearable. Paradoxically it is also argued that drugs and alcohol are a source of short-term enjoyment. When the time-frame is short (as is often the case with people living hand to mouth), then the long-term effects do not matter so much. The immediate or existential joy is what matters. It also blots out the pain.

The challenge is to think systemically and retroductively in terms of the meanings ascribed to intervention. This study is about demonstrating the value of systems thinking for solving complex problems. It unravels the coils and making policy suggestions rooted in praxis. In order to explicate CSP, the case study is an appropriate means to demonstrate process and issues. Examples stress the value of creating links across sectors and disciplines in order to enhance health, education and employment opportunities for all citizens, including those who are striving for recognition and self-determination (Diagram 1.3). Given the context and issues, the policy goal was to address power, empowerment and governance needs of people marginalised in terms of conceptual, geographical and cyberspace/time. Social systems are addressed in terms of culture (the way people think and act), structure (institutions and interactions) and power (based on means) (Ackoff and Pourdenad 2000: 216) and policy (based on decisions as to who gets what, when, why, how, why and in whose opinion) (Diagram 1.4).

The study demonstrates a praxis approach (CSP) that stresses the value of creating links across sectors and disciplines in order to enhance development and health through capacity building. The barriers to achieving health, education and employment outcomes for some citizens living in Alice Springs are due to social–cultural, demographic, geographic, economic and political factors. The challenges are a result of both present and past policies. The impact of colonisation on Indigenous people needs to be considered in all analyses of the current status of social health in Alice Springs (Menzies Annual Report 1999) Alcohol misuse is an effect of a legacy of colonisation and marginalisation and a cause of social ills. Higher mortality and morbidity rates in Alice Springs and the Northern Territory are outcomes associated with violence and road deaths as well as diseases directly and indirectly linked with alcohol and the associated poor nutrition (as a result of both spending money on alcohol as well as the unavailability and very high cost of food in remote communities). The ramifications of the abuse of alcohol need to be understood as being

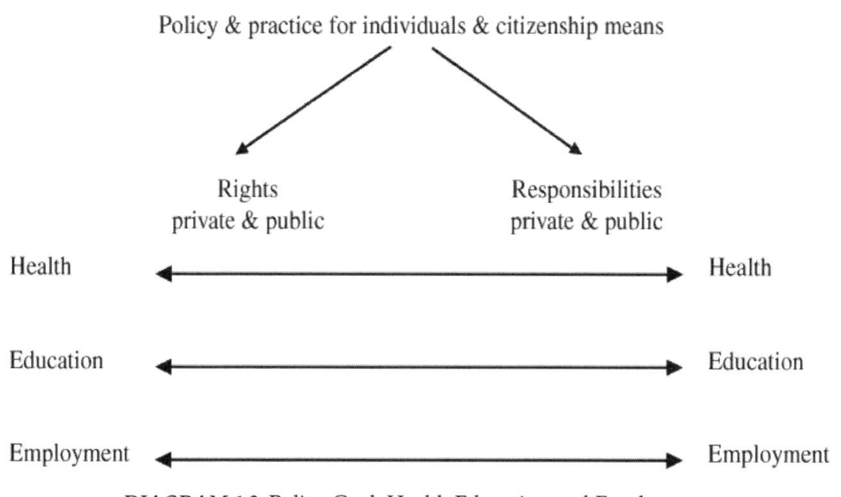


*DIAGRAM 1.3* Policy Goal: Health Education and Employment

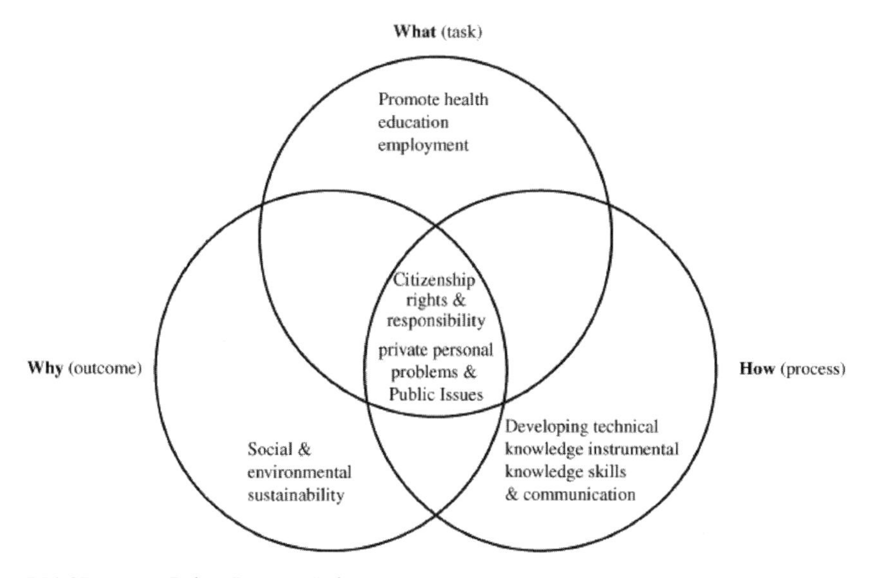


*DIAGRAM 1.4* Policy Process: Enhancing Participatory Governance and Policy Outcome

systemic in their causes and effects. Alcohol misuse is the immediate cause of many social ills. The causes and effects become a cycle of damage to individuals, families and communities resulting in ongoing and intergenerational poverty because of the modelling of behaviour to the younger generation and the sense of cultural loss, meaninglessness and dependency on welfare. The cycle is one of marginalisation, alcohol misuse and further marginalisation. *A broad-based systemic approach* has been advocated by recent NT public health policies in line with the World Health Organisation's Approach (1995), but the policy needs to be translated into practice. In a paper entitled 'Alcohol policy and the public good', the WHO concluded that programs are only effective if they are broad-based and if they address the wider social context.

Much could be learned about different ways of seeing and perceiving through engaging in a dialogue that celebrates the diversity of cultures. There is a need to also shift planning and design away from predetermined ideas and to co-create designs that are functional because they are relevant. Services need to reflect the values of the users and for this to occur the users need to participate in and decide on policy design and governance (Diagram 1.5). The essence of the problem is that services do not really meet the perceived needs of Indigenous people who are treated as consumers, rather than as participants in the development process.

Development is not necessarily going to make much sense unless it can be framed in a mutually acceptable way. Also learning some of their ideas can refocus the direction of development away from 'the market rules' philosophy and to consider the liberative potential of another framework for living. This does not mean that any framework has all the answers, but some answers that are worth considering. This requires

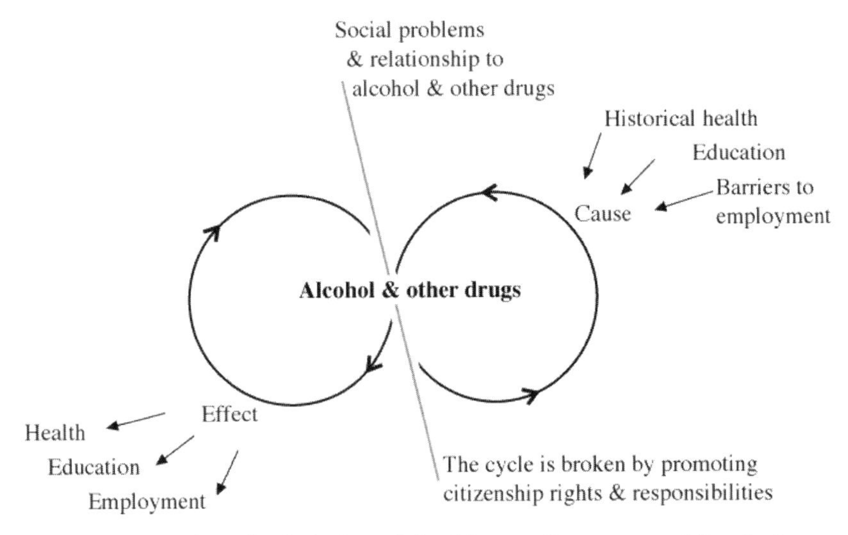


*DIAGRAM 1.5* Breaking the Cycle through Participatory Governance and Developing Citizenship Rights and Responsibilities

ongoing dialogue and political will amongst all the parties. Framing standards of living needs to be considered in terms of perceived and expressed need. The way data is collected and who collects the data will influence the quality of the data. The links between current social problems and the history of colonisation cannot be forgotten. The results are spelt out in alcoholism, cultural despair and a sense of real powerlessness.

## 1.2 *The Audience and the Relevance of Critical Systemic Thinking to Planning and Policy: Implications for Accountable Policy and Practice*

The book is aimed at professionals and students (at undergraduate and postgraduate level) in public policy in the public, private and volunteer sectors. For instance, academics and professionals in social studies, development studies, cultural studies, human services, management, governance, health, education, employment, crime prevention could find the book relevant as a source work to stimulate ideas for social planning and policy. Practical examples are used to exemplify theory and policy suggestions are made. The design demonstrates the value of critical systemic thinking and practice and although the details refer to a particular place and time, this holistic praxis could be of relevance elsewhere.

By thinking critically and systemically and applying a range of praxis tools to enable us to hold in mind more than one culture, ideological framework or discipline at the same time, it is likely that we will move closer to nurturing the social and environmental capital required for current and future generations. The implications of multiple factors are analysed and policy suggestions are made. Development responses do not, however, have to be at either end of the continuum of approaches that are economic rationalist or socialist in orientation, as neither of these models is sufficiently systemic in nature. Potential ideas to address social and environmental justice need to take into account multiple variables from many sources. This requires open communication and diversity management to include stakeholders that represent multiple interest groups, so that multisemic solutions can be achieved. Multiple methods were used so that both qualitative data (to address subjective and intersubjective meanings and perceptions) and quantitative data (to address the socio, demographic and epidemiological challenges) could be addressed. The World Health Organisations' Healthy Settings approach (that resonates with the Indigenous concerns about land, a sense of place and wellbeing) is one such approach. Strategies need to take into account *historical, socio-cultural, political, economic and environmental variables.*

The study analyses the implications of the social, political, the economic and the environmental in terms of the development context and challenges. These indicators are historical, environmental factors, sociodemographic and economic factors. On the basis of analysis and discussion participatory design features for management, policy and program suggestions are made.

When people are excluded from participation in governance they cannot co-create the design of their own futures. In order to communicate and participate in conceptual, geographical and cyberspace numeracy and literacy are vital. But so is respect for the personal knowledge that people have developed as a result of their lived experiences. This is relevant to individuals, organisations and the cultures of interest groups. It is vital to be able to identify:

- Levels of data, information and knowledge narratives across interest groups, sectors and disciplines and the relationships across these,
- Sources of knowledge narratives and to be able to
- Co-create and manage data, information and knowledge narratives appropriately at each level.

Underpinning all these levels are values, assumptions and emotions, that need to be acknowledged through critical systemic thinking, in order to understand their implications for personal and public praxis.

A sense of being left behind is often expressed in terms of reactions of closure to options and opportunities. Competent thinking and practice involves theoretical and methodological literacy with a view to developing co-created solutions for people and the planet based on respect for diversity, the jump lead of creativity. Barriers are distrust of the other based on fear, greed and contempt. When we slip from co-creation to zealotry or cynicism, these are more likely emotions. When identities are defined in opposition and based on competition (encouraged by the market rules paradigm) there is less scope for co-created, sustainable solutions.

I realise that to some systemic thinking could be considered authoritarian, or potentially so. But critical systems thinking doesn't need to be! It can play a useful role as it takes cognisance of (all) scientific knowledge narratives, as well as personal knowledge and lived experience. But most importantly it also maintains the belief in rational enlightenment. The value of 'sweeping in' and acknowledging other narratives is considered a means of testing ideas through falsification (as per Popper) using respectful dialogue (as per Habermas) to communicate across diverse viewpoints. Dialogue that is inclusive of diversity is the jump lead of creativity and an essential aspect of democracy. CSP attempts to avoid the dangers of the worst versions of post-modernism and modernism through applying ecohumanisic principles that are mindful of weaving webs of understanding across self, other and the environment. CSP is discussed in more detail in chapter 3.

I explore rights and responsibilities and ways in which life chances could be redressed at a local level through (i) enhancing representation of citizens and through (ii) establishing links across community councils and a wide range of GOs and NGOs and (iii) by situating analyses of life chances within the context of relevant global benchmarks. Self determination is analysed in terms of the different ways in which it is defined by stakeholders: ranging from user pays for services and rent to land rights. Striving for self-determination and citizenship involves mediating the struggle of user pays (associated with economic rationalism of human services) versus landrights (associated with Indigenous liberation movements) by co-creating solutions that are based on mutual respect and trust to achieve access to a sense of place and well-being. Nationalism is a response to the history of dispossession and the current feelings of social marginalisation.

At a personal level the study strives to describe the processes for addressing co-creation (the challenge of all boundary workers) and (i) both private and public wellbeing. The purpose of this work is to give a voice to the participants (informants and respondents) who identified social, political and economic needs. The policy suggestions are based on the identified social needs in terms of (ii) people's perceptions and in terms of (iii) secondary statistics. A complementary approach to methodology is applied (as per Jackson 1991, Flood and Romm 1996) and with due regard to accountability (Romm 2001).

Systemic approaches to social policy enable us to hold in mind multiple variables. Systemic thinking requires holding in mind the following:

- Various sectors (public, private and third or voluntary),
- Range of disciplines and associated models for addressing social problems but also,
- Dimensions of the social, historical, political and economic context and the considerations of the local, national and international context,
- Personalities, emotions, assumptions and values of individuals and interest groups and the interplay of social dynamics in and between organisations and across sectors and disciplines.
- Identifying the paradoxes and the implications for policy and practice and then finding ways to address the paradoxes through dialogue and negotiation.

Theoretical and methodological literacy is required to understand the values and assumptions underpinning the different models of social policy and welfare delivery (McIntyre 1996). The way people think about key concepts such as health, education, employment and crime prevention impacts on the way in which policy is developed and implemented. This is increasingly challenging as welfare is redefined in increasingly residual terms to the extent that we can talk of 'post welfare' (Jamrozik 2001). Unless 'we get the approach right' then resources (time, money and people skills) will continue to be wasted because development processes can either enhance or retard social capital, the basis of trust in a community. The very notion of social capital is a construct and it means different things to different stakeholders (White 2002).

The way people think about key concepts such as development, education, employment, health and citizenship impacts on the way in which policy to address social and environmental issues is developed and implemented. Our values and assumptions shape: (i) approaches to problem solving and influence the way we define social problems, (ii) processes for addressing social problems, (iii) development outcomes and (iv) evaluation of development outputs and outcomes. We need a high degree of theoretical and methodological literacy (McIntyre 1996) to address the questions who gets what when how, from whom, why and to what effect and in whose opinion. We also need to know that the answers to these questions depend on who is asking the questions and the values that they hold.

There are many areas of knowledge and diversity management unfolds the complexity by asking questions. A CSP to development, involves questioning and thinking holistically across social, political, economic and environmental factors. CSP employs multiple methods to work across rather than within boundaries to give wide angled insights. As such this critical systems praxis (CSP) builds on and extends the tradition of ideographic, holistic case studies[18] undertaken by social anthropologists and sociologists working from functionalist or interpretivist perspectives (and with varying degrees of identification across self, other and the environment). CSP stresses the need to design human services as open and responsive to biodiversity. Multiple variables need to be borne in mind and this requires diversity management, based on the co-creation of shared meanings. An understanding of webs of meaning across the 'hyphens' of 'self-other' (Fine in Denzin and Lincoln 1994) and the environment is essential for responsive design. A map of knowledge narratives provides a

[18] See for instance the work of sociologists Lynd, Whyte and Becker of the Chicago School and the pioneering studies by Malinowski and later by Mead and Geertz.

means of illustrating that we can think, learn and practice within single frameworks or discourses or we can engage in dialogue that enables us not merely to compare frameworks in an adversarial manner, but to see the value or potential of the other, so that co-created meanings can be enriched to respond creatively to specific challenges.

We need to work across disciplines and sectors and realise that working with, rather than within knowledge areas, requires managing knowledge and understanding that information is based on 'either or' as well as 'both and' definitions of data. Churchman's contribution was to stress that the design of policy (and other) research is vital for determining the extent to which it can make a difference to addressing real, complex problems. Working in multidisciplinary, multisectoral teams is not easy. Power and conflict are associated with different levels of status associated with different professions and different positions. Personalities and politics also play a role. Managing group dynamics and communicating in such a way that participants from diverse disciplines are brought successfully into the process is a communication challenge.

The study analyzes the systemic nature of problems, the impact of changes and the flow on affect that they have. It is as much an attempt to discuss the challenges that I faced trying to address the issues of being the observer whilst I was observed, as of being affected by the research process and the social context in which I was undertaking the research. CSP is about working the links between the research and the research issues. It is also an attempt to share what is usually regarded as behind the scene issues. It is an attempt to teach and to share the ethical and research dilemmas by using discussion and reflection on this case study. Through dialogical vignettes I explore the issues that I faced as a researcher. Axial areas for power struggles internationally and locally are social space, conceptual space, geographical space and cyberspace. Being marginalised means that one's stories are not considered relevant to mainstream knowledge, that one is unlikely to be given much opportunity to make public decisions, one's life chances in terms of health, education and employment are lower and one is unlikely to be sufficiently skilled to use digital technology. CSP argues that conceptual space must be shared as a starting point for problem solving. An ability to address a problem from more than one paradigm or set of meanings could be an essential first step in achieving social and environmental justice, based on co-created meanings that address diversity (see McIntyre-Mills 2001).

A central theme is to show how research process and content is a reflection of the values of the research participants. It is also as much about the way values and power are linked, as it is about emotion and personality. The challenge was to avoid slipping into cynicism or zealotry;

by imposing my own narrative of the way things ought to be and to continue to co-create meaning. At another level the key argument that flows from an analysis of axial themes (of access to geographical, social and cyberspace), is that local and national 'nation within a nation responses' are fed by a sense of exclusion. Tools for ethical decision making could help to ensure that social policy is accountable to citizens who have the least power. To a large extent changes in space and time have been altered by culture, but the ecological imperatives remain. For those without access to the 'cyberspace of cultural change' the imperatives affect their lives but they have very little opportunity to participate in the changed reality.[19]

Although the theory of diversity management encourages the managers of bureaucracies to shift towards matrices and fluid structures in response to the fluid, fast changing economy (where market borders no longer provide buffers and so reactions have to be fast), the reality is that the rhetoric may be used quite frequently, but the practice on the ground remains much the same. The frustration of trying to respond quickly and appropriately is very difficult for staff in the public and private sector who remain hemmed in by the boundaries within which they are required to work. As Toffler (1990) outlined in 'Power Shift', the change from cubby-hole thinking to systems thinking or network thinking will require changes in the way societies are organised. Space for some is no longer merely local and neighbourhood-based because communication through interactive technology has created close links with the national and international domain, but for others space remains the same as before. Two parallel senses of space thus coexist. One for the 'on line connected haves' in the population, another for 'the unconnected have-nots'. Cyberspace has effectively impacted on local space for some people, but not others. Castells (1997) argues that it is through the flows of information in cyberspace that the real power lies, because it is more powerful than that wielded by state bureaucracies. This is partially true, but the power of cyberspace needs to be harnessed to address the needs of local groups in geographical human space. At the time of the research globalisation impacted on the lives of those in primary industries as collapsing commodity prices and on the lives of those in the service industries as a suspicion of the Internet at a time when the millennium bug was touted as a challenge for the future.

[19] Haraway in Zimmerman 1994: 373 dismisses cyberspace as 'community hallucination'. But exclusion from it, means exclusion from yet another space. If one does not participate because one is unable to communicate through this medium, one is more likely to be subject to others decisions than one's own.

## 1.3 *Setting the Demographic, Socio-Cultural, Political and Economic Context of CSP*

The study reflects on an experience of undertaking a study of the life chances of citizens in a remote region of Australia. Indigenous people make up 16–20% of the 27,000 population on any one day in Alice Springs. Alice is essentially a borderland of cultures made up of Indigenous Walpiri, Luritja and Western and Eastern Arrernte, tourists (during the cooler months) and the shifting population of professionals and business people. Numbers fluctuate, because of the tendency of Indigenous people to move from place to place within the region for a number of social reasons (accessing services) and cultural commitments (visiting sites in the landscapes that write their story and sorry business for the bereaved). Paradoxically despite the local transience of some Aboriginal people throughout the region and Alice Springs they are, however, regionally stable. Mobility has been a defining characteristic of Aboriginal culture, which halted for a while as a result of the impact of pastoralism and missionary settlements associated with the colonisation of Australia (Coulehan 1997) and government edicts. A transient population of service providers and users come to Alice Springs on government contracts or for business reasons. The service providers are professionals usually on short-term contracts who move away once they have fulfilled their service agreements. In excess of 20–30% of the population changes every 2 years (according to estimates made by the General Practitioners Association, based on ABS 1996 census data). Alice was described by informants to be 'like a company or government town'. Transience has an impact on the quality of service delivery by professionals who leave just when they begin to have an insight into local and regional issues. It is unlikely that transience will end in the near future because of the changing nature of work that has created a very mobile global workforce. Also the tourist population ebbs and flows with the seasons. In 40°C heat there are few visitors.

Research data from a range of sources (as detailed in the following chapters) underlines that Indigenous people score lowest in terms of employment, health and education outcomes (the pillars of citizenship rights, that are now sidelined in a 'post welfare state', to use Jamrozik's 2001 term to refer to Australia) and the highest in terms of incarceration rates. Self-determination is still a goal for this 'nation within a nation', which is hardly surprising as a result of their feeling of marginalisation. Isolation for some from the rest of the world (as a result of illiteracy and innumeracy) and connectedness for others (through being mobile knowledge workers is a reality). Added to this is the marginalisation

Indigenous people feel from local public spaces and a lack of respect for some Indigenous people in banks, on pavements and in shops. In places where their separate, parallel lives are visible.

At the time of the research Alice had almost twice the number of human services as the national average and yet the outcomes in terms of social health were dramatically below the national average. Planning needs to address the lack of continuity in service delivery because of the high turnover in staff in human service organisations. Also added to this any plans need to take into account the historical and global context and the regional service nature of Alice Springs and the way in which the health services are accessed. The role of digital technology to span the divide could support existing services such as the Flying Doctor Service and similarly the education and health opportunities could be expanded through digital technology. It is myth that all technology continues to be regarded with suspicion by Indigenous people. Cars, radios, television, telephones and particularly computers are embraced as ways to enhance communication. Despite the interest in technology, it has not as yet succeeded in bridging the gap between remote desert communities and Alice. Sustaining and using technology in a meaningful way (based on the communicated values of stakeholders) remains a challenge. For example, treatment services in town appear to be used by Aboriginal people only once they are already at a chronic stage of their illness. Accessibility to services is always a problem in terms of cost, distance and sense of appropriateness. These are important factors when vast geographical and social distances need to be covered by poor, remote users with a different cultural sense of need.

The study addresses citizen's perceptions of the quality of their lives. They prioritised the need for programs to address local issues within the context of broader social, political and economic shapers at the national and international level. The research data identified needs that differed across age and interest groups. The issues include: (i) Social and environmental issues pertaining to the well-being of people. (ii) Poor education and health outcomes amongst the Indigenous population. (iii) The high unemployment rates of young people and Indigenous people, in particular. (iv) High levels of incarceration for Indigenous people, particularly young people. (v) Domestic violence, the lack of safe accommodation for women and children. (vi) Alcohol misuse and related issues. (vii) Self-determination of Indigenous people in order to achieve well-being.

The social, political, economic and environmental indicators of need demonstrate the majority of those with limited life chances are Indigenous.

In focus groups issues of well-being were raised, such as attitudes to people using public spaces and services and the need for self-determination in terms of the recognition of Indigenous identity (language and land) and the need for some separate specific services. The research investigated the social, political and economic life chances of the citizens of Alice Springs. Life chances were examined in terms of both qualitative and quantitative indicators derived and adapted from the World Health Organisation Healthy City/Environment literature on indicators. An integrated approach to addressing the identified areas of concern was advocated. The research explores rights and responsibilities and ways in which life chances could be redressed in part through: (i) enhancing representation and listening to citizens of all age groups and backgrounds in decision-making processes and through (ii) establishing links and partnerships across a wide range of government organisations, non-government organisations and the business sector to redress inequalities in life chances (in particular perceived access to resources and services).

The municipal area (of about 410 square km) serves many of the people who come from remote areas to use its services. Alice Springs needs to be considered in regional terms rather than merely in terms of its municipal boundaries. If we consider the region to include a 500 km radius of areas which it serves then the population rises to 48,318 and if one includes the international tourists present on a daily basis in this region as a whole the number rises to a further 4,000 (approximately).[20] If one adds the number of Aboriginal citizens moving from Western Australia, Queensland, and South Australia to visit family the number rises. Alice Springs is a regional service centre where the residents and service users are characterised by a high level of transience. Transience is a defining characteristic of both the Indigenous and non-Indigenous population albeit for different reasons. Transience is in fact as much a product of the mobile global workforce impacted by the changing nature of work as it is the way of life of the Indigenous people who move across the landscape once again, now that limitations on their movement have been lifted since their being recognised as citizens with the right to move and not have to carry documentation as they once did. Alice Springs is characterised by a diversity of cultures and lifestyles.

The following groups are demographically significant in Alice Springs: migrants of English and non-English speaking backgrounds, Arrernte people, the traditional owners of Mparntwe (Alice Springs), other Indigenous groups, the visitors from surrounding rural areas who follow their songlines to Alice Springs, who visit relatives and who seek urban services, the local and international tourists and, the pastoralists and miners from outlying areas. Special interest groups are represented

[20] ABS Regional Statistics Report 1362.7 and ABS Profile for 500 km radius of Alice Springs.

by different age groups and the level of physical and mental health. The young and the elderly and the disabled are specific lobby groups as are gender specific groups, such as support groups for men and women or parents of young children and adolescents. The population overall is young and mobile, without extensive family support. Demographically there is a small, but growing group of elderly people, despite the fact that some tend to leave at retirement age. The growth in Indigenous young people is however demographically significant. The identity of the different groups within the population is particularly interesting. To be a Territorian one is required to have lived in the area for more than 25 years, which is little when compared with the 40–60,000-year Indigenous connection with the place. (This is considered to be in a different category!) Migrants are welcomed if they can speak English and if they are going to make a 25-year commitment.

Some locals consider that they are in touch with the world and have a global frame of reference (because they are computer and Internet literate, for instance). It is well known that Alice was built as a result of a communications system, namely the Overland Telegraph and it is a tourist destination and thus has a global reference point. Another international link is the result of a joint Australian and American Strategic Military Base[21] for surveillance and intelligence. Others resented the way in which their small town was changing as a result of tourists and those with different cultural backgrounds. A cosmopolitan atmosphere prevails in areas visited by tourists. Tourists although essential are regarded as outsiders by some, but many realise that they are essential to the local economy. Tourists, contract workers and short-term workers for government, the private sector and the military have specific recreational and service needs. The ABS census data for 1996 indicated that 49% of the local population enumerated different addresses 5 years previously whilst 32.6% enumerated the same address 5 years previously. This means that many people tend to be mobile and long-term stays in one home are not very prevalent.

[21] Some ambivalence to the Military presence by some local people centres on the possible threat to Alice as a result of the large military installations. This has been expressed by the Peace Group. The significant contribution to the economy by the military presence offsets these disadvantages as far as many residents are concerned. After the research had been completed and the process of writing up and analysis was all but over, the world changing bombing of the American symbols of power (commerce and defence) and the civilians from many countries were killed by the deliberate use of planes as weapons of destruction, the role of Pine Gap has escalated.

The locals who have lived in Alice Springs for generations, however, have a sense of pioneering the development of Alice. Paradoxically 'pioneering' is seen to have destroyed vernacular streetscapes as far as the aesthetically and historically minded are concerned. In terms of Aboriginal history and a sense of place it has led to native title, land rights claims and claims against the destruction of sacred sites. The Aboriginal population has a sense of being the custodians of the land. They have an existential sense of place. Self-identity is linked with place. Class, culture and gender continue to define life chances today. Indigenous people are statistically more likely to be unemployed and destitute than other groups. But it is undeniable that class positions cut across Aboriginality. The life chances of Indigenous people can be plotted along a continuum from ownership of business and property interests, employment, underemployment to destitution. Alcohol, its availability and usage are an effect and a cause shaping the life chances of citizens. Alcohol and poverty play a role in property crimes in Alice Springs. At the time of the research Alice Springs and the Northern Territory are a focus of international attention because of mandatory sentencing and its particular impact on Indigenous young people and their families. This law is a leitmotif of a class and cultural divide. Property laws were implemented in such a way that the judiciary is subject to the laws of the Territory. These limited the right of the judiciary to assess each case and to determine punishment that fitted the specific case.

At the time of the research, for those on low incomes or without incomes, the social health outcomes did not compare well with other parts of Australia and internationally. Young people and Indigenous people experience the highest levels of unemployment. Poor education outcomes, high levels of youth suicide, high levels of youth incarceration as a result of mandatory sentencing laws (that apply to adults and young people), high levels of domestic violence, high injury rates along with high morbidity and mortality rates (associated with the highest abuse of alcohol and some other drugs in Australia) and poor nutrition define the different life chances of some citizens.

For others who have medium to high incomes, the standard of living compares well with other parts of Australia and internationally. The range and quality of facilities and the unique arid landscape for outdoor recreation provide attractions for those who come to Alice to work and to set up businesses. Overall the employment levels were high and overall the median level of weekly income compared favourably with other parts of Australia.

International and local travelers swell the numbers in Alice Springs during the cooler months to experience the 'borderland of cultures' or 'frontier town'. The commodification of culture by tourism is yet another issue which Indigenous people articulated as an interest in aspects of culture (real and manufactured) that can be commercialised for the tourism industry. Aboriginality is divided into that which is saleable and acceptable and that which is of no commercial value and 'a nuisance'.

Indigenous people at that time scored lowest in terms of employment, health and education outcomes and highest in terms of incarceration rates. A diverse population of Aboriginal people in Alice Springs have overall health and development indicators that differ in terms of life chances from the rest of the population. The health and development indicators for Aboriginal people (such as morbidity and mortality rates) are at levels way below those for non-Aboriginal citizens, despite the delivery of community services at almost twice the rate at which services are delivered at the national level.

A transient population of service providers and users come to Alice Springs on government contracts or for business reasons. The service providers are professionals usually on short-term contracts who move away once they have fulfilled their service agreements. Some stay on and provide a sense of continuity and a sense of history. The commonly cited reasons for moving are the climate, the lack of schooling, the need for a more diverse lifestyle and the need to rejoin families elsewhere. Transience has an impact on the quality of service delivery by professionals who leave when they begin to have an insight into local and regional issues. It is unlikely that transience for some sections of the population will end in the near future because reasons are not only climatic but a product of the changing nature of work that has created a very mobile global workforce which is employed on contract rather than in terms of long-term careers. It is also a result of a long-standing Aboriginal cultural practice to move across the land which was halted when pass books were required by government authorities and has now resumed as a social and cultural expression of identity.

> There is a constant spending of resources to stem what I sense is an irreversible trend for the bulk of the people to stay for relatively short periods. I advocate working with the shifting populations and demographics rather than trying to change them. (Bret Galt-Smith 1999, personal communication).

For some, who are long term residents of Alice Springs with a local rather than a national or international focus, the changes away from a small town to a larger town is disappointing. Those who are not professionals and who do not value a connection with a wider sense of place and a wider sense of values (who are less interested in the local potential offered by globalisation and computer networking) mourn the change from a sense where local place and local culture is all encompassing. 'The Centre' continues to have a distinct cultural meaning not withstanding the changes. For those who are long-term residents with professional and business interests that expand a local sense of place, the potential for change is more exciting. These different perceptions need to be borne in mind when planning for development interventions.

Another challenge is for many citizens to recognise their shared history of colonisation and migration and sense of place. This can be summed up by quoting an informant, Olive Veverbrants, who stresses that the intersection of cultures needs to be acknowledged more openly and embraced as valuable and relevant to people in their current lives:

> The descendants of the First Australians, explorers, pastoralists, the miners, Chinese market gardeners, Afghan cameleers, telegraph operators, police officers and the first magistrate are here in Alice Springs. We are tied by kinship and the same is true of many families, if people would only recognize this. Aboriginal people do recognize kin very widely. I am one of the people in town who can trace her family tree to include many cultures and many classes. In 1892 my Chinese grandfather lived in Alice. Hong Street is named after him. He worked with my grandmother, an industrious Western Arranda woman who spent 10 years at school in China learning to speak and write Cantonese. Perhaps she is the only Aboriginal woman to have that experience, I wonder... In later life she married an Englishman. Her children, my parents believed in the value of education. All my sisters and I were sent to a Catholic boarding school in Alice from the remote mining fields where we lived. We received a book each year at Christmas. All my family is educated, I have a daughter who is an opera singer, not famous but nevertheless sings in the Sydney Opera House. I have visited Canada and Scandinavia where I have family, England, Belgium, Portugal and I have learned about many ways of life. All these cultures have enriched my life. Barriers do not help human beings.

The story of Alice Springs past, present and future is one based on the contributions of many people from many places and as such is a meeting place of many cultures. For some, like the members of the Stolen Generation,[22] coming to grips with culture is a daily issue. As is achieving a sense of self-determination, rather than determination by 'the other'. Aboriginal Australians, who are small in number and density in other parts of Australia, have a strong cultural presence in this remote town as artists and as interpreters of the landscape in terms of their dreaming stories, but have minimal socio-economic opportunities. The Arrernte people are the traditional owners of Alice Springs, who are no longer in

[22] National Stolen Generation Workshop (1996) Alice Springs, NT *The stolen generations*. Darwin, NT: The Stolen generations Litigation Unit of the North Australian Aboriginal Legal Aid Service.
National Inquiry into the separation of Torres Strait Islander children from their families (Australia) 1997. *Bringing them home* (Commissioner Wilson, R.) Human Rights and Equal Opportunity Commission.

the majority, because other Aboriginal groups have moved into the area and compete for resources as have Australians of non-Aboriginal descent, migrants from a range of European, American and Asian destinations and International Tourists who are attracted to Central Australia, because of its mystique, landscape and culture. The divisions in the population expressed in terms of the different life chances of people in terms of age, culture, race, access to employment and the level of education and health outcomes. As a result of the past approach of emphasising an infrastructural approach to development, recreational sporting, commercial and cultural facilities are of a high standard.

Gaps in service delivery need to be identified not merely in order to determine how best to respond to the existing policy environment and to formulate new policy suggestions, but to develop opportunities for better civil governance at the grass roots. This means creating opportunities for the non-elected leaders in the community to influence policy decisions and to keep the elected leaders accountable. The overall challenge is to address the priority needs which together with other agencies across government, non-government, the third (or volunteer sector) and business through facilitating strategic links. These sectors need to respond to globalisation and thus need to review their organisational structures (so that they become more open to changes), human and environmental resource base, management styles to take into account the wider systems in which they operate.

In this monograph of a participatory action research approach in a remote town in Australia, I attempt to apply a modified form of the World Health Organisation's Healthy Cities approach, with specific emphasis on power, empowerment and governance to address the needs of people marginalised in terms of conceptual, geographical and cyberspace/time. Social systems are described in terms of culture (the way people think and act), structure (institutions and interactions) and power (based on means) (as per Ackoff and Pourdenad 2000:216) and policy (based on decisions as to who gets what, when, why, how, why and in whose opinion). A CSP to development, involves thinking holistically and in terms of the links across social, political, economic and environmental factors. CSP employs multiple methods to work across rather than within boundaries.

Thinking is bound to disciplinary bases and socio-political environments and these bases are used for defining competencies for practice. Compartmentalisation and the isolation of social problems lead to simplistic approaches to problem solving. Associated with compartmentalisation is an unwillingness to share in paradigm dialogue that often occurs when political agendas are at stake. This book builds on previous arguments (McIntyre-Mills 2000), through a specific case study. The axial content themes of access to geographical, social and cyberspace and the implications of exclusion are a central thesis with specific reference to age, gender, cultural factors and levels of ability. Technocratic solutions, power plays across and within the social and natural sciences and the limitation of social responsibility were challenges faced in this learning experience, as was the rationalist social policy. In this action research thinking tools for systemic development were applied. In Australia like elsewhere the welfare state that was paradoxically set up as a contract between capitalist markets and the state is in the process of being dismantled. The gaps between rich and poor are growing wider. The governance systems set up to administer the new order of global markets have been supported by a new grand narrative: 'the market rules because it must and it is the only solution'. This zealotry flourished partly because postmodernism (when taken to extremes) creates cynicism and the belief that there is no truth. This has dire results for social and environmental justice. Similarly the zealots believe that they have the ultimate truth and use competitive language and governance to impose their meaning on the world. The challenge however is to realise that neither the zealots nor the cynics have all the answers. The zealots use top down decision-making and competitive language and force to back up their decisions. Power and knowledge (as per Foucault and Gordon 1980) are indeed linked. The cynics have abrogated responsibility because they have lost a sense of direction and thus cannot contribute to designing or problem-solving.

The theme tools for ethical thinking and caring to redress the impact of the market economy (commodification of body and mind) is central to the discussion. The focus on participatory design for the future is topical because thinkers and practitioners such as Banathy have been working in the area of intelligent design of systems for a long time and has recently launched the AGORA project (Banathy 2001,[23] Banathy, Jenlick and Walton 2001[24]) that uses the Internet for participation. The approach is problematic, only insofar as it creates space mainly for those who are able and articulate enough to take part in the project on the Internet. Intelligent design also needs to include a diverse group to co-create meaning and so considerable work needs to be done to enable the

[23] Self-guided evolution of society: The challenge: The self guided evolution of ISSS 45th International Conference International Society for the Systems Sciences. July 8th–13th 2001.

[24] The Agora project: Self guided social and societal evolution: the new agoras of the 21st century 45th International Conference International Society for the Systems Sciences. July 8th–13th 2001.

most marginalised to have access to conceptual, geographical and cyber-space. Diversity remains an important jump lead of creativity. Training through networked computer hubs in neighbourhood houses, schools, libraries and youth spaces, for example, where equipment is sponsored by the public private and volunteer sectors could be a starting point to redress this problem and to make participatory decision making via the Internet more accessible. Ultimately the power of the social agent who interprets the input remains in a privileged position. This monograph is not meant to be the last word, however, only a means to encourage systemic outcomes. Silenced voices throughout history have had the potential to contribute to social evolution. Gender, age, culture, race and class have been and continue to be used as a basis for discrimination. If the grand narratives exclude these voices historical evolution will continue to be slanted towards economic rationalism. This grand narrative is based on the belief that markets are natural systems and that it is rational to allow the natural systems to prevail. In fact the market is a constructed choice made by human beings and it is in the short-term interest of some, but not in the long-term interests of any given the non of the environment.

The fact that people (particularly young people and Indigenous young people) perceive that they cannot control their life chances or that 'a real job' is unlikely to be obtainable (because Community Development Employment Programs or social security is currently the norm) and that they have too many external constraints, results in directing a sense of control over their body (hence the emphasis on drugs and alcohol to help them cope and the ultimate act of despair, namely suicide). The programs for prevention to address these indicators of social marginalisation and despair need to redress real and perceived social marginalisation through integrated development across sectors and in so doing build social capital and environmental capital across all interest groups. Development concepts need to be defined in culturally acceptable ways if a sense of well-being (associated with self-determination) is to be achieved. By thinking systemically and using a range of tools to enable us to hold in mind more than one ideology, discipline or framework at the same time, it is likely that we will move closer to nurturing the social and environmental capital required for current and future generations. The implications of multiple factors are analysed and policy suggestions are made. Potential ideas to address social and environmental justice need to take into account multiple variables from many sources.

This requires open communication and diversity management to include stakeholders that represent the range of diverse interest groups so that multisemic solutions can be achieved. The World Health Organisation's Healthy Settings Approach (that resonates with some of the Indigenous concerns about land, a sense of place and well-being) is one such approach that could contribute to improving the life chances of marginalised citizens. Strategies need to take into account historical, socio-cultural, political, economic and environmental variables. If appropriate systems solutions can be introduced then the positive flow on effects will be multiplied. If we persist in fine-tuning systems that produce inequality and ill health then we will make the situation much worse. Essentially this is a case study of a community system and the ways in which actors have tried to shape the direction of the system.

## 1.4 *A Mandala to Heal Divided Thinking: Narratives for Understanding Culture and the Tension Between Totalising and Critical Approaches*

Indigenous people the world over use the landscape and the seasons as reminders or their shared history and their own personal connections with their place. Oral culture needs to be recorded somewhere and the landscape provides a space for shaping and holding their stories/(or Indigenous holons?) that are told, sung and painted in varying degrees of complexity for those who will understand and appreciate them. The landscape has also provided a useful analogy for social, cultural thinkers, historians, ethnographers and travel writers who attempt to understand culture. Travelling from the urban developed centre to the wilderness remains a theme shared by many ethnographers and travel writers. I share a personal journey with my maternal great-great-great-grandfather George Thompson. In his words

> ...I was also desirous of penetrating, should circumstances admit of it, into the countries beyond...; and ...gratifying the ardent desire of exploring unknown regions, and of contributing, however humbly, to the enlargement of geographical science, and to a more exact acquaintance with the character and circumstance of the native.... Such were my objects. (Thompson 1827: 1)

Like early ethnographic travel writers, I too was drawn to cross boundaries and to learn more, but also to earn a living. Like Thompson I need to express humility and to stress my inability to cover all the transdisciplinary (Jackson 2002)[25] issues raised with much expertise. Instead I

[25] Key note address Systems Thinking: managing complexity and change, International Systems Sciences 2002. 46th Annual Meeting, Shanghai. August 2–6th.

try to map the stories and opinions of the people who shared their stories and points of view with me and to attempt (through exploratory conversations with the many participants) to gain a sense of their similar and different points of view. My goal was to ask questions and to suggest a process of transcultural (McIntyre Mills 2000), transdisciplinary and trans sectoral work in an attempt to address complex issues.

Culture is a map of rules for living and the extent to which it is seen as static and unchangeable varies. For some, culture is a fortressed island. For others it is connected or interconnected with other maps through travelling physically, mentally and emotionally to other places. Power and choice come into the creation of dominant maps or narratives. It is one thing to choose to revise a map or a structure, because one is a privileged traveller, it is another to have a map torn up by a more powerful group of people who rewrite the maps or impose their stories of how things were or should be. 'So cultures can be seen as isolated, connected by choice or privilege or merged by force. They can also be seen as sharing a space as frontier, fringe or finger populations', according to Joe Watkins (in conversation with a group of Indigenous and non-Indigenous participants at the Sharing the Space Conference at Flinders University July 2002).

Creating spaces for shared maps or structures and different structures of meaning requires dialogue based on respect for diversity (in the sense used by Romm and Flood 1996, based on the work of Habermas) and a sense that power is shared in the process. This is the generative dialogue of Frieire (1982) that is as vital for two-way education and participatory action research. It is what Banathy (1996) in the management space calls participatory design. It is a delicate process of creating and finding lines of shared meaning or interpolations that follow the lines of topography. It is about finding parts of the map with similar topography and creating lines of communication for shared understanding. It is about recognising that some parts of the map need to remain isolates spaces, whilst people make decisions about the value of communicating. The power differences can mean that people perceive that dialogue is dangerous. Isolation, however is problematic in the long term (Cox 1995).

Following Churchman (1971) I believe that knowledge is purposive and that objectivity is fragile because it always requires judgement and judgement entails a value-laden decision. Making judgements requires constant self-reflection. Recursiveness in thinking shapes the world and the world shapes thinking, in the sense used by Giddens. But in the final analysis the mind in this millennium may be able to shape the world, probably in unanticipated ways that do not address complexity (Bananthy 2000). We have created pollution and the potential of a nuclear holocaust by means of science. By means of social science we have created subjects that wriggle beneath the thumbs of experts. We are part of our subject matter and we can shape it through thinking and dialogue and action. How we shape it is a matter of extreme concern and it is a matter of morality more than pure ingenuity. Unfortunately technical knowledge outstrips moral astuteness.

We as critical systems thinkers and practitioners[12] are mindful of the contributions made by many theorists and many disciplines. Critical systemic praxis is based on the idea that we need to work across disciplines and across sectors to solve problems. This is another approach to thinking and practising. Instead of working in isolation or within paradigms, we can work with frameworks.

The aim is to promote an understanding that how and why some issues are defined as social problems by some interest groups and not others is vital for development that is not merely efficient and effective in terms of a discipline or a sector, but is systemically effective and systemically efficient. Understanding socio-cultural frameworks associated with different life chances is a starting point for integrated development. Values and assumptions shape the way in which social problems are defined. An analysis of these could highlight similarities and overlaps and thus ensure better communication for problem-solving and planning.

Some people think there is only one absolute truth. Some think that truth is based on shared understanding that is constructed through debate and some cynics think there is no truth, they are extreme postmodernists. The way we think shapes the way we see social issues or the way we construct them. This in turn impacts on the way we address them in our management of issues and the way we develop policy. Critical praxis entails respectful communication, co-creation of meaning, generative dialogue and listening to the other. We consider assumptions, values and emotions. We compare (how are things the same) and contrast (how are things different). We are open to other ideas and our map of reality is wide enough to accommodate different ideas and competing ideas. We ask what is the nature of reality from the point of view of this individual stakeholder or interest group? It can be very important to understand this when you are managing or developing social policy of who, gets what when why and to what effect. You see each person/group has a perspective of reality (an insider perspective), but at the same time outsiders have an etic perspective. Our challenge is to understand both insider and outsider views and to try to establish common ground. Critical thinking looks at issues from a number of angles: it considers the social, the political and economic factors, it considers the way the problem is defined and in whose opinion. Critical thinking requires ethical considerations and making decisions. It is one thing to represent diverse opinions, but we also need to be able to use critical thinking to come to policy decisions. Critical thinking is about the pursuit of truth and enlightenment in a bid for greater social and environmental justice for a sustainable future. It also requires working the hyphen across self and other and the environment. Thus it can be used to build a sense of the value of relationships. The expert is less vain and more humble and interested in what other people have to say. The key point is that boundaries are reworked and reconsidered.

Boundaries between actors can be addressed by considering their frames of meaning and finding ways to establish some common denominators through iterative dialogue. Dialogue helps to establish links across knowledge areas. The characteristics of critical systemic praxis (drawn from Jackson 1991) are complementarism, co-creation, emancipation of thinking, critical reflection and 'sweeping in' (as per Churchman 1971,1979,1982 who follows Singer), commitment to the essence of the enlightenment belief in the capacity of human beings to think rationally about their world (see Valero-Silva 2000). The methodology of critical systems thinking or heuristics, as Ulrich[26] (2001) uses the term in his article 'the quest for competence in Systemic Research and Practice' is about 'sweeping in' ever more aspects of the research process. In this research I sweep the relationships across embodied social actors and their environment. Emotions are part of the sweep expressed in prose, poetry and art. Values and assumptions are also part of the research process and are opening acknowledged. This is vital for honest research, in the sense of holistic representation.

The logic of unidimensional social science instruments is insufficient to represent the whole story. It needs to incorporate multiple meaning systems. The challenge is how do we work with multiple meanings and encompass more than one way of seeing the world. We need to unfold assumptions and values when we talk about policy. This is useful for developing theoretical literacy and enriching our thinking through contemplation. How do we shift from seeing the world through

[26] Critical thinking per se is as an approach that has been associated with the Frankfurt School that disbanded in Germany during the Second World War and reformed at the University of Columbia in New York, but today it is much more that just the legacy of this group that stressed that the focus of thinking needed to be both on human thinking and on the structures of society that shaped their behavior. Criticisms of the Frankfurt School have stressed that they focused on criticism and not on making practical suggestions in the real world. Others have said: 'well criticism and debunking is practical and can have a real impact on life'.

one set of lenses or based on one set of assumptions? How do we achieve reflection through arranging multiple lenses into a stained glass window on the world?

Extending Bateson's (1972) ecology of mind can enable us to operate within one type of knowledge narrative, compare narratives or co-create narratives for contextual problem solving. Concepts such as 'commodity, work, environment, sense of place and identity' are interpreted very differently by those with different life experiences. The different cultural values provide ontological insights that can be of value in developing co-created futures that do not veer towards extremes on the socio-economic continuum of possibilities. Many factors need to be considered and praxis tools can help us to hold in mind multiple variables and attempt to manage the diversity through critical systemic praxis. The design process can be enhanced when tools are used in culturally sensitive ways to add heartfelt conversational depth (Banathy 1996) and creativity to brainstorming, focus groups, nominal groups and the Delphi Technique (Flood and Romm 1996, McIntyre 1996). Improving our ability to co-create design is vital if the quality of social policy is to be improved. The key pitfalls when thinking about, doing and recommending policy decisions in the area of development is to think that we are accountable to only one set of values, or one interest group. If we can remember that we are ultimately accountable to one planet and one systemically linked ecology (of which we are part) then it becomes less an issue about control and more about finding ways to work with one another in terms of all our interests, that are ultimately linked (Zimmerman 1994, White 2001).

Co-creation makes it possible to draw out the liberative potential and recognise the 'bad news' of adopting a univocal notion (Flood and Romm 1996). For instance: modernist approaches assume that asserting 'the right solutions' could control causes and effects and achieve a desirable order, but was not mindful of the harm caused by top down, one eyed views. At their worst, postmodernist approaches to development assume that recognising differences and diversity undermines the possibility of rational truth and also the possibility of justice. Chaos theory (as per Prigogine 1997) recognises that life is dynamic. Unlike deep ecology that is about the primacy of nature and can be used as yet another master narrative, ecological humanism (McIntyre-Mills 2000 and which this monograph develops) is about facilitating and co-creating as an ongoing process that respects the web of life. The book develops a systemic approach to public policy issues.

The orientation is not so much a destination as a staging post that provides tools for further mapping. It has been reached via (i) a reading of: primary and secondary texts in sociology, women's studies, social anthropology, cultural studies, via community development thinkers and practitioners, policy work on citizenship, via a range of systems theorists and the work of Bateson on 'the ecology of mind', (ii) The critical systemic thinking of Romm on paradigm dialogue and co-creation of meaning (who draws on Habermas), (iii) Critical thinking for management and governance (as per Flood, Romm and Jackson); (iv) My own application of critical thinking and caring to problem solving and participatory planning, via the strategic work of community development workers such as Saul Alinsky, (1972) and the development work of Robert Chambers (1983).

Methodologically I am influenced by empowering ethnography and participatory research and in a loose sense by the Popperian notion (1968) that through falsification we are able to assess probability. Testing through rational dialogue means that if an idea can stand up to debate then it has a greater likelihood of being true. Diverse points of view are important for testing and for debate to occur. Humility is required in order to test knowledge, not arrogance. Humility is a competency, as is respect for diversity and for personal knowledge. Diversity is the jump lead for the future. It is required in order to challenge all grand narratives that attempt to silence the creativity that is rooted in being different.

This CSP builds on and extends the tradition of ideographic, holistic case studies[27] undertaken by social anthropologists and sociologists working from functionalist or interpretivist perspectives (and with varying degrees of identification across self, other and the environment).[28] CSP stresses the need to design human services as open and responsive to biodiversity. Multiple variables need to be borne in mind and this requires diversity management, based on the co-creation of shared meanings. An understanding of webs of meaning across the hinge of self, other and the environment is essential for responsive design. A map of knowledge narratives provides a means of illustrating that we can think, learn and practice within single frameworks or discourses or we can engage in dialogue that enables us not merely to compare frameworks in an adversarial manner, but to see the value or potential of 'the other', so that co-created meanings can be enriched to respond creatively to specific challenges.

[27] Systems Research and Behavioural Science vol. 18, number 1 vol. 46.1.

[28] See for instance the work of sociologists Lynd, Whyte and Becker of the Chicago School and the pioneering studies by Malinowski and later by Mead and Geertz. Although these case studies are different in that they did not build into them a concern about praxis.

If we consider information to be derived from both 'either or' thinking and 'both and' thinking then we will be more reflexive in constructing narratives. Co-creation always occurs within the context of social, political and economic factors and power and empowerment for the long term. The sustainability of biodiversity is the ultimate challenge. This is explained in *Diagram: Mandala of knowledge narratives*. This is a meta narrative for the spiritual, religious nature of knowledge, based on assumptions, or in other words, faith. (Churchman 1971: 243). The detail is expanded in the following chapters. Data can be defined as Bits (in terms of computer language, binary oppositions that can be computer read/interpreted as technical information) and as Logons (continuums from wave theory)[29] and as units of energy that resonate as a continuum of life in all living systems, as per Simms (2000). Energy and ripple effects are the focus of systemic analysis. The analogy of energy webs is fine provided they are seen as pulsing in the same way that electrons pulse around the human body or any other body (organic or inorganic) and each of these is in turn linked.

The challenge when collecting/reading data is to be open to variants. The grammar of meaning is contextual and the text is no longer static, it is the wave of data that pulses through networks. Recognising fluidity and dynamism is central. The mandala is a metaphor for the endless process of the dialectic that 'sweeps in' and unfolds issues. It is argued that this is the essence of a systemic approach that strives for a multi-layered and complex truth.

Simplistically, structuralist anthropologists such as Levi-Strauss (1987 as per the Savage Mind) found that in many cultures categories of binary oppositions provide a basic structure for thinking and practice in a society. Human beings think in terms of day/night, light/dark, right/wrong, raw/cooked, civilised/uncivilised, left/right, male/female. But it is also assumed that data can be defined as being both dualistic, that is 'either or' for sorting, categorising and patterning as well as 'both and' for including, synthesising and making patterns or webs of meaning. The human mind is capable of both forms of thinking and both are functional. Left hemisphere thinking has been characterised as traditional and qualitative and right hemisphere thinking as modernist and quantitative (McLuhan and Power 1986).

In fact it is merely a matter of cultural emphasis, because all human beings have the capacity to do both. Binary oppositional thinking if applied to drawing a distinction between self and other and the

[29] See Bradley, R. (2001) Bits and Logons: Information processing and communication in social systems 45th International Conference of the International Society for the Systems Sciences.

environment has some 'liberative potential' (as per Gouldner 1971), in so far as it can lead to the ability to think dialectically and critically in a loosely Popperian sense and to establish whether ideas can stand up to testing. If ideas can stand up to testing they have a higher probability of being accurate. This ability to separate self from others and objects, enabled dualistic thinking to break out of 'the charmed circle' (Popper in Zhu 2000). But on the other hand, the negative potential of dualistic thinking was to objectify the other and to pave the way for commodifying the other when science serves the economy in a narrow profit sense. The other and the object can become means to ends, not ends in themselves. Fortunately human beings also have the ability to think in terms of the links across self and other and the environment. These links have been interpreted as sacred and the spiritual essence and energy of the continuity of life has been the realm of spiritual healers, philosophers and the sages across many cultures.

The connectedness or spirit of life is as well known to Indigenous peoples of many cultures, as it is to the major religions such as Buddhism, Taoism, Hinduism and Christianity, despite the mistranslation of the early writings about the role of people as custodians (servants) of their universe not dominators who could and should master the environment.

The negative potential of emphasising only the energy of continuity is that it does not allow for space to reflect on actions and thought. It can also lead to what Zhu (2000) refers to the dark side of idealist thinking that places mindfulness at the centre and allows it to absorb the universe. The values of the thinker will determine the nature of divine will or rightness. Once again the dominance of one narrative can be seen to be potentially harmful. So the challenge is to harmonise styles of thinking through intersubjective sharing, or co-creation of meaning (Banathy 1991,1996, Romm 1998, McIntyre-Mills 2000, 2001, Zhu 2000) in order to achieve intelligent and participatory designs.

The first question in response to what appears to be a normative statement is why? Why is co-creation better than imposing a decision and is co-creation always better? The answer is that in the long run if we can work the hyphens across self, other and the environment and consider the systemic implications of our praxis we are less likely to make decisions that will lead to conflict or damage to the environment. It is thus based on ecohumanistic values that are pragmatic in so far as there is a realisation that co-operation and living in harmony is essential for caretaking our planet. This approach is different from, but not contradictory to Gu and Zhu's (2000) construct of an integrated objective, subjective and intersubjective reality that requires multiple methods in order to address complexity and a particular emphasis on values.

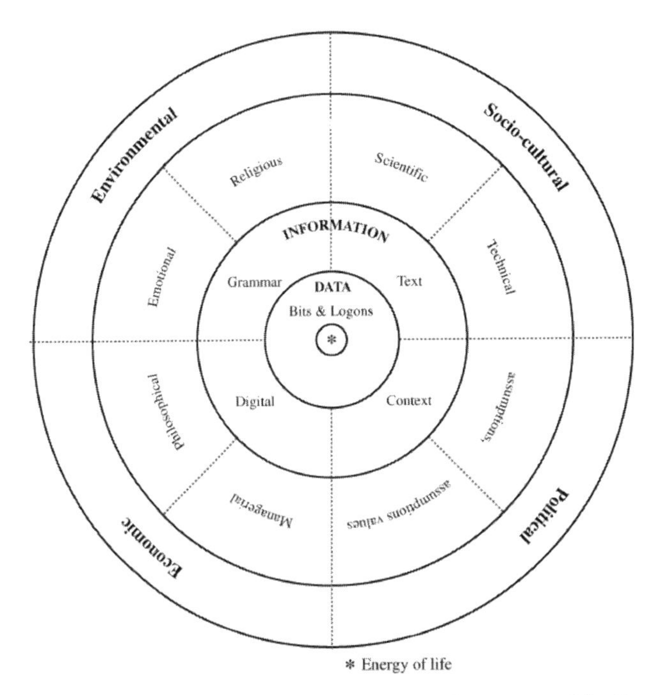


*DIAGRAM 1.6* Mandala of Knowledge Narratives Explains some of the Complexity

Knowledge narratives can be explained in terms of a mandala, symbolic of a meta systemic view of knowledge, rather than as separate categories (Diagram 1.6). A mandala is symbolic of a complex whole. Data can be defined as bits (binary oppositions that can be computer read/interpreted as technical information) and as Logons derived from telephonic communication research. Bradley defines information as the 'minimum uncertainty beneath which the fidelity of a signal cannot be compressed', (2001: 68). He argues in mathematical terms this is a space time constrained hologram.

There is yet another definition of information from the physics of living systems as the basic unit of energy required for life from conception to reproduction. As Simms (2001), a physicist argues, information is required for energy, a determinant of living systems. Thus it is possible to argue that units of energy resonate as a continuum of life in all living systems. If we consider information to be the potential to carry energy to enable work (Simms 2000), then we can accept that information can be derived from both 'either or' thinking and 'both and' thinking and we will be more responsive to diversity and its value in sustaining creativity.

Energy is the basic unit of co-creation may or may not lead to consensus, but the process of communication could construct new narratives and build webs of understanding that can help to re-wire the closed/compartmentalised ways in which we think. Troncale (2001, personal communication, ISSS workshop) stressed the importance of understanding linkages as a means to address closed thinking leading to oppression and bigotry. Energy flows, it gives life and is radiant. Radiance is the difference between meaning that generates hope and goodwill and meanings that are imposed and limit creativity and good will (Churchman 1982). Thus decisions (derived from the Latin, to cut off) options (as pointed out by Churchman 1982: 55) should not be taken lightly. Information could be argued to be the most basic unit of energy to shape life, whether we use religious, mythical or science narratives.

Energy is possibly the basis for all communication in and between the organic and inorganic life (if I understand Simms 2000 correctly). Communication to make responsive adjustments is thus integral to the universe. Open systems sustain energy flows. Closed systems lead to entropy (Flood and Carson 1998).

Reason (2002) makes clear connections across 'justice, sustainability and participation'. This makes systemic sense. Social and environmental justice need to be seen not merely as rights not responsibilities, but as essential for life as we know it. It is from this starting point that I move towards a discussion of the way that welfare policy has changed and to include a critique about the way that the relationships across self, other and the environment have been couched. Other definitions need to be 'swept in', as they have shaped knowledge and practice. The either/or thinking and the space/time definition of information are indeed relevant because information and the way we understand it, shapes our perceptions of the world and our positions in interactions. But we can re-construct this information through critical thinking, in order to steer the direction of change (Habermas 1984). The structure or map/text/language and grammar (in the sense used by French/Algerian postmodernist/structuralist thinkers such as Derrida in Young 2001) needs to be seen as part of a whole system of possibilities. To decide in the Churchian sense 'to cut off' some of these stories from the whole is to limit our understanding of one another, our environment and ourselves (Diagram 1.7).

Co-creation of meaning may or may not lead to consensus, but the process of communication could construct new narratives and built webs of meaning may or may not lead to consensus, but the process of communication could construct new narratives that can help to re-wire the closed/compartmentalised ways in which we think. Troncale (2001, personal communication, ISSS workshop) stressed the importance of

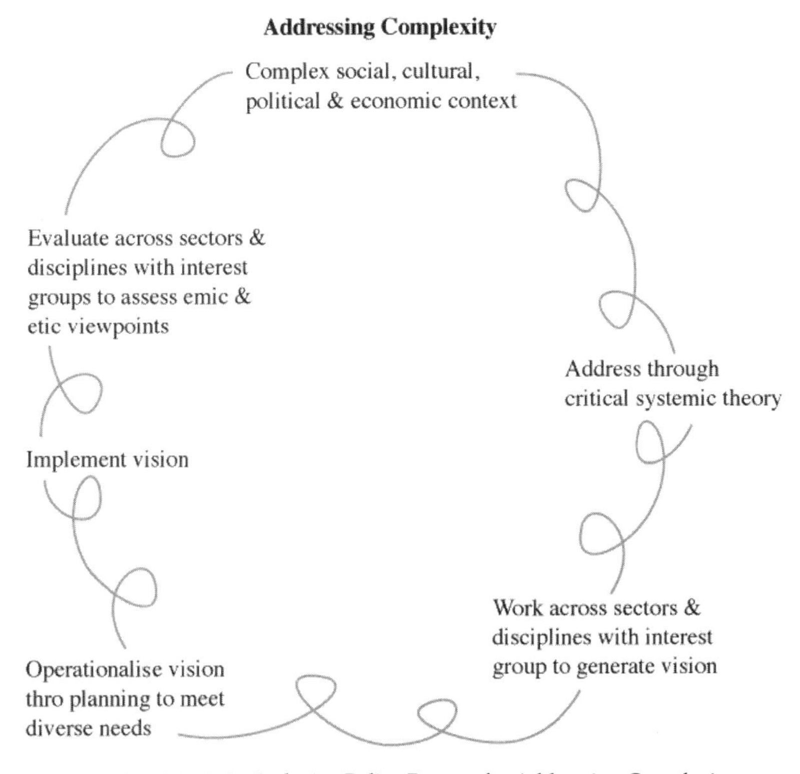


*DIAGRAM 1.7* An Inclusive Policy Process for Addressing Complexity

understanding linkages as a means to address closed thinking leading to oppression and bigotry. Information shapes life and thinking, whether we use religious, mythical or science narratives. Psychological narratives appear to confirm this as well: 'What we know about the world becomes projected upon the world' (Banathy in Romm 2001). In terms of this research I follow Simms definition as it provides a generic basis for understanding across social, political and economic disciplines.

If I can make a contribution it is through demonstrating how tools for ethical thinking and caring can help to show how marginalisation feels to the other and how systemic tools can help us to think about a shared future. I place myself alongside the other. At the time of the research I was not in a powerful position, albeit the position of writer enables me to have a voice and to interpret the research experience. This is about how participation engaged me and affected me as well as a story of the research and an analysis of what could be done to address the issues raised. Because I am a teacher by trade and by nature, it is also an attempt to share what I have learned with others. If it is accepted that health, education and employment are the pillars of citizenship rights in a socially just society, then in a post welfare state (Jamrozik 2001) we need to consider the extent to which citizens can achieve health, education and employment. Further, we need to consider to what extent citizenship continues to be a relevant concept if the nation state abandons the notion of rights and stresses responsibility. By identifying and describing the social, political and economic dynamics and the way in which the status quo is maintained and challenged, I hope that not only will a contribution be made to social justice but that some of the processes and barriers for achieving rights and responsibility in a socially and environmentally just society (the goal) can be analysed.

As such the conceptual, practical and policy contribution of the approach serves to support the social movement for social justice and democracy through improving governance in the public, private and voluntary sectors. The research was undertaken at a time when some locals debated whether or not to fly the Indigenous flag in the town (after the Olympics) where it was flown with national pride. In this context Indigenous people from remote areas were regarded by some with ill-disguised contempt in shops, banks and pavements, where parallel lives become visible.

This monograph addresses epistemology (the question about how we know what we know), the combination of multiple methodologies (qualitative and quantitative) that were used reflexively to consider and triangulate the data narratives (in a constructivist sense). I also strive to use qualitative research data alongside quantitative data from a range of sources to build a composite picture. But most importantly I was a participant observer, living in Alice Springs and experiencing the social dynamics from the inside and with my own narrative of that experience. The monograph addresses ontology (the nature of reality, my reality and the reality of multiple stakeholders) and the way that theoretical perspectives can assist in shaping research and the way research can reshape theory.

This approach is ecological in so far as it considers the importance of holding in mind multiple variables and multiple meanings when addressing social issues. At one level the case study attempts to show how existing structural hierarchies impact on interest groups and how they are interpreted. At another level it is about how systems can be understood in terms of analogies from the natural sciences. Structures shape thinking and thinking shapes structures, but ultimately it is assumed that critical thinking and practice can reconstruct reality.

Through using multiple methods and reflection on these methods, the aim is to surface assumptions and values that can help to sustain the enlightenment agenda. As a critical systems thinker I follow Popper (in so far as the process of knowing can only be based on attempts at falsification through testing. In this instance this implies reflection and the reliance on multiple sources of data). Further (as per Foucault and Gordon 1980) the layers of reality of all the stakeholders need to be comprehended and taken into consideration in building a mandala or meta narrative of knowledge. Stories need to be told irrespective of whether they can meet the criteria for rational communication. Worldviews differ, but without sincere attempts to listen to all narratives then we have little hope of establishing rational meaning (as per Habermas 1984) and we cannot hope to achieve rational democracy. The problem of course is that many successful exercises of co-created meaning are not in the interests of everyone, only some people. This is where the process of surfacing assumptions and values comes into play.

The issue of power (associated with different worldviews on politics, religion, morality and aesthetics), to use Churchman's (1979a,b) categories, is perennial and not one that can be set aside. The might–right issues (Flood and Romm 1996) need to be surfaced, whilst mindful of the strategic implications for the stakeholders. Because critical thinking draws on Habermas (who is concerned with rightness) and Foucault (who is concerned with power and its impact on what is perceived to constitute knowledge) the difficult tightrope is acknowledged.

Romm (2001) raises the point that inductive research that makes wide sweeping conclusions would do well to embrace the Popperian humility (rooted in the belief that science is at best tentative in its conclusions, based on attempts to falsify hypotheses). Scientific testing cannot cover all possibilities. Thus science can only make probable statements. The process of science and the search for truth is thus rooted in ongoing dialogue that tests out ideas. Romm (2001) demonstrates cogently that accountability and responsibility go beyond merely research ethics and social policy recommendations; they go to the very heart of the way a research project is designed or framed. This is why Churchman (1982) following Singer stresses the need to consider not only the technology of research, but also the ethical reasons for framing it in particular ways. Churchman (1982: 134) stresses the value of the dialectical tools for honing in on layers of meaning through 'unfolding' and 'sweeping in' a host of social, cultural, political, economic and environmental considerations. He stresses the value of both personal and public responsibility for issues and the need to strive to address the ideal, not merely the most practical and immediate solution that is merely a short-term quick fix and that does not get to the heart of the issue. It is about avoiding limitations and being open to the contributions that can be gained through a wide range of perspectives. It is also about building a sincere dialogue based on trust, so that co-creation of shared meaning is possible. Imposition of ideas (as truth that is not open to testing) and denial that truth is possible are equally problematic for empowering human beings within their society and within their environments.

Ontologically critical systemic thinking is not convinced of a totalising calculus of life. It does, however believe in the enlightenment agenda. It is influenced by the healthy scepticism of Foucault. Von Bertalannfy (1975) is a founding father and one of the widest systems thinkers, concerned with achieving understanding across the sciences. He has been accused (by soft systems thinkers such as Checkland [1990]) of confusing systemic thinking (using holons or heuristic tools to comprehend complexity) with the belief that there is in fact a calculus that can pertain to all living systems. Totalising theory is problematic, simply because knowledge for progress is a matter of construction and debate. Critical systems thinking and practice differentiates itself from soft systems thinking (a form of engineering) and follows in the footsteps of critical thinkers and enlightenment practitioners who have a healthy sense of humility and a belief that the closest we can get to truth is through dialogue. In this sense critical systems thinking follows both Habermas (who embraces Popper in so far as testing through intersubjective dialogue) and Foucault (who stresses the need for critical dialogue and narrative discourses to establish knowledge). Stafford Beer (1974) although working in the area of cybernetics, also pushed forward systemic approaches to achieving socially just outcomes, another specialist offshoot of systems thinking. Bausch, follows Ludwig Von Bertalanffy's (1975) approach that consensus is possible across maths, physics, biology, psychology and sociology. This totalising calculus enables human beings to see themselves as potentially the authors of their futures. A totalising calculus sets aside the notion of analogies being applied from the natural to the social sciences, in order to enable us to understand the social world with more insight. Instead, a calculus of all life not only models, but is considered to be representative of laws that enable us to engineer our futures with a sense of certainty, rather than to design potential directions. The nature of knowledge becomes less intersubjective and more objective. The challenge is for critical thinkers and practitioners to be part of the process, as citizens this is vital for democracy and social justice in order to ensure that we remain subjects, rather than the objects of planning!

Critical systems thinking and practice differentiates itself from soft systems thinking (a form of engineering) and follows in the footsteps of both critical thinkers and enlightenment practitioners who have a healthy sense of humility and a belief that the closest we can get to truth is through dialogue. In this sense critical systems thinking flows from many strands of thought in order to address in a complementary way (as per Jackson 1991, 2000 who draws on Habermas 1984) the three interests of human action are: (i) instrumental and technical, (ii) strategic and practical, (iii) communicative and emancipatory.

# 2

# *Participatory Design and the Heart of the Process*

## 2.1 *Introduction and Rationale*

Definitions that are owned and that reflect the needs across interest groups can form the basis of conversations and practice that 'have radiance' (Churchman 1979a,b, 1971) and power to transform. Radiance is the difference between meanings that flow from self-confidence and a sense of dignity and identity, to meanings that are imposed. Values are at the heart of the definition of spiritual well-being and what constitutes development. Unless the initial definitions are owned by specific interest groups and shared to develop a co-created sense of citizenship rights and responsibilities then the process is meaningless and without resonance or radiance. Harmonious meaning is the goal of ethical systemic thinking and practice. By unfolding and sweeping in many considerations, it is more likely that ethical outcomes can be achieved. Systems thinking resonates with traditional and advanced forms of complex thinking that holds in mind multiple variables and the relationships across the researcher and the researched.

Poverty is defined not merely in monetary terms, albeit this is a crucial aspect, but in terms of exclusion from participation in the fabric of social life that is vital for well-being and even more importantly from acting as design participants for the future. I am also aware that resistance to the current approaches of professional control (paradoxically expressed as welfare) encourages withdrawal into nationalism.[1] Language, identity

[1] Current and past professional control approaches of Indigenous people comprise for instance: the control of physical movement across the land, the control of the right to be parents (as per the Stolen Generation), the control of alcohol, control of definitions of citizenship until 1962 and at the time of the research, control through property laws and attempts to control language tuition.

and nationalism are linked (as the early identity theorists explained) but power and knowledge (as per Foucault and Gordon 1980) are also linked and for the first time in history we have the ability to design out futures (as per Banathy 1996, Laszlo 2001), our bodies, minds, social and environmental systems. These implications are addressed in relation to culture, life chances and strategies for the future. Personal knowledge in the form of cultural maps is revised to take into account a diversity of circumstances, because culture is a means to respond effectively to one's environment. Cultural maps give meaning and can also provide the means social, political and economic to respond to circumstances. Marginalisation can be addressed through the politics of resistance. Culture can and does play a part in resistance; for instance the culture of economic rationalism is used to justify much behaviour. Language is needed by Indigenous people to engage in the wider world and to maintain their cultural heritage. One language cannot achieve these multiple goals. Learning a widely understood language such as English should not be a prescription to abandon Indigenous language.

In Strehlow's (1958:27) dated, somewhat patronising, but nevertheless relevant words:

> Above all let us permit (sic) native children to keep their own languages—those beautiful and expressive tongues, rich in true Australian imagery, charged with poetry and with love for all that is great, ancient, and eternal in the continent. There is no need to fear that continued knowledge of their own languages will interfere with the learning of English as the common medium of expression for all Australians. In most areas of Australia the natives have been bilingual, probably from time immemorial ...

If however the wider world is seen as a threat, then mother tongue language can be perceived as a means to protect oneself from the dangers of the outside world. When the map provided by a culture fails in terms of providing a voice and a hearing in the wider world, then cognitive dissonance occurs. Depression is the result of marginalisation and a sense of cultural despair at one's own impotence. The return to country movements, native title, land rights and sacred sites are part of an attempt to preserve another knowledge narrative essential for cultural identity as it is currently defined, but openness to diversity for re-invigoration is also necessary. This is important for all meaning maps. It applies as much to dominant narratives as it does to the narratives that are in the process of being silenced by powerful groups. On the one hand, if we are not open to attempts at falsification of our maps through testing, then we lose touch with others and the changing world that our cultural narratives and our interpretations are based on. On the other hand, if ours source of resistance is rooted in our culture then protecting it from colonisation or contamination is vital. It can be a rational decision and pragmatic but we also need to realise that isolation can be disempowering and it is likely to lead to stasis not growth. It is vital that we preserve the links across self, other and our environment if we wish to have a stake in designing out future evolution.

Substance abuse including alcohol, cigarettes and other substances (such as petrol that is inhaled as an intoxicant) and the associated violence is a problem, but the way it is defined to be merely a cause and not a part of a systemic web of cause and effect is in itself problematic. Better systemic designs to address poverty are possible when research is premised on the dignity and value of the narratives of all the participants and if it does not disempower 'the other'. This means that the participants need to resist being objectified by so called objective research. Silencing of the other through drawing the boundaries of research to exclude values and emotions are part of the process of maintaining power and knowledge, just as the refusal to participate is a means of cultural resistance. Critical Systems Praxis can help to re-work boundaries by recognition of the webs of life that need to be respected across self, other and the environment of all living and non living things. Buber (1965) conceptualised the I–thou relationship, based on respect for the self and other. That same respect needs to be extended further to achieve eco-humanism, not merely for social and environmental justice, but for our survival as humans on this planet. This requires finding the answers through co-creating meaning. Diversity is the jump lead for creativity and for problem-solving. Diversity of life is sacred for that reason. Diversity is thus also pragmatic and rational, besides being socially and environmentally just. Respect for Indigenous cultures the world over can be rooted in the 'liberative potential' (as per Gouldner 1971) that they contain for environmental thinking and the reminder that material wealth is not the only way to achieve worth and social status. In ideal terms, Indigenous people perceive that they do not merely respect the environment, they are the environment. They do not gain status by saving commodities for the future, but by sharing all resources (often at the risk of sanction) with significant others.

This should not however blind us to the negative potential of naïve romanticism about traditional systems suited to economies where storage of food commodities was limited by technology and climate and where reciprocity was a means of personal and group survival. Nevertheless reciprocity and maintaining connections remains a tried and tested form of social interaction (as per Malinowski 1922, Mauss 1990,[2]

[2] Mauss, M. (1990) *The Gift*, translated by Halls, W.D., foreword by Mary Douglas. Routledge, London, New York.

Strehlow 1997[3]), necessary for mutual survival. It is mirrored by symbiosis in the animal and plant world. The negative potential of extreme forms of nationalism (often in response to brutality) that can lead to antithesis before achieving synthesis could be avoided if a co-created meaning in the interests of self, other and the environment is the basis of governance. Co-creation requires humility and strength, but most of all it requires trust, based on the belief that it is in the interests of all people to design our futures together. Pooling, linking and allying[4] energy and ideas for the future can best be achieved through life-long learning networks that can help to create learning communities. For this to occur mutual respect and trust has to be generated across self, other and the environment.

## 2.2 *The Principles of PAR*

The research principles underpinning the research was to improve the quality of life of all citizens by working according to the World Health Organisation's Ottawa Charter, namely to address health and development and by working in partnership across sectors and disciplines. Recognising that development is the outcome of shared definitions and shared goals, but it is also the recognition that empowerment requires the right to difference and diversity. Sustainable development is by, with and for people within their environment.[5] Ensuring that development is sustainable and the health of human and environmental resources is a priority. The scope of the research was socio-cultural, economic, political and environmental health and thus the quality of life of citizens.

Data that reflects the insights of different interest groups is discussed through Participatory action research (PAR) and provisions are made to accommodate diversity in development decisions. Action research for development provides a way of redressing the imbalance between research in ivory towers and action in the real world.[6] The spiral process of data gathering through listening and learning whilst

[3] Agencies of social control in Central Australian Aboriginal Societies Occasional Paper 1 1997.

[4] See Moss Kanter, R. (1989) on this concept in the context of management in the private setor. 'Becoming PALS: Pooling, Allying and linking across companies' The academy of Management Executive. **3**(3): 183–193.

[5] Adapted from Chambers (1983) Rural development: Putting the Last First Wiley, New York.

[6] Described so well by Rees (1993) as the challenge to avoid 'doing without thinking' and 'thinking without doing'.

implementing projects involves feedback that is ongoing, iterative and incremental. This is the basis for integrated development strategies outlined in the figure. Project planning and policy design based on PAR[7] strives to give voice to all participants, not only the facilitator(s) of research and development. This ongoing approach to planning and evaluation enables participants to determine for themselves the concepts, the boundaries or parameters of development, the content, the process and the outcomes. The involvement of local citizens in the responsible and active pursuit of a socially and environmentally sustainable environment is the goal. The initiatives of many people internationally to address their local environments has evolved into a social movement.[8]

Communication skills, counselling, advocacy and negotiation, networking and lobbying skills are vital to all stages of PAR. The focus on intelligent participatory design (Banathy 2001, Banathy, Jenlick and Walton 2001). It included a diverse group to co-create meaning to enable the most marginalised to participate. Diversity remains an important jump lead of creativity. Silenced voices throughout history have had the potential to contribute to social evolution. Gender, age, culture, race and class, for example have been and continued to be used as a basis for discrimination.

The enlightenment agenda of critical systems thinking (as per Habermas and Ulrich in Flood and Romm 1996: 180) strives for the following communication tests, that I have modified to suit a transcultural context, with due regard to the following: (i) Foucault's concerns about power and knowledge (and what constitutes knowledge) and (ii) Churchman's (1979) concerns about decision-making, meaning literally in Latin 'to cut off' areas of possibility and (iii) his worry that we should always recognise the enemies of wisdom within and without, namely politics, religion, morality and aesthetics that both motivate and subvert our thinking and practice; unless we remain ever vigilant of the light and

[7] The process uses repetitive, spiral feedback loops for problem-solving. This differs from linear or straight-line thinking in that it is not based on an expert thinking in terms of causes and effects, in order to solve problems. Instead it is based on participation in thinking and implementation and 'learning by doing' by, with and for people.

[8] For instance the healthy cities/environment movement has a website, World Health Organisation (WHO) and supports many participating cities the world over, including Australia. Similarly the United Nations Summit on the Environment led to Agenda 21 that has been translated into a number of local projects internationally and with considerable support from municipalities in Australia.

the shadow in our thinking:

(1) **'Do you understand what is being said?'** (Language, dialect, mode of communication, nuances of meaning [intended and unintended], tone, pitch and body language may influence our understanding)
(2) **'Is the speaker sincere?'** Is the person you are interviewing interested in the topic or are they busy and keen to say anything so that they can get back to an activity they find more relevant? Are they concerned that there could be a hidden agenda? Are you in a position of authority? What is the social, cultural, political and economic context of the conversation?
(3) **'Is the speaker's point acceptable to you?'** The issue of values comes into play. What are the benefits of understanding why the speaker has framed his/her argument in a particular way? This could be very important for a study of attitude to management or attitude to a particular social policy. This is where the rationalist approach is modified to link with the work of Foucault on the need to understand different stories or discourses of reality, in order to understand multisemic reality. Decision-makers often decide what can be called knowledge and truth and what can be called myth, lies or mere idealism. The implications of these powerful decisions marginalise and silence 'the other'.
(4) **'Do you agree with the speaker's use of information and/or experience?'** What is the source of the information? Can it be checked out in more than one way? We need to realise that some of these points require unfolding levels of meaning. The work of Ulrich on critical unfolding and the work of Foucault on narrative dialogue is relevant to this part of the process as is listening to the other (and considering the hyphen of self and other, as per Fine in Denzin and Lincoln 1994). Our thinking and practice is a product of our upbringing, our gender, our age, our culture. And miraculously the same is true for every one else too! And we need to think through the implications of this for policy and management if we wish to work respectfully with people and if we wish to be able to hold multiple variables in mind and if we wish to be mindful of our embodied, socialised thinking. Enlightenment in the sense of spiritual holism can help us to travel further when we genuinely respect another person and their feelings and their dignity, simply because we are human and we worship the sacred webs that make us part of one human race and one planet. Compassion for the other is a process of worshiping the sacred. It is a moral activity as much

as a religious activity. We would do well to understand the Greek concept of hubris and the implications for life, as we know it. The implications of changes to the web of life and the planet are as yet unknown.

(5) **'Are the power dynamics such that some narrators are given more status than others?'**[9] How does power shape the conversation and the decisions made?

## 2.3 *PAR for Planning and Problem Solving*

The implications of recursiveness (derived from Giddens 1991) at each stage of the cycle are detailed in chapter 3. As participant observer, I reflect on citizenship rights and responsibilities and the way in which these concepts are linked with experience and life chances. As such it goes beyond the current debates concerning so-called 'victim mentality' that tend to limit social analysis to an individual or at the micro level. This critical approach considers broader systemic issues. If appropriate systems solutions can be introduced then the positive flow on effects would be multiplied. If we persist in fine tuning systems that produce inequality and ill health then we will make the situation much worse.

The act of undertaking research framed in a questionnaire, designed in terms of a cultural model which has often used research as 'a mask for non action' to quote Chambers (1983), makes undertaking research a challenge amongst people who have been regarded as 'a honey pot' for researchers who come to Central Australia. The answers to many of the problems are believed to be already known by aboriginal people as the following insights demonstrate:

> *Family and community issues*: 'We understand, but what can we do ... people bring in grog to camps and we find it hard. *Socio-cultural, economic and political issues*: "People get sick as a result of the cost of food in remote communities, then they come here." We know what the issues are'. *Power dynamics associated with race and class*: It is about 'black fella white fella', but it is also about the greed of 'black fellas who forget about where they came from and what they suffered'. 'Desert people were ignored, in the past the whites sometimes got lost, starved and died of thirst. Desert people have got something to offer, if people would only listen.'

In this way the research design was iterative and reflexive, in order to attempt to avoid setting myself up as the expert judge. A composite, multilayered picture was built by means of a range of data sources and it is intended as a resource to which others can add—a living document for

[9] An addition to Habermas cited in Flood and Romm's (1996: 180) framework.

local learning and reflection and to give insights/ideas to others elsewhere. Multiple perspectives provide reality checks. I turn the lens of accountability on myself in chapter 3. Even though the brightness of the light is uncomfortable, it is appropriate that an attempt is made to include myself in the picture, with as few edits as possible. The overall approach was one of PAR for planning and problem solving (Diagram 2.1).

PAR is learning, reflecting on the new insights gained and implementing them in a series of feedback spirals. This ongoing approach to planning and evaluation ideally enables participants to determine for themselves the concepts, the boundaries or parameters of development, the content, the process and the outcomes. In reality, structural limits imposed by post welfarism (Jamrozik 2001) will make this goal increasingly difficult, but increasingly important. The role of active citizens, versus passive consumers and the wisdom to know the difference remains vital for democracy. The involvement of local citizens in the responsible and active pursuit of a socially and environmentally sustainable environment is the goal. Reflection is not merely a tool for research but also a tool for living. It enables

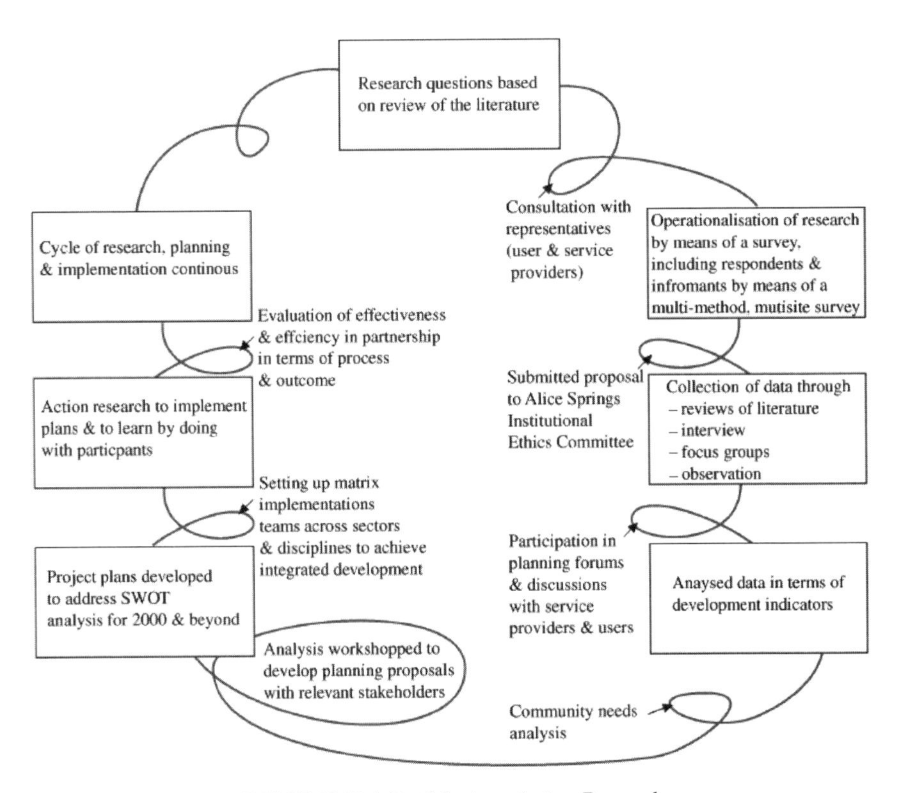


*DIAGRAM 2.1* Participatory Action Research

us to peel away the layers of meaning and the layers of misunderstanding and assist us to cultivate mindfulness. The data itself and the process are the subject of reflexive critique and constructivist reflection (for examples see chapters 3, 4, 5 and 9). It is much harder to be accountable as a researcher when organisations use closed paradigms and communication, not open communication systems. But demonstrating PAR can be of value, even if it leads to small changes, such as the agreement to a memorandum between organisations that have been at loggerheads. In this instance some of the key organisations that contributed as informants, had previously communicated via lawyers, the liquor commission, a magistrate's court and the high court.

Participatory design requires an ability to think critically based on theoretical and methodological literacy of available statistical data. Policy researchers need to be able to apply appropriate and a complementary research method to PAR to establish needs (normative, perceived and expressed through service usage).

## 2.4 *The Aim and Focus*

The aim of the project was to undertake a systemic approach to social planning and policy to: (i) Identify social, cultural, economic, political and environmental needs of the citizens in Alice Springs and regional service users. (ii) Provide a comprehensive analysis of needs. (iii) Assess the extent to which existing services are appropriate to the needs of the range of interest groups that make up the community of Alice Springs. (iv) Identify the factors that impact both positively and negatively on the community. (v) Assess ways to work with the private, public and volunteer sectors to achieve change. (vi) Identify gaps in the efficiency and effectiveness of the delivery of services. (vii) Make recommendations for planning to meet any outstanding needs based on listening and learning from a range of citizen interest groups, in the community of Alice Springs. These include gender groups (men and women), age groups (infants, children, adolescents, young people, middle aged and elderly), overseas and Australian-born ethnic groups with both English and non-English speaking backgrounds, pastoralists, town and town–camp-based Indigenous people and people from remote communities, as well as the needs of tourists and visitors.

The focus is on how different interest groups perceive their community and one another in terms of access to resources and participation. The research addresses indicators for participatory planning purposes[10] and is

[10] The indicators have been adapted to accommodate a constructivist approach from Southern Community Health Research Unit (1989) in *Healthy Cities: Research and practice* by Davies, J. and Kelly, M. Routledge, London.

structured in such a way that social planning indicators indicative of social well being, life chances and social capital are addressed:[11]

The research addressed (i) Social policy, service delivery, organisational management and the need for integration across sectors and disciplines to address the development challenges as linked systems and (ii) The need to establish dialogue between citizens and government and to build in the opinions of citizens to inform policy and planning.

In order to assess the needs of the diverse demographic groups, the multisite, multimethod design research design comprised an integrated approach that aimed at complete coverage and representation.The enquiry comprised:

(i) An interview administered by fieldworkers to the following units of investigation: household members (living in formal and informal structures) and non-household units (hostels, the hospital and alternative accommodation for people with special needs) across Alice Springs. Households and non-household units were sampled in the Metropolitan area of Alice Springs.[12] The purposive sample[13] aimed to ensure that each

[11] With reference to research literature and original research.

[12] There is a total of: 5173 separate houses, 1168 semi detached or town houses, 1077 flats units or apartments and 1127 miscellaneous other dwellings according to the 1996 Census (some of which are Town Camp Dwellings). 371 households were included in a purposive sample to ensure that each of the 7 areas were represented in terms of the number of people who live in each area.

[13]

| Area | Number of residents | Refusals per area | Households sampled per area |
|---|---|---|---|
| • Area of Golf Course Estate and Eagle Court | N = 1890 (7%) | 5 | 26 |
| • Area of Saddadeen, Eastside and Hidden Valley | N = 4927 (18%) | 7 | 68 |
| • Area bounded by Wills Terrace, Railway Line, Todd River, Heavitree Gap | N = 3253 (12%) | 33 | 46 |
| • Area of Larapinta West of Bradshaw Drive | N = 4925 (18%) | 3 | 68 |
| • Area of Gillen and Bradshaw | N = 3341 (12%) | 6 | 45 |
| • Area outside the Gap | N = 3016 (11%) | 1 | 42 |
| • Area bounded by Larapinta Drive, Stuart highway, Lovegrove Drive to Town Boundary | N = 5339 (20%) | 8 | 76 |
| Total | 27092* | | 371 |

* The total is based on ABS census and it should be 26691 or 26754 with refusals/absent at the time of census.

of the 7 areas were represented in terms of the number of people who live in each area.[14]

(ii) Focus group discussions with a broad range of the users of human services were organised through other Human Service Organisations at a total of 18 selected venues. These were representative of age, gender, cultural background and different life chances associated with physical and mental health. The groups comprised advertised public forums and specific discussions with government and non-government human services and with advocacy groups addressing domestic violence, disability, legal issues (for example from land rights to mandatory sentencing, sacred sites, stolen generation issues).

Focus group discussions were held with residents of town camps and with migrants with limited English language skills. Because an English language interview schedule would have been inappropriate for those from different language backgrounds and a formal interview per se would have precluded conversation, so a conversational interview style used prompts. A purposive sample was achieved with the help of professional facilitators through the Migrant Resource Centre and Tangentyere Council (an organisation responsible for the administration of town camps through housing associations). At town camps I was introduced as one 'who would like to talk about things that matter, such as health, transport, recreation, education or any other thing'. Group conversations were followed up with a series of conversations with those who wished to talk or show me specific things (gardens, housing, communal facilities) or introduce me to others with specific needs (for instance: access to resources and services, social justice concerns, employment and health). Networking out from people met at these conversations (snowball sampling) helped to establish issues that they considered important. Initially I was accompanied by

[14] Household members were asked to describe the needs of their members in each household. The numbers in each household were recorded on the cover sheet. The interview schedule for each area was colour-coded and each area had a clearly delineated map for purposive sampling purposes. Because of transience a complete sampling frame was unavailable and thus households were selected on the basis of a map of each area prepared and enlarged by the Department of Lands, Planning and Environment. Five fieldworkers worked according to the map to select houses/units/hostels in each area. Purposive sampling was used to ensure a spread in age, gender and life chances. If research in an area was conducted during the day on one occasion it was ensured that research was also conducted during the evening or weekend to ensure that informants were not skewed in terms of employment or child rearing choices. Monitoring enabled problems such as interview style to be addressed.

a staff member from Tangentyere Council, until relationships with members of the community developed and they acted as facilitators of conversations and directed me to meet specific people and see particular places. I joined in activities where I knew that I could meet people with different needs (for instance, play activities of mothers and children, participant observation in public places, youth centres, schools, libraries, facilitating a painting workshop to develop a wall mural depicting things that matter with young people, joining a youth justice coalition, acting as an advocate on behalf of a town camp that wished to address alcohol and domestic violence issues). In this way conversations could take place alongside activities. Similarly, for example through the Migrant Resource Centre I attended a range of activities for different interest groups and documented detailed descriptions at the end of each day.

(iii) Participation in (a) planning forums for health and education, (b) Participant observation in a range of human service contexts also provided qualitative data for thick description (as per Geertz 1973), (c) interviews conducted with a number of human service managers, (d) Interviews with key informants from the Indigenous community to identify the needs of local people and those from remote communities, (e) The concerns of pastoralists were detailed in a conversation with a consultant (Monica Bradley 1999, personal communication) who was working on a range of social and economic policy concerns. Phone interviews were organised with pastoralists via her network. My conversations with informants centred on the services that they needed to access in town and how services could be improved. These were followed up with conversations at the Cattleman's Association, School of the Air, the hospital/medical services and services to remote communities, such as Waltja.

(iv) Review of available literature, documentary research and Australian Bureau of Statistics census data.

(v) Iterative reflection on key findings involving self and others was achieved by using tools such as dialogical vignettes, painting and poetry.

(vi) Governance project using generative dialogue with Indigenous participants of a housing association.

## 2.5 The Indicators

The life chances of people are examined in terms of both qualitative and quantitative indicators derived and adapted from WHO healthy settings literature. Research data comprised both *quantitative* data (statistical indicators) and *qualitative* data (meaning and perceptual indicators).

Primary data and secondary data were used to build indicators comprising:

(1) A local and regional demographic profile, because the municipal area of Alice serves a wider area.
(2) Social indicators of well being and life chances:
Employment, unemployment, literacy and numeracy, food security, transience and homelessness, incarceration
Socio-economic status (type of dwelling, ownership of dwelling, type of transport, attendance, level of education)
Perceived social involvement in the community
Perceptions of access to public facilities based on access indicators (infrastructure, attitude, information and communication)
Political attitudes to local government and state government
Sense of safety indicators
(3) Physical and mental health status indicators
Mortality and morbidity and indicators of poor coping behaviour (Domestic and community violence, drug and alcohol use, suicide, road accidents).
(4) Environmental Health indicators
Housing, water, sanitation, refuse removal, pest control.

When considered together the indicators provide a basis for analysis and planning. Research was undertaken by working in partnership with a range of community stakeholders. This aimed to ensure that the process was inclusive and stakeholders will be able to contribute. This participatory process provides invaluable insights into the way in which participants perceive their needs.

## 2.6 *Ethical Considerations*

Co-creation and iterative reflection on the design and process included the submission of the design to the Alice Springs Institutional Ethics Committee to assess the co-created research parameters that we had set (as widely as possible to include the perspectives of all interest groups to assemble social, political and economic indicators of need). The process included a review of the research in order to ensure that the following issues were addressed: (i) ownership of the data, (ii) recognition of the research input, (iii) ensuring that the research is culturally sensitive, (iv) due respect for Aboriginal people's requirements in the research design, process and outcomes. The research was assessed to determine whether it could add to new knowledge, whether it could benefit or harm people and whether piloting had occurred and whether the process could develop governance capacity. Reviewers made assessments to ensure that personal and specific questions were neither invasive nor ethically unacceptable. In this research conversational prompts were used to address issues in a less directive manner. Conversations and interviews were used to build up a picture of perceived needs across socio-economic levels, age groups, cultural backgrounds and groups with specific needs. This open-ended approach complemented the data derived from a more traditional questionnaire aimed at establishing a few specific issues.

### *2.6.1 Ecohumanistic Tools for Enhancing the Accountability of Social Policy Research*

Habermas (1974, 1984) argues that the ability to develop rational meaning between human beings is essential to our humanity. He follows the rationalist approach, based on the enlightenment idea of science that is the bases for modernist thinking. The post modernist thinkers have argued that the enlightenment project was arrogant and suppressed some voices and some forms of knowledge. The work of Foucault is very important in this regard too. Dialogue and narrative are placed at the forefront of both so-called critical rationalist and post modernist thinkers. Habermas kept his faith in enlightenment and the ability of human beings to construct rational worlds, while Foucault (1980) stressed that power and knowledge are often linked. Nevertheless human beings have the capacity to deconstruct narratives that are biased. The modernists have been criticised (quite appropriately) for being arrogant towards people with different frameworks about the world by virtue of their gender, culture, class or political position or perceptions on the environment and the nature of life. As practitioners and thinkers we have the chance to become more accountable to all the stakeholders and to the environment of which we are an integral part. That is the challenge. The work of the critical thinking and practicing is still evolving. It is about applying the tools of thinking and practice to improve their interventions as accountable co-researchers and policy-makers.

Power relations are at the center of interactions and Habermas has tried some routes to address power, as has Werner Ulrich and many other practical thinkers like West Churchman (1979), Checkland and Scholes (1990), Stafford Beer (1974), Jackson (1991, 2000), Flood and Romm (1996). They have argued that if you try to do something about social problems you have to look at a number of areas: social, cultural, political, economic and environmental.

But no single theory or theorist can possibly find all the answers. Critical thinking is not about denigrating others (criticising one another for our shortcomings). Analysts concentrate on some aspects of power, but not all areas, particularly power over nature. Habermas appears to deny the value of narratives and myths, unless they are perceived as rational through a process of communication. He sees appropriate communication as based on establishing intersubjective or shared meaning. His validity claims (1984: 123) do not take into account that across different contexts (social, cultural, political and economic) it can become more challenging to create shared meaning. Also Habermas dismisses subjective reality and mythological reality in this work as false consciousness. But the challenge is to accept that the myths of the first and the so-called third world are equally challenging and need to be debunked through critical reflection. What is comprehensible? What is right and in whose opinion? What is the appropriate tone and the appropriate non-verbal communication? Once again it is contextual and thus the need for ongoing dialogue. The closest we can get to truth is through respectful and ongoing dialogue (McIntyre-Mills 2000, 2001) to understand what is important and why. Even the strongest proponent of positivism, Popper, believed that openness to other points of view was vital for progress towards 'truth'. This is not nihilistic and it follows in the footsteps of humanist rationalists who show humility, respect for others and compassion for one another's shortcomings. The determination of what is rationality is also linked with positions of power. What constitutes knowledge is often a decision based on power. But nevertheless to abandon the enlightenment commitment does not serve any purpose. We need to take on the positive potential of modernism and try to avoid the past mistakes, by gleaning lessons from post-modernism. Extreme forms of post-modernism lead to nihilism and can also feed oppression. The same can be said for modernism, if the dark side of totalising control is allowed to take over. The Jungian admonition (in reference to human minds) is always to recognise our shadow, or as Gouldner (1973) said, to always recognise the bad news. Walking the tightrope is possible through co-creation of meaning. I have always been concerned about the way that time is spent breaking down an argument or a practical suggestion without trying to find what Gouldner would call 'the liberative potential' within it and use that to develop some useful changes with in a specific context. The battles raged between those who think capitalism alone is the basis for all problems versus one or more other factors (such as social structures, human thinking, psychology and physical embodiment) that need to be considered without lapsing into one-eyed solutions.

As critical systems thinkers and practitioners we need to be mindful of the contributions made by many theorists and many disciplines. CSP is based on the idea that we need to work across disciplines and across sectors to solve problems. This is another approach to thinking and practising. It is based on the assumption that of the three options(isolationism of paradigms, non-commensurability and commensurability),the latter makes the most sense.

Romm and Flood (1996) argue that the critique of commensurability has been made along the lines that you can't compare apples and oranges. The answer they stress is why not? This is because it depends on what aspect one is comparing: social aspects for instance some people may prefer one to another, they may find some easier to eat or they may find they are allergic to oranges but not apples. The exercise can be extended to discuss apples and oranges in terms of economic trade or perhaps agricultural considerations, for example. The entire critical systemic perspective is built on the argument that we live in a complex world and that we cannot try to impose neat categories without accepting some ambiguity and necessity for contextualisation.

Romm and Flood (1996) discuss the way purists work alone in isolation, those who engage in critical discussion and say that things are incommensurable and those (thank goodness, who are prepared to be open to the ideas of others). Closure is the false consciousness of those who are so arrogant that they believe they have all the answers.

The issue of methodological isolationism, commensurability and incommensurability (as per Flood and Romm 1996) is central to the work of policy-makers and managers who need to be able to work with complex situations. Following Bateson (1972) I have argued elsewhere (McIntyre-Mills 2000) that the following distinctions are important for systemic problem solving: simplistic or rote learning (Level 1) which means thinking and working critically only within one framework. Level 2 learning involves being able to think about or reflect on a framework and Level 3 learning is holistic, interdisciplinary and transcultural, because we rethink taken-for-granted notions of development pertaining to health, education and employment. These are the pillars of citizenship rights in a socially just society. So much professional thinking is bound to disciplinary bases and socio-political environments and these bases are used for defining competencies for practice.

I fear a complete abandonment of trying to test for truth. Through co-creation, instead a kind of unfolding process as per Ulrich (1983) and Romm (2001), could be more effective for sustainable solutions, because if we do abandon truth then (as in the most extreme forms of post-modernism) we can throw out the baby of rational social justice along with the bath water of a healthy scepticism about absolute positions, to use a well worn analogy. What are the policy implications of throwing out the absolutes of the need for mutual respect and the need for social justice? Asking is there such a thing as social justice is the post-modernist downfall. It becomes no better than the most conservative of positions (Flood and Room 1996, McIntyre-Mills 2000).

Habermas does have shortcomings, but if we accept that he thinks that dialogue is vital for rationality, then it is possible to combine that 'liberative potential' (as per Gouldner 1971) with that of Foucault on the need to listen to narratives (as discourses that reveal layers of truth). The work of Ulrich, Churchman and Singer on critical unfolding provides a useful bridge between what Romm (2002) calls foundationalists and non-foundationalist thinking.

Isn't it worthwhile to create webs of shared meaning across divides? That is what co-creation, unfolding and dialogical critique can do for heuristic thinking categories. We need to avoid reifying these categories. This work uses a process of unfolding to see the various constructs of truth in order to get closer to an enlightened understanding and by no means abandoning the search for truth (Diagram 2.2). If we go the route of trying to find strands that link positions that strive for finding common denominators by virtue of our shared humanity and shared planet that we can bind together the rifts between those that (i) think there is no ultimate truth (ii) those who believe that the closest we can get to truth is through dialogue (McIntyre-Mills 2000), and (iii) those who believe in that great rational (and potentially oppressive truth). If all three positions accept that there is indeed a common foundation based on our humanity and our shared planet, then perhaps it is possible to find that ultimate healing synergy or dependent origination idea (as per Buddhist philosophy and many indigenous ways of thinking). At the outset we need to ask ourselves: where we stand when we undertake development?

The diagram summarises our positions on social justice and as such can be used as a tool to enhance governance through extending our understanding.

(i) Tools for surfacing assumptions and values are vital in design conversations within multiple arenas. In essence all tools stimulate the unfolding of ideas (as per Ulrich, Churchman) and the sweeping in (as per Churchman and Singer) of considerations through questioning and thinking in complementary, not merely competive ways. The use of tools that foster the inward and outward gaze are invaluable, because they focus on the quality of the individual mind, the collective process of co-creation of a vision and the process of working together to make it

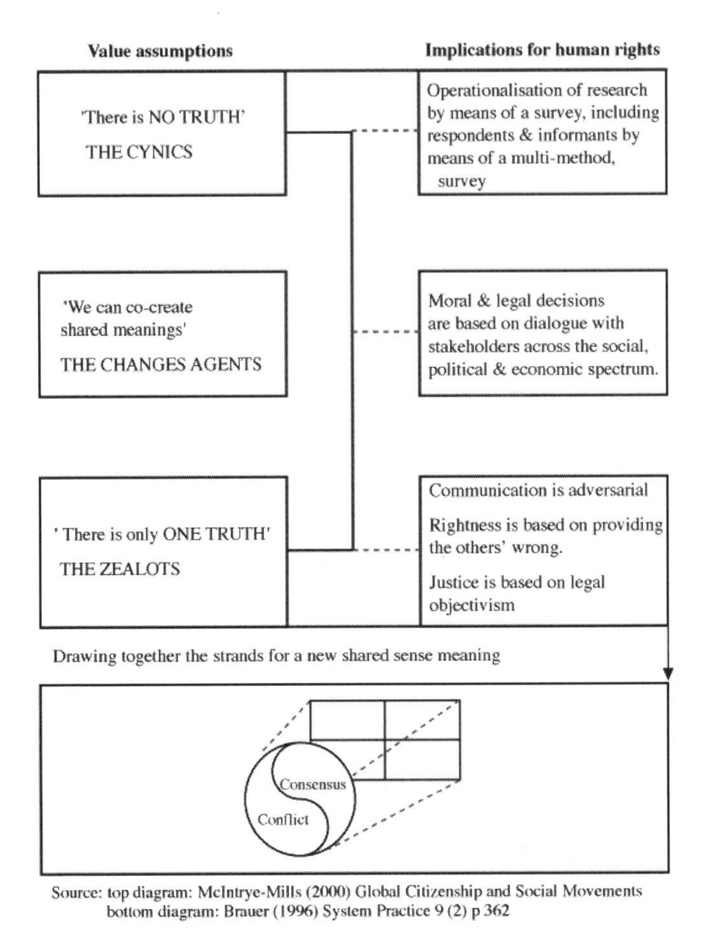


Source: top diagram: McIntrye-Mills (2000) Global Citizenship and Social Movements
bottom diagram: Brauer (1996) System Practice 9 (2) p 362

*DIAGRAM* 2.2 Where do we Locate Ourselves in Terms of Assumptions and Values?

possible. Development can be guided by tools to achieve greater accountability, for instance attempting to implement participatory design, mapping systems of communication and communication breakdown, critical heuristics of unfolding discourses in conversations about vision, history and planning across sectors and disciplines and working out strategies for co-creation and strategies for conversations in arenas where those with limited power are more likely to be heard.

In a general sense as human beings do we believe that there are absolute rights and wrongs, given that we are faced with the absolute of sharing one planet and sharing similar needs irrespective of apparent difference? Development in the past (an in many instances today) can be guided by:

(a) The idea that there is one truth, zealotry based on the fundamentalist or modernist belief that there is one truth and one way of doing things.
(b) The idea that truth is based on co-creation and falsification/ testing. The more humble and post-positivist belief that truth is obtained through testing out ideas with other people and that if they can stand up to testing then they are closer to that elusive multilayered truth, or
(c) No truth, or only constructs, as the most extreme post-modernists have argued. This is problematic, because we can throw the baby of rational human rights and respect for human beings and their dignity, out with the bath water of skepticism.
(d) There is yet another approach, suggested by C.West Churchman (1982), who draws on the work of Singer, namely that epistemology and technical thinking of science per se, is not adequate to meet the challenges posed by development. A greater striving to understand ethics and human meanings and values is essential. This is why we need the techniques of 'unfolding' and 'sweeping in'. These are dialectical tools that hone in on issues and draw in a range of considerations. Churchman's perfect monad or whole system (see Churchman 1971: 41,[15] 1979, 1982) is based on the assumptions that 'all things contain all things'. If we go the route of trying to find strands that link positions and that strive for finding common denominators (by virtue of our shared humanity and shared planet) we can hold in mind the ideal to bind together the rifts. The challenge is to avoid slipping into cynicism and zealotry when undertaking social policy research. This tool can be of some assistance in achieving the goal of systems thinkers, namely to hold in mind multiple variables and not be overshadowed by just one big idea. Light and dark are analogies for wisdom and ignorance throughout the ages in religious thinking. To quote Churchman (1979: 212):

> Suppose we consider, not the rationality of holism, but its spirituality. Holism traditionally says that a collection of beings may have a collective property that cannot be inferred from the properties of its members.

[15] Drawn from his reading of the Greek Philosopher, Anaxagorus, who assumed that 'all things contain all things'.

The approach used in this book is based on the belief that meanings must be co-created in order to address mutual long-term interests of everyone within the ecosystem. If all three positions can accept that there is indeed a common foundation based on our humanity and our shared planet, then perhaps it is possible to re-construct that ultimate healing synergy of the 'dependent origination idea' (Koizumi Tetsunori 2001) who refers to Buddhist philosophy and many indigenous ways of thinking. A range of complex issues needs to be addressed if social and economic sustainability is to be achieved.

The methodologies need to be in line with the culture of the community. Poetry and art are transferable tools of aesthetics that enable us to focus our thinking at a tacit level, that helps us to understand phenomena more explicitly when we unpack the meaning. It enabled me to explore reflexively the extent to which I was considering all potential factors and to explore emotion. Poetry allows for emotion that can be analysed by the researcher. Poems act as portals. They draw together the themes. Poetry compresses, distils and represents the webs of connection, the paradoxes, the duality and the transience of identity and role over time and space. These are useful, reflexive and analytical tools that enable the researcher to reflect on 'the hyphens' across 'self', 'other' (Fine in Denzin 1994: 70) and the environment. They can also help to improve a sense of where the perception of self and other(s) and their relation to the environment begins and end.

Poetry seems to be useful when analysing key themes because it allows for associations to be made as elisions and slippage across and between categories. It can be used to show hybridity and can compress without merely summarising. New configurations and new meanings emerge from the elisions. Poems like conceptual diagrams provide awareness of linkages. Traditionally the personal notebook of the ethnographer was considered a useful tool where the researchers own perceptions could be located. But the reality is that as we are 'part of our own subject matter' (as per Habermas) and the location of emotions and perceptions need to be confronted and reflected upon in the main body of the text (and not in parenthesis), in order to ensure that narratives are co-created and not imposed. These self-reflective tools are an aid to prevent slippage into cynicism or zealotry.

Painting is meaningful as is story telling about a sense of place (see Roux 1991 in Denzin and Lincoln 1994: 327 on the relevance of other methods). This is important where literacy, language and cultural concepts can be barriers. Expressing oneself in the way one is most comfortable is vital for meaningful communication. For example in line with the PAR approach I include a conceptual representation of feelings and perceptions at that time co-created with a range of Indigenous stakeholders and non-Indigenous lobbyists. Introducing alcohol controls had been requested by Indigenous lobbyists (versus the private sector) at that time, whilst mandatory sentencing of minors, along with issues of self-determination, health and land were other strong social justice concerns raised during the research experience.

The Gap pictures in Figure 2.1 is a physical feature of the MacDonnell Ranges, but it is also a means to document oral stories about social, political, economic and conceptual gaps. The gap is said to be paradoxically bridged and created by the cocktail of culture, commerce, religion and politics and numbed by alcohol, consumed by Indigenous and non-Indigenous alike. It has been a thirsty place of broken spirits (Hauritz 2000) where spaces and places were maintained through exclusivity and law. The shackles of the past and the impact of colonisation and social control are balanced by the hope of reconciliation, transformation and the re-birth of the caterpillar as yeperenye (the butterfly) in the future;

*FIGURE 2.1* A Composite from Discussions with a Range of Stakeholders: Caterpillar Dreaming of the MacDonnell Ranges and a Systemic Analysis

and the hope that young people will feel a sense of freedom to participate in their futures. The painting was co-constructed in a number of contexts with young people and conceptualised in part by the lobbyists concerned with youth justice. As a conceptual representation it draws together the issues that mattered at that time to the participants. Painting is a daily activity for many in Alice, particularly members of the Indigenous population. Paintings were used as a means of educating people about public health and other issues and to convey policy messages. Arrernte people describe the landscape in terms of ancestral maps. Their links across self, other and the landscape are confirmed through religion and ritual. They do not begin with dualisms in their ontology. From this perspective they do not conserve the land they are the landscape. Their birth mother identifies a particular item of fauna and flora that she sees when the foetus moves within her as an ancestor.[16] This identification is not without political intent, according to Stehlow[17] and can be a means of securing resources and counterbalancing dominance of some social narratives.

(iii) Extending Bateson's (1972) ecology of mind can enable us to operate within one type of knowledge narrative, compare narratives or co-create narratives for contextual problem-solving. Co-creation always occurs within the context of social, political and economic factors and power and empowerment for the long term. Concepts such as 'commodity, work, land, sense of place and identity' are interpreted very differently by those with different life experiences. Sharing in hot desert climates is a necessity, not merely as a means to gain acceptance. Where commodities are scarce and where abundance is shared, webs of reciprocity (as per Mauss 1990) are built. Building strong social links meant survival in desert communities. Culturally giving (and not merely receiving or storing) can be a means to build power. The different cultural values provide ontological insights that can be of value in developing co-created futures that do not veer towards extremes on the socio-economic continuum of possibilities.

(iv) Tools for working with the community, such as De Bono's tools continue to be useful for working with members of the non-Indigenous community. For example his thinking hats can be used in narrative to

[16] Strehlow, T. (1978) *Central Australian religion: Personal monototemism in a polytotemic community.* Australian Association for the study of religions.
Strehlow, T. (1997) Occasional Paper Strehlow Research Centre.

[17] Strehlow Centre. (1997) Occasional Paper. Page 119.

highlight the role feelings and emotions play in cognitive decision-making if it is possible to achieve generative dialogue in a context of shared understanding and political will. They are of particular relevance for problem-solving with a group of people who share similar goals. The analogy encompasses the elements of information and feelings (both negative and positive). The thinking hat method can be usefully applied to strategic planning and participatory design. Each hat tends to shape the focus of thinking. By wearing more than one hat in a problem-solving exercise thinking can become more creative. More potential ideas for problem-solving and decision-making can be generated. If participants can realise that their thinking is influenced by the sort of hat they are wearing and the sort of hat other stakeholders are wearing the exercises can serve as a useful introduction to meta theory or thinking about thinking. It can help people to leap beyond 'mind traps' in the sense used by Vickers. The exercise is quite useless if used in a naïve way to shift thinking that is motivated by political strategy.

(v) Similarly the sociological lenses tool is useful to help support dialogue that is aimed at problem-solving. What are the assumptions and values underpinning our approach? How can we avoid losing part of the picture? The answer lies in putting the lenses together in a stain glass window and reflecting on the whole (Diagram 2.3). (see McIntyre 1996 and McIntyre-Mills 2000.)

Indigenous participants tended to think in terms of wholes and so there was little need to prompt the conversations using this tool. It was however useful to mention these lenses to enable non-Indigenous participants who operated in terms of parts, not wholes within specialised organisational bureacracies to think beyond a partial picture. It was also a useful prompt in focus groups to enable people to consider an issue in terms of many variables and in terms of more than one set of assumptions.

(vi) Dialogical vignettes (Scott Hoy 2001) and poetry are tools that can be used to explore positioning as a researcher and as a responsible citizen who has feelings and values. Vignettes enabled me to explore reflexively the extent to which I was considering all potential factors and to reveal the layers of identity as researcher and as participant. Vignettes of dialogue reveal the layers of identity as a researcher and poetry allows for emotion that can be analysed by the researcher. Vignettes and poetry also compress, distill and represent the webs of connection, the paradoxes, the duality and the transience of identity, role and space. These are useful reflective and analytical tools that enable the researcher to reflect on the hyphens across self, other

**Assumptions filter the way we define social problems**

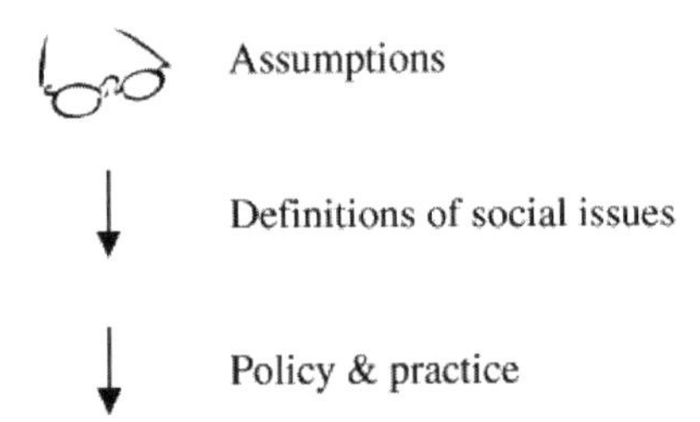


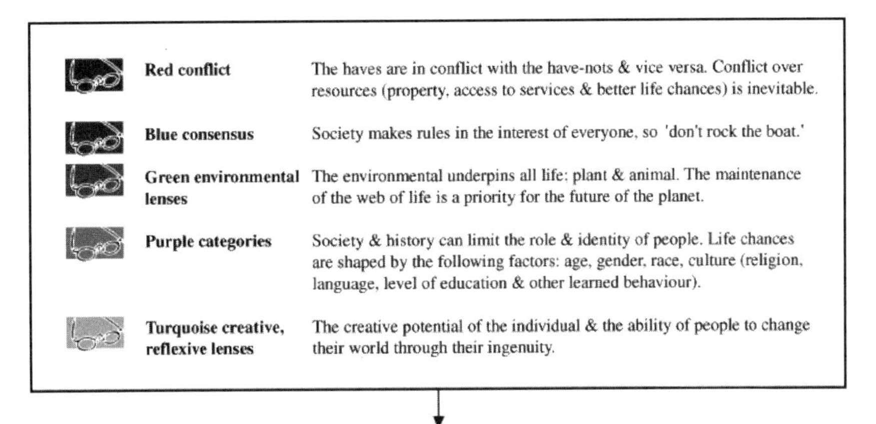


**By arranging these into a window, the wider picture can be seen**

*DIAGRAM 2.3* Sociological Lenses

(see Fine in Denzin and Lincoln 1994) and the environment. They can also help to improve a sense of where the perceptions of self and other(s) begin and end. Last and by no means least they inspire sociological imagination and acknowledge the link between biography and history (C.Wright Mills 1975).

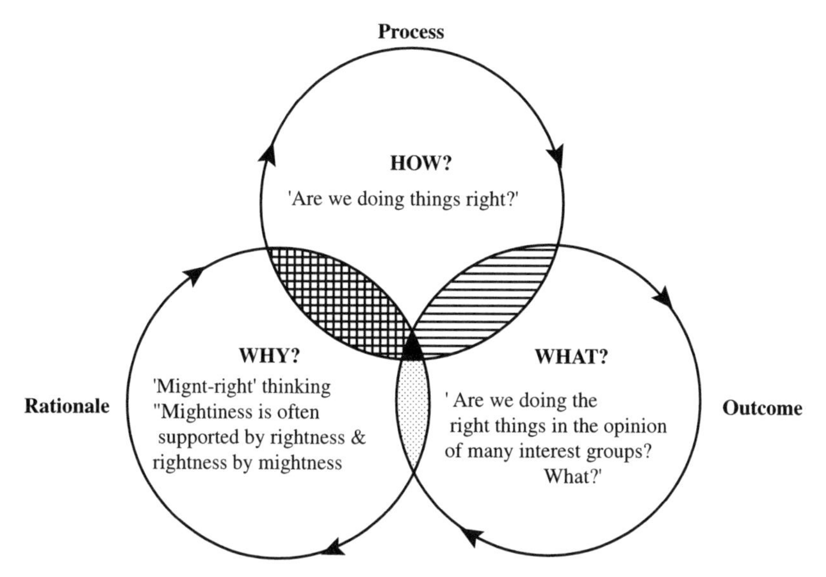


Source: adapted from Flood & Romm 1996 p Xiii

What are the implimentations of the questions for professional decision making?

a. It guides thinking
b. It enhances inclusiveness in decisions
c. It helps to inspire reflection on decision

*DIAGRAM 2.4* Managing Diversity

(vii) Triple Loop Learning (Flood and Romm 1996) ask the question: are we doing things right (process) and are we doing the right things (policy content) and are we doing the things we are doing for the right reasons? (in so doing they too draw on the work of Ackoff, Argyris and Schon). Who decides what, why and how is also important for Indigenous decision-making, because consanguineal (blood) and affinal (social) ties are important as are power and personality factors (Diagram 2.4).

Managing diversity through awareness of different areas of knowledge is a vital tool in reassessing governance. In the Diagrams 2.5 and 2.6 approaches are contrasted for heuristic reasons and are purposefully simplified.

We need to work across multiple arenas and not to be limited in one area of intervention. So often detailed research is undertaken in order to solve problems defined in narrow terms to fit a managerial compartment, rather than in response to a multifaceted problem. The possibilities for problem-solving are limited by the fact that people work in one sector and

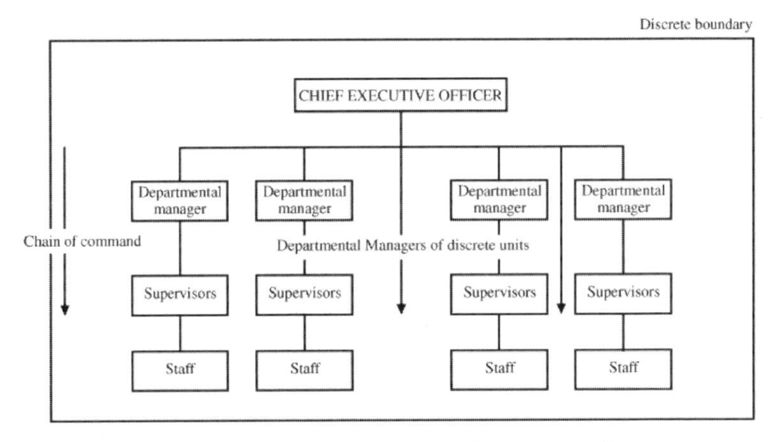


*DIAGRAM 2.5* Closed Communication and Segmented Management

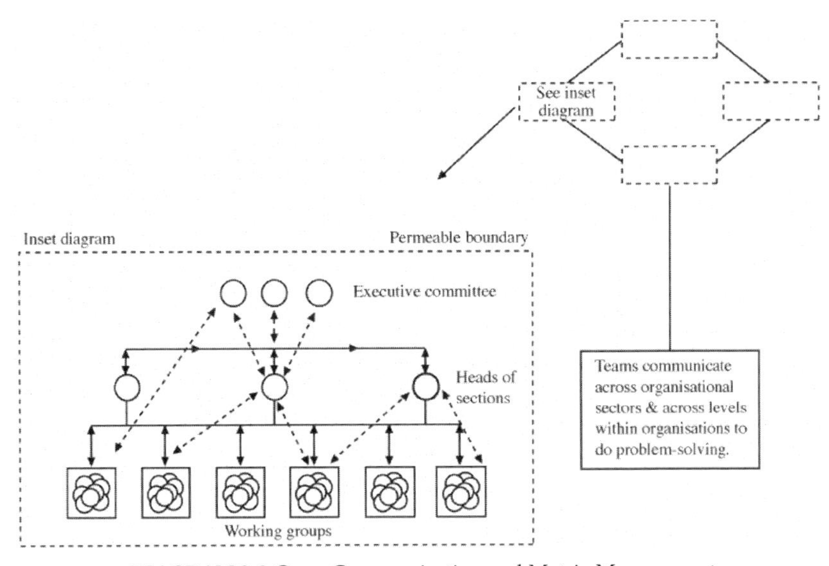


*DIAGRAM 2.6* Open Communication and Matrix Management

in isolation. Detailed work the size of a postage stamp can only be of benefit if the wider implications are considered. Perfecting a wrong system rather than reworking the system is of little benefit.

It is pointless to refine a system that is not working, as stressed in the recent work of Acoff and Pourdenad (2000). Unless we re-assess by asking about these key questions a great deal of resources will continue to be wasted. Organisational structure (flat, hierarchical, matrix teams, networks), organisational dynamics (leadership, communication style and power base) impact on the kind of dialogue that is possible and the way that problems are conceptualised and addressed.

Modernist approaches were about asserting 'the right solutions'. Postmodernist approaches to development were merely about recognizing differences; ecological humanism is about ensuring that the balance of human, plant and animal life or biodiversity is maintained for the benefit of all and in the interest of sustaining life for the future.

Many factors need to be considered and praxis tools can help us to hold in mind multiple variables and attempt to manage the diversity through CSP. The design process can be enhanced when tools are used in culturally sensitive ways to add heartfelt conversational depth (Banathy 1996) and creativity to brainstorming, focus groups, nominal groups, the Delphi Technique (see McIntyre 1996). Improving our ability to co-create design is vital if the quality of social policy is to be improved. The key pitfalls when thinking about, doing and recommending policy decisions in the area of development is to think that we are accountable to only one set of values, or one interest group. If we can remember that we are ultimately accountable to one planet and one systemically linked ecology (of which we are part) then it becomes less an issue about 'power over' and more about finding ways to work with one another in terms of all our interests, that are ultimately linked (Zimmerman 1994, White 2001).

Co-creation makes it possible to draw out the liberative potential and recognise the 'bad news' (Flood and Romm 1996), for instance: Modernist approaches assume that asserting 'the right solutions' could control causes and effects and achieve a desirable order, but was not mindful of the harm caused by top down, one eyed narratives. At their worst, post-modernist approaches to development assume that recognizing differences and diversity undermines the possibility of rational truth and also the possibility of justice. Chaos theory (as per Prigogine 1997) conceptualises the energy of life as dynamic. Deep ecology is about the primacy of nature and can be used as yet another master narrative, whilst ecological humanism is about facilitating biodiversity for the benefit of all, in the interest of sustaining life. Indigenous people the world over identify closely with nature for both spiritual and instrumental reasons. Compartmentalisation and the isolation of social problems lead to a mechanistic approach to problem-solving, based on trying to improve specific problems without considering whether they are a result of wider issues. There is unwillingness to share in dialogue across sectors and disciplines when narrow agendas are at stake (McIntyre-Mills 2000). This study builds on the previous arguments, albeit by means of a specific case study. The axial content themes of access to geographical, social and cyberspace and the implications of exclusion are a central concern of this monograph. The hope for the future lies amongst those who are prepared to co-create meaning and believe that the closest we can get to truth is through the process of sharing ideas to address contextual, systemic issues. Better governance of human services is achieved through close participation and sharing, based on a realisation that multiple variables need to be considered and that holistic, integrated decisions based on multilayered or nested systems (that have multiple, weblike feedback systems) can lead to order or to chaos. Dealing with complexity is the challenge.

Better governance needs to be based on the realisation that:

> ... the previous period was characterised by a prevalence of hierarchy, certainty, bureaucracy, class, centralisation and the state. Now we can find the following: complexity, dynamism, fragmentation, uncertainty, diversity, multi-identity, decentralisation and confusion (Kooiman 1993 in White 2001: 243).

I believe that currently we find the characteristics of *both* the previous and the current period in human systems, for instance: internationally, globalisation has been paradoxically translated into openness or closure depending on the degree of faith in 'the market rules' knowledge discourse. Social problems in terms of this economic rationalist approach are increasingly individualised and citizenship models emphasise the responsibility of individuals and families. Many current human service and social policy models in Australia (like elsewhere) are non-systemic (psychological, medical, education, crime prevention and economic). Class issues are also once again relevant in Australia given the growing disparity between the rich and the poor (Jamrozik 2001: 97, 114, 116).

CSP attempts to achieve generative communication (as per Beck in Banathy 1996) in a divided town where co-creation was particularly difficult for social, cultural, political, economic and environmental reasons. The praxis strives to shift governance away from segmented organisations that served different class/cultural groups and an acknowledgement of the value of trans-sectoral communication and a shared local agenda to address health, education and employment opportunities.

Generative communication is essential for change. The research found locally that when Indigenous citizens feel that they cannot take responsibility and that they have no rights or minimal rights then they direct their sense of anger inwards (high rates of mental ill health and suicide) or express their despair at being marginalised as indicated by the high rates of domestic violence and injuries linked with substance-abuse.

Violence and physical prowess was always a recognized part of Central Australian[18] society, but it was based on a rational structure of law, admittedly rules are not always followed, but the likelihood of acting irresponsibly and with unusual violence is heightened through substance abuse.

The programs for prevention to address these indicators of social marginalisation and despair need to redress real and perceived social marginalisation through opportunities for sustained conversation for integrated development across sectors and in so doing build social capital and environmental capital across all interest groups.

Participatory planning to co-create meaning is discussed in the following chapters and the outcomes of non-systemic thinking are detailed. I reflect on (i) the content, (ii) the brokerage process of the research, (iii) the assumptions and values that underpinned the approach, (iv) the way that knowledge is constructed and (v) how being imbedded in the social fabric renders one necessarily part of one's own subject matter. The study strives to demonstrate critical systems praxis (CSP) at a community level using tools to address indicators of need. Qualitative data in the form of dialogue is used as a vehicle for exploring complexity and surfacing diverse perceptions to provide a nuanced and multilayered approach to policy design. I have applied CSP to working not only the hyphen between the researcher and researched, but also the environment. The value of a specific case study is that it allows a detailed discussion of how CSP can be applied. It also gives a detailed analysis of the significance of understanding the linked nature of social problems. The contribution of the extended and sustained use of a case study gives a detailed understanding of how CSP works in the field. As such it could enable practitioners to learn how to apply analytical tools so that they could design effective policy and be effective practitioners. Policymaking is situated and articulated within the local, national and international context. Technology could be used in a liberative manner to assist the most marginalised people. The case demonstrates a complementary use of tools to facilitate dialogue and problem solving. The reader is taken through the

[18] 'Fights were the acknowledged means of settling disputes not only between individuals but also between groups of individuals, as long as the settlement of the dispute did not involve the death of the offender'. Strehlow Occasional Paper (1997: 21). He concedes that wife beating was acceptable behaviour in the past but that as a result of changed ideas about women's rights marital violence is no longer accepted by women and that males disapprove 'persistent wife beaters', implying that violence is still considered acceptable by males in some contexts. This was written in the 1950s but was relevant at the time of the research.

process of planning and implementing public policy in context. The social, cultural, political, economic and environmental challenges are 'swept in' (to use Churchman's term) not brushed aside. The stakeholders with diverse points of view need to be included at all times in interactive designs. Knowledge narratives of participants could be employed as a means to enhance creativity and improve practical decision making. This is important in a world where 'managing diversity' effectively is vital for the sort of governance that will enable democracy to be sustained.

# 3

# Globalisation, Citizenship and Critical Systemic Thinking for Policy Development Through Participation, Observation and Research

## 3.1 Interactive Policy Design via Communities of Practice to Address Current Development Challenges

Individualism is a particularly Western ideal. Being part of a society is however as important for many non-Western and Indigenous societies as it is for so-called Western society. To be human is to live through others. Trust and reciprocity are the basis for getting things done in all societies. Distrust and competition are also options, but they need to be considered carefully, because they too bring reciprocal outcomes.

I am well aware that the Western universalising and totalising approaches, even those that present themselves as being open, dialectical and mindful of sustainable social justice can still be considered problematic, if the praxis does not allow space for Indigenous otherness (Foley 2002). This is important, it is argued not because of a desire to deny holistic systems, but because of the fear born of colonisation and its impact. Perhaps geographical colonisation will be followed by conceptual colonisation. Cultural constructs of difference are important, in order to retain the diverse kinds of linkages across self, other and the environment that have been developed for centuries in isolation. But this does not mean that because biodiversity is vital, there is no space for co-creation. Rethinking ways to make public policy more systemically responsive to diverse needs is an important area of co-creation.

According to Jamrozik (2001), paradoxically, welfare has been the means by which capitalism was provided a human face. He argues that welfare has always been considered a contract between citizens and the state; it has indeed varied as a result of social and political shapers, but it is driven by economic considerations. Keynesianism (the approach that moderates the markets) has been largely set aside and competition of the market has been allowed to rule, without much constraint (Cox 1995). The winners are those with capital or on large salaries, until they lose them and the real winners are those with vast amounts of capital that can be used as a shelter when times are fraught. The medium to small shareholders are also vulnerable. The real losers are the casual labourers and the long-term unemployed that have inappropriate skills and cannot access educational opportunities. Very often age/gender/culture/race/language and religion become class proxies. Power differences reside in access to resources and access is determined in terms of the categories that will be used to include some and exclude others.

The reality is that low-income, low-skilled workers in Australia (most often defined as being of non-English background or Indigenous, or young and unskilled and if you are also female and disabled then the layers of disadvantage just become greater). These disadvantaged groups will also have to compete with other low-skilled low-income workers internationally. Standards of living and standards associated with the rights of being Australian citizens could be increasingly challenged. What effects would an approach to address health, education and employment have on citizens? Citizens could be active contributors aware of both rights and responsibilities in a safe environment where social justice is considered important and where social capital is considered alongside cultural, economic and environmental capital. This is a far-sighted approach to welfare and the wellbeing of the state and its citizens, rather than a short-term horizon that considers only short-term gain and maximising profits. Spending is on the prevention of social and environmental ills, not on addressing problems through spending on social control and addressing the end results of poor social health and marginalisation.

Welfare has been used politically in positive and negative ways as emphasised for instance by Rowse (1998) in his work 'White flour, white power' in which he gives a historical analysis of the way in which missionary handouts shaped dependency and eroded a sense of self-reliance. This is supported by Pearson (1999) who writes that welfare can be positive or negative depending on the social policy values on which it is based. He also stresses that personal as well as social responsibility is required to break out of dependency.

According to Jamrozik the post-welfare state appears to do more social monitoring downwards of those drawing 'social welfare benefits' and less monitoring upwards of those making the decisions and controlling policy (Jamrozik 2001). Those drawing 'the dole' are perceived to be using up the tax resources of people who have worked hard and for long hours (Jamrozik 2001). There is little sympathy from overstressed workers for those who are unemployed, even though they could be one step away from being unemployed. As benefits are cut, an ever higher proportion of the welfare dollar comprises professional salaries and in Alice Springs there are still many professionals despite the decrease in the past few years (since 1996).

Residual welfare is limited to the necessities of life as opposed to universal welfare that addresses quality of life and can be used to promote participation in society and the basis for trust in one another. When we also consider that the frameworks professionals apply are inappropriate (because they are too compartmentalised and the time frame is short, not long-term) the problem deepens.

It is a fact that overall although Australia has a demographically ageing population; Indigenous Australians have a life expectancy 20 years lower than the average. The ratio of dependents caused by an ageing population and a population of shrinking or declining births and fewer people working than ever before is a worrying social and economic scenario. Immigration intake is currently limited based on arguments about the scarcity of resources (environmental and social) and that we already have enough challenges ahead without the burden of further immigration, but policy has begun to address the implications of the population on economics. To sum up, in the future the challenge will be to do problem-solving that takes into account the social, political and environmental factors that after all shape the context of welfare policy.

In some areas such as Alice Springs welfare is one of the main industries. But the health and welfare outcomes of Indigenous people remain woeful, because the approach to welfare (the policy and practice) is inappropriate and unsustainable.

Let us consider the context of welfare work at the individual, local, national and international level. The challenge is to address the hyphens across self, other and environment and shift welfare from passive to active participatory processes. The importance of the spiritual, the emotional, the psychological and the biological need to be considered systemically with work at the interpersonal level, group dynamics and interorganisational dynamics as all are important factors for governance. By this I mean the need to address interdependency and links. The Indigenous philosophy of ubuntu, for instance stresses that we are people through other people and that we are well to the extent that we live in harmony with ourselves, other people and the environment. This value and assumption is relevant to policy and practice.

The impact of economic rationalism and a residual versus universal welfare policy needs to be considered systemically. Fewer people are employed in Australia in full time positions and working longer hours, which impacts on family life. More people are in casual and contract positions and are prepared to move in order to work. Alice Springs in particular attracts mobile professionals keen to progress their careers. I will assume that we are talking about a 'post welfare state' (to use Jamrozik's term 2001) that has implications for employment, governance, welfare education and practice. In Australia, like elsewhere, the welfare state has already been trimmed back increasingly since 1996 when the Coalition Government came into power, but the changes can also be linked with the opening up of our markets and the floating of our currency, and these changes occurred during the era of the Keeting Labour Government. What has been derogatively called the Nanny State no longer exists. Hand holding (as it has been parodied from the cradle to the grave) has now been replaced with freedom to sink or swim with an increasingly limited welfare wings to support us. According to Jamrozik (2001) the reality is that Australia has fallen further behind social democratic welfare states being more on par and moving ever closer to the USA and Canada than the Scandinavian welfare states. He argues that in model welfare states such as Sweden have also had to cut back as a result of increased demands by an increasingly diverse and changing population and that the shift in welfare is from status entitlements that are universal and based on for instance being sick or being a parent or a widow to more residual entitlements based on proven need for this context according to Jamrozik (2001) the determinants of need can be increasingly narrowly defined by government.

Finally Jamrozik (2001) sums up the issue as a future in which we face an ageing population and low rates of immigration, thus raising questions of sustainability and dependency. An added concern is that the baby boomers are about to retire and to draw down their superannuation, thus depleting the reserves available for funding welfare in the future.

Added to this is the fact that demographically Indigenous young people are growing and their educational outcomes do not match employment opportunities. In Australia it is a fact that in 1996 there were proportionately more people aged 16 years and more receiving benefits than there were in 1966. In 1996 10.8% of the population were on benefits in 1996 24.8% of the population were on benefits (see: page 114 in

Jamrozik 2001)[1]. ABS 1996 data show that poorer households spend a higher proportion of their disposable incomes on essentials such as housing, maintenance of housing (including running costs of power, water), transport and food than do households with higher incomes (Jamrozik 2001: 116). This is not sustainable. There are more people on welfare and more people spending a higher proportion of their disposable income on the necessities of life. It is also a fact that these lower income households have to pay GST on their purchases and the user pays approach (that applies increasingly across health, education and recreation and transport) means that there is less and less disposable or discretionary income after having paid for basic necessities, such as rent, food and transport (Jamrozik 2001). Privatisation of essential services such as power, water and the telephone will according Jamrozik bring even further changes and in the developed world where these services are taken for granted, standards of living will change for those who cannot afford the increased costs associated with privatised resources, or they will need to be innovative. New forms of power will need to be generated, water will need to be distributed, conserved and recycled in more sustainable ways and cooperatives will need to replace privatised distributors of services. If not, people will be facing stark choices as to whether they will eat or remain cool in summer or warm in winter. Increasing numbers of citizens will be merely consumers who cannot pay their power bills not citizens who have a right to a healthy and safe home setting.

As a bridge from the known to the unknown, community of practice provides some of the thinking tools that can help participants *leap outside* (as per Banathy 1996, 2000, Romm 2001[2]) the emic (or taken

[1] It is a fact that currently Australia has higher dependency ratios of people making use of government services than ever before, in other words the proportion of people on welfare is higher than before (Jamrozik 2001, table 5.11, p 98). The proportion of the population on welfare spend more of their social wage on necessities than before because of GST. Costs are higher as services are de-regulated (see Jamrozik 2001, table 6.6, p 116). All Australian families with dependant children have less disposable income than ever before and the Gini co-efficient (0 = complete equality and 0 = complete inequality) shows that the gap between rich and poor is growing (see Jamrozik 2001, table 5.10, p 96).

[2] Romm, N. (1998) The process of validity checking through paradigm dialogues 14th World Congress of Sociology, Montreal.
Romm, N. (2001) Our responsibilities as Systemic Thinkers 45th International conference of international Society for the Systems Sciences, USA, Asilomar.
Romm, N. (2001) *Accountability in social research: issues and debates*. Kluwer/Plenum, London.

*TABLE 3.1*
*Characteristics of a COP*

| | |
|---|---|
| Assumption | The more variables we sweep in and the more issues that are unfolded, the more likely we are to achieve sustainable social and environmental justice |
| Arenas | Local, national and international groups, organisations in the public, private and volunteer sectors |
| Task | Individual and group support across sectors and disciplines that focus on social, political, economic and environmental factors |
| Process | Participatory design through conversation using open ended matrixes that can be used for lobbying locally in and across organisations and through social movements |
| Rationale | Personal and public need |

for granted) and engage in creative practice that focuses on rights and responsibilities.

A modified form of community of practice (COP) is a management approach that spans arenas, based on educational disciplines or specialisations, sectors (private, public and volunteer) and geographical space (Table 3.1). It also makes use of technical/instrumental, strategic and communicative or interactive knowledge (as per Habermas 1984). COPs provide a vehicle for addressing complex inter-related socio-demographic, economic, technological and environmental challenges. The approach provides a way to work systemically with knowledge categories, rather than within categories. Webs of meaning (as per McIntyre 2000) can be developed across disciplines and with due consideration of all the stakeholders in co-creations that are meaningful to all the participants.

Interactive design needs to address multiple challenges. Education and jobs need to keep pace with the needs of young people with limited learning outcomes. The public sector will have to address this challenge in partnership with the private and volunteer sectors, because the current resources for government are increasingly residual in the era of what has been termed the 'post welfare state' (Jamrozik 2001).[3]

Globalisation may have winners, but the losers will need to be assisted to obtain services and skills. Currently, the gap between rich and

[3] The definition is based on a shift from universal quality of life projects, to residual welfare to meet basic needs. This is because of the challenges posed by the way that the market functions and the way it is currently viewed and constructed, and in part because Governments make policy decisions on who gets what, when, where and why, but also because, more people currently draw wellfare benefits than ever before.

poor in Australia (among other countries) is widening and as benefits are cut, an ever higher proportion of the welfare dollar comprises professional salaries. The public sector workforce will therefore become older, more diverse in terms of gender and culture and there will be fewer staff overall (Paul Case 2001).[4] It is also possible that there will be more contracted staff and more outsourcing if current trends continue. This is debatable and it depends on whether there is a realisation that continuity is a necessary aspect of quality services. In order to address the need for managing the knowledge base of the public sector, creating digital databases is useful for teaching people how to use them confidently and contribute to them, and is the next hurdle.

As people have fewer religious or national, or ethnic boundaries people feel lost unless they can integrate these changes into their lives in a meaningful manner. The impact on well-being cannot be underestimated. Technological challenges are posed by biotechnology and nanotechnology will impact on the ethics of *being* human and will have particular relevance to the health sectors. Social and cultural changes that face this generation are accelerating in response to many challenges and the technological changes are dramatic (as per Banathy 1996). This means that people's identities need to accommodate change. A globalised, workforce that moves, in response to the dictates of the market to find work, faces the loss of a sense of place, a sense of continuity in their work, home and neighbourhood. A sense of continuity, becomes a characteristic that has to be achieved in new innovative ways.

Social and environmental sustainability in terms of natural resources, soil and water, in particular will be key shapers of quality of life. Social sustainability involves meeting other needs such as: Affordable power and service costs, housing and transport.

## 3.2 *Governance, Management and Social Policy*

'Survival; is not mandatory, it is our choice' (Paul Case 2001), is the mantra we need to bear in mind as we develop designs to address current social and environmental challenges! Good governance is increasingly hard to deliver and it will depend on the ability of the players to become innovative and to work effectively across organisational sectors to enhance designs, plans and operational delivery. One way to enhance governance is work across sectors and disciplines. Digital databases for

[4] Speech by Paul Case, Commissioner for Public Employment PSA Conference 14 August 2001.

knowledge management across sectors and disciplines are also a useful starting point for knowledge workers who need to address complex problems that require integrated responses by teams of specialists who are increasingly adepts at pooling knowledge. The narrow compartmentalised approach to addressing problems is being replaced by a realisation in many areas of research and practice, that in a complex, changing world, integrated responses are needed. Designing healthy settings in homes, neighbourhoods, and places of work, places of recreation and places of learning can positively and negatively impact health, education and employment. One of the key challenges for the future is building and sustaining the rights and responsibilities of citizens through developing social and environmental capital, not only economic capital. The triple bottom line (people, the environment as well as economic considerations) will be vital.

When we also consider that the frameworks we apply are sometimes inappropriate (because they are too compartmentalised and the time frame is short, not long-term) the problem deepens. We need to be mindful that we are citizens, not merely customers and that the social and environmental contract is deeper than any commercial contract. But even more important is the idea that as the boundaries of the world fall to market forces, people need to be seen as citizens of the world, not merely as members of nations. If money can move and trade is open, then it is logical that people need to be free to move as well!

Learning needs to enable the participants to design for the local and the global. Participants need to be able to make sense not only of their own field of practice, but learn to engage with others in other fields of practice. Identity based on one set of assumptions and values associated with one's own learning and experience needs to be expanded to negotiate with other frameworks of meaning. This involves becoming boundary workers across subject areas, sectoral areas and cultural areas. Skills will include transcultural understanding, knowledge management and management of diversity. These are not entirely new skills and they can be developed by building on existing competencies in conflict resolution, communication skills (based on respect for the other) and one's enthusiasm to learn from one another in a range of contexts and via a range of media. Imagination is helpful not merely in trying to predict the future and then attempting to plan rigidly for it, but in being able to anticipate the best route (as per Banathy 1996, 2000).

Health, education, employment (as per Jamrozik 2001) and participatory governance are the pillars of citizenship and quality of life. Participation in decision making and in designing systems to address these factors is essential for the future well-being of all people. Critical thinking is required to apply diversity management and systems thinking strategies to improve governance across a range of human service organisations. Ackoff and Pourdenad, sum up the issue that we have to stop doing the wrong things well and start to think about what we should be doing and how we can begin to do these new things better.[5]

> Many, if not most, social systems are trying to do things right, but are not doing the right things. Examples are drawn from transportation, health care, law and order, and education ... Recall Peter Drucker's observation of the difference between doing things right and doing the right things. This distinction is fundamental. The righter we do the wrong thing, the wronger we become. If we make an error doing the wrong thing and correct it, we become wronger. Not so when we pursue the right thing. (2000: 216)

The aim is to promote an understanding that how and why some issues are defined as social problems by some interest groups and not others is vital for development that is not merely efficient and effective in terms of a discipline or a sector, but is systemically effective and systemically efficient. Understanding socio-cultural frameworks associated with different life chances is a starting point for integrated development. Values and assumptions shape the way in which social problems are defined. An analysis of these could highlight similarities and overlaps and thus ensure better communication for problem solving and planning. Understanding values and building trust are the starting point for development. The seven guiding principles for sustainable development defined by the United Nations International Council for Local Environmental Initiatives are: 'Partnerships, Participation and Transparency, Systemic Approach, Concern for the Future, Accountability, Equity and Justice, Ecological Limits. These principles provide the framework for organising, implementing and evaluating'.[6] The sustainable development approach requires management of diversity and open communication system as discussed below that is sensitive to the emotional content of the participants. The research strives to provide strategic direction for social planning and policy for 2000 and beyond.

The principle of the social policy research is to improve the quality of life of all citizens by (i) working according to the World Health Organisations Ottawa Charter, namely to address health and development and by working in partnership across sectors and disciplines. (ii) Recognising that development is the outcome of shared definitions

[5] Many of these concerns have been raised by Jamrozik (2001) in his discussion of the post welfare state and I also raised these concerns about needing to think across sectors and disciplines (McIntyre-Mills 2000).

[6] ICLEI International Council for Local Environmental Initiatives (1998).

and shared goals, but it is also the recognition that empowerment requires the right to difference and diversity. Sustainable development is by, with and for people within their environment.[7] (iii) Addressing sustainable development in terms of human and environmental resources is a priority.

Policy will be discussed in depth in chapters 7 and 8. The following summary provides a context for understanding the argument as it unfolds, but it also provides a brief overview.

A systemic approach to public policy issues pertaining to health, education and employment (with reference to accessible, linked examples) is developed to carry forward the theoretical arguments throughout the book (Diagram 3.1). Specific suggestions are made that have been drawn (in part) from the Healthy Settings Movement (HSM) that has enjoyed widespread support internationally. Specialisation needs to be balanced by integration. An ecological approach to planning health and development is needed. In Indigenous societies health is defined as living in harmony with oneself, other people and one's environment. Beginning with the Ottawa Health Charter of 1986, the life chance links between health and development in terms of physical, mental and spiritual well-being have received increased recognition in recent years (and there has been a retrieval of some of the wisdom of Indigenous cosmology) as a result of the official ratification of the World Health Organisation. At the United Nations Conference on Environment and Development (1992) delegates in Rio developed Agenda 21 which provides 'loose guidelines' not (to use its terminology): 'a recipe book' on: 'social and economic dimensions, conservation and management, strengthening the role of major groups through partnerships and means of implementation' (United Nations Conference on Environment and Development 1992).

A healthy environment approach can be envisaged as a means of encouraging people to see the interconnections across issues and to find ways to work in partnership to address problems. The problem is that there are often bureaucratic power bases and personal value bases to be challenged when duplications and tensions are discovered. The promotion of a healthy environment needs to be placed as a central assumption of planning. The challenge is to find ways to encourage existing organisations (that tend to work separately in bureaucracies) to work together. The Healthy City/Environment Movement (HCM) is not merely about

[7] Adapted from Chambers (1983) *Rural development: Putting the last first*. Wiley, New York.

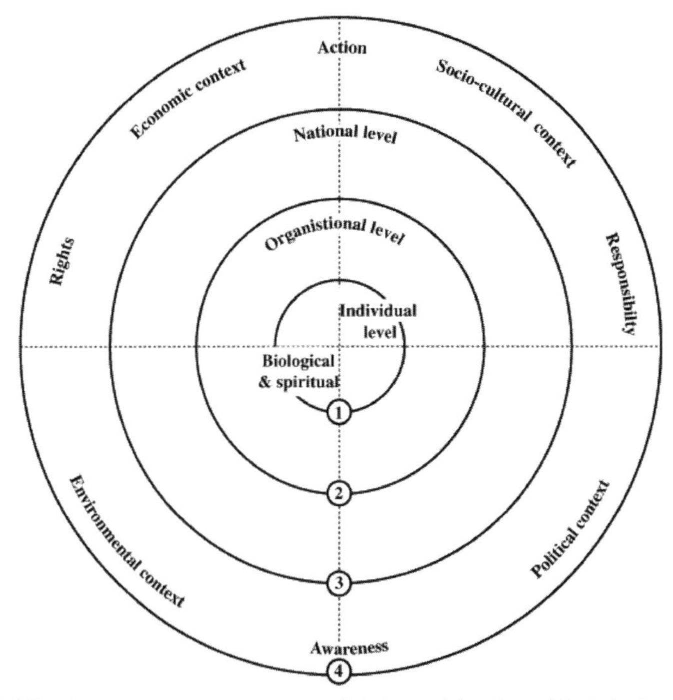


1. At the individual level
   – psychological, physiological & spiritual considerations

2. At the organistional level within the local community
   – interest groups competing for resources & cuts to services

3. At the national level
   – the changes in the economy & nature of work (shifts to permanent options)
   – cuts to welfare state

4. At the global level
   – globalization impacts on the way of life at the local level
   – international benchmarks set by the United nations
   – social & environmental movements for change

*DIAGRAM 3.1* Mandala of Complexity to be Held in Mind when Planning

experts finding technical solutions, it is about citizens working together to bring about sustainable health and development by, with and for people and their environment locally, nationally and internationally. There is considerable potential for the public sector to work closely at the local level with a wide range of relevant organisations. The application of global indicators of social, political, economic and environmental well-being developed in relevant health, environment and human rights charters would be an effective starting point for Participatory Action Research programmes to achieve integrated development.

In order to build trust in society and to establish a sense of shared future a concerted practical effort needs to be made to redress the social health and well-being outcomes. According to the HCM, life chances, life-style and the quality of the built and natural environment in which one lives are linked. By promoting health and development through integrated policies, much could be done to eliminate sickness and social problems.[8] Systemic thinking for integrated development is central. Understanding the links within and across sectors (e.g. education, health, housing, recreation, crime protection) and disciplines (e.g. primary, secondary and tertiary education, public health, engineering and architecture, recreational studies and criminology) is required. An analysis is of value only if it is used for problem-solving to address identified linked problem areas. Shifts in thinking need to be attempted through education and in depth discussion to develop an enhanced understanding of issues and to develop enhanced problem-solving. The uses of specific tools for systemic thinking are detailed in the planning recommendations. Integrated development means working with many disciplines rather than within just one.

To sum up the contributions of Critical Systemic Praxis (CSP) includes: (i) seeing the social, political, environmental and economic implications for citizens. (ii) Processes for intervention and addressing diversity such as: (a) Diversity management approaches to improve governance, (b) Healthy settings approaches through participatory action research, so that there can be shared learning that spans disciplines and sectors and (c) Participatory planning and design. CSP is developed to address poverty to improve governance and to enhance participatory democracy through improved understanding of the way values, definitions, concepts and issues impact on the life chances and learning by people.

[8] According to the Healthy City Website: 'Health for All' Declaration (1976) at the Alma-Ata conference was adopted as the guiding premise for the WHO for the 21st century. The concept was about achieving development outcomes by using a broad systemic approach to health promotion and development, rather than a reactive, vertical issues-based or problem-based approach. The Healthy City/Environment Movement follows the Ottawa Health Charter of 1986. It emphasises an integrated multidisciplinary and multisectoral approach to community development, public health and planning. This means that many disciplines (engineering, architecture teaching, nursing and medicine, psychology, sociology) need to address issues together, because one discipline alone cannot address interrelated social issues. It also means that there needs to be cooperation by representatives across many sectors, such as education, health, youth affairs, policing, the legal sector who need to work *together* not in isolation. Specialisation needs to be balanced by integration. An ecological approach to planning health and development is needed. The Ottawa Charter underlined that health is determined by the quality of the living environment.

## 3.3 *The Complex Policy Context of Postwelfarism in a Remote Region of Australia*

We need to do integrated development in both rural and urban areas using healthy settings approaches so that people who have few resources can be safe in their homes and access essential services and achieve social mobility. Elderly people in remote areas should not have to choose to be warm or to eat, nor should students or single parents, or the children of low-income parents. The cost to tax payers would be the same for addressing morbidity and mortality associated with unsafe living conditions as it would be for the prevention of ill health associated with responding to the effects of deregulation and privatisation through sustainable technology, infrastructure and organisational management. Citizens in the public, private and volunteer sectors could either support health promotion, education and accessible pathways to employment and create new jobs through research and education or taxpayers can pay tax to support social control measures through increased incarceration.

This social policy proposal outlines a way in which organisations within the private and volunteer sectors could respond to identified needs together with all levels of government. A participatory and integrated approach to facilitation, management, planning and development is proposed (see Diagram 2.1 Participatory action research). Membership of the healthy cities or community of practice networks (see Wenger 1998) and mentoring by partners could facilitate the implementation of projects, rather than operating in isolation. Healthy settings in homes, neighbourhoods, schools, public places and organisations, work environments, accessible transport routes and communication systems within a sustainable natural environment make up healthy settings. By concentrating on these settings, we can build healthy environments and thus improve the quality of life of all citizens. At a recent Australian Pacific Healthy Cities Conference,[9] delegates from all over the world, for example from isolated parts of Mongolia to Sydney, Australia explained how they could benefit from this sensible approach. The key point made is the value of working in partnership to achieve sustainable outcomes for all. This monograph provides guidelines for any organisation within the public, private and volunteer-based sectors that wish to undertake systemic planning and outcomes. The policy and planning recommendations are based on

[9] Australian Pacific Healthy Cities Conference: From Global Policy to Community Experience: The First Twelve Years of Healthy Cities in our Region, June 2000, Canberra.

the identified social needs. Indicators showed that the quality of life varies greatly and thus so do the life chances and social health outcomes of citizens. Healthy settings include public spaces (neighbourhood living areas and town camps, parks, shops, work environments, places of learning, libraries, schools), private spaces (domestic environments). Examples of healthy and unhealthy settings are detailed in following chapters.

The challenge is to enable citizens to be independent, not to merely survive, as losers in a globalised economy through residual welfare to meet survival needs through minimal social welfare payments. The issue of citizen's access to geographical, conceptual, social and cyberspace need to be addressed. The class divides today intersect with categories such as culture, gender, age and level of education. It remains true to say that those who can access education and health resources are better placed to achieve better life chances through gaining full-time and ongoing employment. Without access to these citizens are vulnerable and have to access the crumbs from the economic table: casual work or work for the dole. The primary aim is to benefit human service practice and outcomes across the public, private and volunteer sectors in order to promote social and environmental justice. By using a case study and problem solving as a focus, complementary methodologies and critical thinking have been used to draw out creative possibilities from a range of disciplines. The case study addresses a dis-ordered system. I allude to the analogy of chaos, but avoid the use of social entropy, despite the temptation to use it and instead (for the sake of clarity) refer to the way in which the dynamics of power operate in order to maintain the status quo. If the real definition of equilibrium, as Bailey (2001: 55)[10] suggests, is the complete dissipation of energy and not a state of harmony as some sociologists have mistakenly believed, then this is indeed a case study of social policy to re-energise praxis. Development responses do not have to be at either end of the continuum of approaches that are driven by the market (economic rationalist) or controlled by the state (socialist in orientation), as neither of these models is sufficiently systemic in nature. Potential ideas to address social and environmental justice need to take into account multiple variables from many sources. This requires open communication and diversity management (Flood and Romm 1996) to include stakeholders that represent multiple interest groups so that multisemic solutions can be achieved. The World Health Organisations Healthy Settings Approach (that resonates with some of the Indigenous

[10] Bailey, K. (2001) Towards unifying science : Applying concepts across disciplinary boundaries in *Systems Research and Behavioural Science* **18**: 41–62.

concerns about land, a sense of place and well-being) is one such approach. Strategies need to take into account historical, socio-cultural, political, economic and environmental variables.

Also at the time of the research the policies at national and Territory Level (which are implemented in a context of a move away from universal to residual welfare and a global capitalist system in crisis) were in many instances contradictory. At a National government level policy is shifting from universal to residual welfare. The example of the shift of funds from Medicare to the 300-dollar rebate is a case in point.[11] The argument is that if more middle class people move from public to private health cover, then more resources would be available for those who need it. In reality Baum (1999) argues that has not been the case in the USA where the public health systems are far worse than in the UK where a public health service is maintained (Baum 1999). Public health policy in the Territory has received much government support as its mission for 2000 and beyond. But the planned holistic service delivery faces the challenge: How do you move service delivery away from public bureaucratic delivery to competitive privatised delivery which somehow has to become integrated and holistic despite the competitive tendering contracts from government? The shift from specialised programs to generic programs in reality means that specific services (in gender and cultural terms) will not necessarily receive as much funding as previously. The rhetoric of prevention and health promotion can only be implemented in an environment where organisations are prepared to work across sectors in an integrated manner. The NT has the highest rate of population growth (2.9%) of any jurisdiction in the nation, according to Reed (Oct 1998). In line with the 'Planning for Growth' policy outlined in speech by Treasurer The Hon Mike Reed MLA (Oct 14th 1998) the centralisation of services will allow savings in the public sector that can be used for improvement of services. The review needs to be undertaken within the context of the Budget Speech on 28th of April 1998 that indicated the process of Planning for Growth: 'Major reviews are in progress in education, health and PAWA and Correctional Services all oriented towards a better service at a lower cost'. At the time of the research it was argued by NT government that savings would be achieved by economies of scale by bringing together like services and allowing organisations to concentrate on core business. The implications of privatising basic services such as Telstra and the PAWA could have challenging implications for the remote areas surrounding Alice Springs in terms

[11] In Touch: Newsletter of the Public Health Association Inc, vol. 16. No 1 February 1999. Editorial by Fran Baum.

of user cuts and access to vital services. ACOSS cited (Nov 4th 1998) that all-social welfare services show an increase in the number of people using services and an inability to cope with the needs of clients. In particular Centre Link has been targeted as an organisation in need of reform. The socio-economic context reflects the way in which welfare states have moved from universal to residual welfare rights. This is reflected in the ministerial statement of the Treasurer The Hon Mike Reed MLA on 14th of October 1998. He discussed the review of Government that would involve 'planning for growth' which involves maximising effectiveness and efficiency in services, outsourcing, trimming to core business. The centralisation of information services for NT GOV is one example of cost cutting. The welfare state is in the process of being dismantled in most parts of the Western World. The trend towards privatisation of services and outsourcing is likely to continue as government's face the pressure to compete in global markets and trade barriers are unlikely to be restored as markets move from a global to a regional focus. Within this broader context at the local level, small human service agencies respond to the centralising and devolution policies of commonwealth and state government, by narrowly guarding their resources and trying to justify their separate existence, which means considering their own self interest in order to justify a separate service position.

Within this broader context the challenge of conducting a needs assessment is augmented by two opposite tendencies at local and state level. A push towards centralising services at state government level and a pull by many agencies at local level to 'protect their patch' is evident. This local approach to compete for turf was highlighted by service providers at the CARPA[12] Conference hosted by Aboriginal Congress. Fragmentation in the delivery of services is due to a number of reasons:

- Fear of mergers and cuts in services.
- Competition for scarce funding and specialisation of services to meet specific needs, different values underpinning service delivery, such as harm minimisation versus control and zero tolerance, some services are geared to primary health care and others to curative care.
- The political values of players and the personalities of players.

Networking committees that act like matrices are needed for joint working and problem-solving. They allow for agencies to work together but to maintain some autonomy. The local government bodies in Central

[12] Central Australian Rural Practitioners (CARPA) Conference.

Australia need to liase with the many community councils in order to address the needs for self-determination and human service provision in terms of the dual concept of indigenous citizenship. Local government was in existence prior to self-government for the Territory in 1978 an additional tier of government has been added where previously only two levels existed.

The question needs to be asked: What are sustainable standards of living and how can they be achieved in remote places using sustainable forms of technology, based on intelligent co-created designs for living? Solar and wind power is important for living in harmony with nature, as is the biodiversity of the fauna and flora. The current way of life of living in competition and extracting resources is not sustainable. The notion of the user pays for services plus the notion of payment of GST for goods and services means that citizens are required to meet a gamut of needs for which they are more and more responsible to pay. As detailed in chapter 5, the rising need for emergency welfare services from non-government agencies such as St Vincent de Paul is an indication that people are not coping. As the savings of citizen's are eroded and they face contingencies citizens become more vulnerable. The most vulnerable are those who are unemployed and who have limited numeracy and literacy and who are reliant on welfare.

## 3.4 *Implications for Social Policy and Governance*

Poverty studies that are directed downwards have emphasised measurement of poverty. New participatory governance studies could consider measuring the waste caused by non-systemic policy and practice. Corporate accountability checks could be introduced. We need to work more in terms of systems and less in terms of compartments. Social policy based on intelligent, participatory design that is based on co-created solutions could maximize the multiplier effects of positive initiatives to empower people, so that more is achieved and so that scarce resources are used effectively.

We need to understand that spending on promotion of social capital (measured in the crudest terms by health, education, employment and participation indicators) and prevention of social ills (measured by incarceration rates, morbidity, mortality, unemployment and marginalisation) is cost effective. This is a pragmatic argument advocating the low road to social justice, because it is in our mutual interest.

We need to think and practice with theoretical and methodological literacy. This is the basis for intelligent social designs and policy.

Theoretical literacy is required to understand the values and assumptions underpinning the different models of social policy and welfare delivery. The way people think about key concepts such as health, education, employment and crime prevention impacts on the way in which policy is developed and implemented. Our values and assumptions shape:

- Approaches to problem solving and influence the way we define social problems,
- Processes for addressing social problems,
- Outcomes and
- Evaluation.

People need to participate at all stages of policy development (because interventions are only meaningful if it is defined to be so by all the participants. Banathy (2001) calls this intelligent design. Internationally one of the biggest challenges is to realise from the outset that frameworks for thinking shape development outcomes. Unless 'we get the approach to planning and development right' then a great deal of resources (time, money and people skills) will continue to be wasted because development processes can either enhance or retard social capital, the basis of political good will and trust in a community.

Human service workers in Alice Springs and elsewhere need to understand the context of welfare at the individual, local, national and international level. We need to be able to ask the questions who gets what when how, from whom, why and to what effect and in whose opinion, with a high degree of theoretical and methodological literacy. We also need to know that the answers to these questions depend on who is asking the questions and the values that they hold. So it is vital to teach students that questions about ontology (which are about the nature of reality) and epistemology (which are about how we know what we know) are very relevant first questions. So as policy makers we need to know where we place ourselves on ontological and epistemological maps as our values and assumptions are relevant to addressing all the factors mentioned in Table 3.2.

According to Jackson (2001) in his incoming speech as president of the International Systems Science Society, prediction and control in addressing complexity, will be a major challenge, because this is the rationale for most systemic thinking. The essential question that needs to be asked is who benefits and stressed that the work of Foucault and Paulo Friere remain important in this respect. He stressed that meaning and motivation in social policy development will need careful consideration through diversity management and concluded that co-creation of solutions in 'a world of multiple values' remains a challenge.

*TABLE 3.2*
*Who Gets What When, How, Why and to What Effect and in Whose Opinion?*

| | |
|---|---|
| Who? | Demographics of the population and of users |
| What? | Services and nature of delivery |
| | Universal rights or residual benefits |
| When? | Circumstances under which services and resources can be accessed. |
| | Moral, legal and political considerations |
| | Historical trends of welfare |
| From whom? | Organizational responsibility for delivery the public (Commonwealth, State, Local Government) |
| | Non Government organisations that are mostly state funded |
| | Non Government organisations that are mostly privately funded |
| | Private sector (business) |
| | Voluntary Support networks |
| | Neighborhood support |
| | Family support |
| How? | Arenas for delivery |
| | Local level, national level and international level |
| | Organisations at the local level within sectors |
| | Inter-organizational cooperation across sectors through\networking and social movements. |
| Why? | Political and ideological rationale |
| To what effect? | Monitoring and evaluation using benchmarks of integration and marginalisation, for instance: health status, educational access and outcomes, employment access and outcomes, governance (participation indicators), marginalisation (incarceration rates). |

Empowerment and emancipation in a world of inequality will need to be the focus of systemic policy and practice. The concept of social capital is relevant to thinking, practice and policy insofar as: (i) understanding poverty in social, cultural, political, economic terms and environmental terms, that is, systemically. In a world where the zealots of economic rationalism believe that their grand narrative has the solutions we will need to think in terms of addressing citizen's rights and responsibilities through creative solutions. We need to think in terms of education for the future and to realise that social and environmental capital are our future, not merely being competitive in a narrow economic sense. The notion of people, the environment and making a living now and in the future is paramount. Such a role for citizens is active not passive, not merely as welfare recipients but as active shapers of their community.

## 3.5 *Reflection on the Theoretical and Methodological Orientation and Tools: Implications for Accountable Policy and Practice*

Modern science developed along the lines of disciplinary specialisation. Working within narrow predefined areas in order to develop detailed knowledge extended the frontiers of knowledge in many areas. In development studies the attempt to apply modern scientific thinking has failed (Hettne 1995). Research and development based on the notion of controlling variables in the real world made up of socio-cultural, political, economic, technological/scientific and environmental variables (in order to engineer changes) was based on an attempt to control or limit causal connections in constructed experiments. New development thinking recognises that models of development thinking that attempt to simplify reality are just that: artificial models. The environment comprises multiple links across multiple causes and effects that are weblike rather than unilinear. In order to bring about changes a multisemic approach is needed. Systemic approaches are not limited to working within bureaucratic structures, but instead explore and use a broader range of arenas, namely: networks and matrix teams that span local, national and international contexts. A brief typology covers aspects of ontology, epistemology and implications for policy and practice.

The Table 3.3 does not, however, give a sense of the way the old approach could take on the terminology of the new approach as rhetoric, while the approach remains the same. One of the terms in particular that has been filleted of its intended meaning is sustainability, particularly by microfinance specialists who use it to refer to financial sustainability. Internationally one of the biggest challenges of development work is to realise from the outset that frameworks for thinking shape development outcomes. Cynicism and meaninglessness are the by-products of perceived or real marginalisation. Hence the vital importance of Aboriginal role models (of athletes such as Cathy Freeman, leaders such as the late Charles Perkins, a civil rights activist or Senator Aiden Ridgeway, a member of parliament at the time of the research) but also local, attainable success stories that are part of the everyday social fabric and alongside real opportunities are required to break the cycle of hopelessness that continues across the generations.

Locally there is also a particular need to see that apparently separate issues are in fact linked and impact on life chances. Life chances refer to opportunities in life experienced by people as a result of a host of

*TABLE 3.3*
*Comparison of Approaches to Thinking and Practice Based on Closed and Open Approaches*

| | |
|---|---|
| Old approach to science: Compartmentalisation | New approach to science: Systemic, integrated thinking and practice |
| Short term horizon | Long term horizon |
| Profit and economic capital | Environmental and social sustainability and social and environmental capital |
| Thinking in terms of the meanings of one culture or one interest group. | Addressing multiple sets of meaning[13] when undertaking development. |
| 'Either or' thinking in narrow terms that is specifically about social or political or environmental issues. | 'Both and' thinking in social and cultural and political and environmental terms. |
| Hierarchical structures for management, communication and program delivery. | Weblike team approaches (matrices) that span sectors and disciplines in order to address issues. |
| Citizenship models stress individual and family responsibility. | Citizenship models stress social and environmental responsibility. |
| Expert driven by specialists working within a single discipline. | Community driven by a range of stakeholders, interest groups and professionals representing multiple disciplines and sectors who contribute to research, problem solving, the development content and process. |
| Individual responsibility for problems. | Social and environmental responsibility for problems. |
| Management stresses efficiency and outputs (number of items of service delivered). | Management stresses effectiveness and outcomes (the qualitative perceptions of the impact of a development intervention). |
| Top down research and development. | Participatory action research based on learning from successes and mistakes. |

demographic, socio-cultural, political and economic factors. An example could illustrate this point:

> If a general practitioner, community nurse, social worker, youth worker, school teacher and psychologist sat together to discuss a young person 'at risk', to use the current label, the frames of reference of the 'professionals' could be very different. The GP could be concerned about the impact of binge drinking on the young person's liver and nervous system. The psychologist could be concerned about the young person's depression and inability to think in terms

[13] Multisemic thinking has been stressed in the development literature (for instance McClung Lee 1988, Berger 1976, 1977, Giddens 1991, Castells 1997).

> of the consequences of their actions and could fear that the young person would commit suicide. The community nurse could be more concerned about the young person's vulnerability to a range of diseases such as tuberculosis and renal disease, whilst the social worker could label the child 'at risk' of becoming a school drop-out, unemployable and on track for a life of 'criminal behaviour', made particularly likely as a result of mandatory sentencing in the Northern Territory.[14] The schoolteacher may simply view the young person as a liability to the class she teaches, because of the disruptive role model he sets for the others in her class. She could also be concerned because the young person models the violent behaviour he sees in the community and that he also conforms too much to youth counter-culture and not enough to the mainstream norms set by Australian society. She may be concerned that the traditional community norms are being eroded by the availability of alcohol in his neighbourhood and may be unaware that Indigenous young people have to make sense of not only more than one culture but they also have to cope with adolescence along with the particular challenges of poverty. A social/cultural psychologist with an understanding of the changing patterns of work may understand the meaninglessness that young people experience, because available work for unskilled workers has decreased, except for CDEP work or social security, which provide almost exactly the same rates of pay. CDEP work can provide a valuable pathway to 'a real job' if linked with training and it can also provide an alternative to those who wish to remain in their communities, but it is also perceived paradoxically as a dead end by many young people and as low status or 'dirty work', that is not and can never be 'a real job'. Thus for some it is a poverty trap, because if CDEP and the 'the dole' provide equivalent rates of pay and if these are perceived to be the only options available, the notion of education for employment opportunities is questionable. These can be demonstrated as follows (see Diagrams 6.1: The context of alcohol use).

To address the issue not only do professionals need to work together, but there needs to be a realisation that the modeling of 'at risk' behaviour by adults and the resultant lack of sleep as a result of domestic violence play as much of a role and the host of factors that lead to marginalised people resorting to the use of alcohol. It can be argued that once people (Indigenous and non-Indigenous) have better life chances they are less likely to use alcohol in hazardous ways. Also the environmental factors such as access to services (transport, safe housing including power, water and refuse removal) are essential in order to improve health and education and then employability. Indigenous connective frameworks traditionally conceptualize webs of factors that shape their health

[14] See Editorial, *Criminal Law Journal* Volume 22 August 1998 page 201. Mandatory Sentencing was introduced on 8 March 1997. This removes almost all judicial discretion. Judges and magistrates cannot ensure that the punishment fits the crime. Mandatory Sentencing applies to a wide range of property offences, and does not distinguish between the most trivial and the most serious.

and well-being and this resonates with the latest systemic thinking that eschews categories and instead considers the ramifications of interrelated or webbed variables that connect across health, environment, education, employment and crime prevention policies. A broader intervention than merely a medical model of casework and counseling, or an education model of supportive learning environment, or a reactive, punitive model such as mandatory sentencing is required if the issues of drug and alcohol abuse, poor school retention rates, unemployment, suicide and domestic violence are to be addressed. If each of the participants in this scenario sees only a slice of the situation, but not the whole, little can be achieved. Added to this, the status differentials amongst the professional participants also shape decision-making (Smith and Williams 1992, Reason 1991), as do the personalities and value positions of the various participants. Assumptions and values shape policy and practice. A broad systemic approach could take a preventative stance to youth alienation by working across a number of sectors and disciplines. Ideally for instance, public health approaches do not try to isolate a single cause. Instead an attempt is made to discover web-like causation and feedback systems amongst variables. Health and development outcomes have not been achieved, because whilst considerable effort is made in terms of resources, the interventions often work against one another. For instance: some people blame individuals for their social health problems and for behaviour that society considers criminal and emphasise the responsibility held by individuals, not societies for social problems. These are explained in terms of what the individual ought to do. Disciplines are used to focus on the psychology of individuals or on the impact of the social dynamics of families on individuals to explain issues. Others lay the blame squarely on society and history and stress that current social structures are to blame and the history of dispossession and social marginalisation is the major current cause of problems. Fragmentation of thinking in terms of set paradigms or models can limit the ability to solve problems.

Also there is considerable antipathy amongst individuals and organisations working from different points of view. This leads to undermining one another at a personal as well as at a professional level. Conflict is used for political and economic advantage, not merely to hone thinking through constructive dialogue.

Diagram 3.2 conceptualises 'Either or Thinking' as causing polarisation of options in problem solving, particularly in relation to crime prevention, based on the opposing assumptions that individuals are free to choose to conform to laws or to break laws and that they must then take the consequences, versus the assumption that society is structured in such a way that for some people, life chances are limited by a host of social,

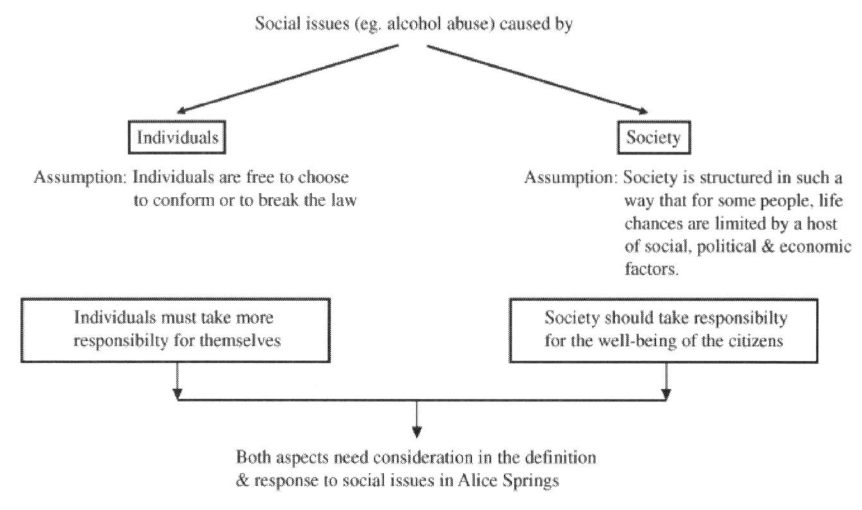


*DIAGRAM* 3.2 'Either or Thinking'

political and economic factors. Even more importantly, the very meaning of responsibility can be different in cultural terms. Alcohol misuse in Alice Springs is the product of many factors. The Indigenous cultural nexus of the norm to share and the norm to use what is available (common to all desert communities, CARPA Newsletter 1998) and the Territory drinking culture across the entire population and the social availability of alcohol (more than seventy outlets in a small town) is part of the problem. The other usual factors such as peer pressure amongst young people, a lack of entertainment options and personal risk taking behaviour of young people, added to modeling the behaviour they see in their own families are factors. Finally, social responsibility and reciprocity are eroded as a result of welfare dependency (Rowse 1998, Pearson 1999).

There is another way. Instead of working in categories, holistic thinking can address interrelated social problems as discussed in chapters 6–9. This does not mean that widely opposing values of social justice need to be jettisoned. Instead there needs to be a way of moving toward a realisation that through ongoing dialogue amongst those with different ideas, the liberative potential of a focus on both the individual rights and responsibility and social rights and responsibility can be discovered. A systemic approach to health and development transcends compartmentalisation and the isolation of social problems from an understanding of the social, cultural, historical, economic and environmental context. Networking, thinking holistically and exploring different ways of looking at the same issue are the most vital aspects of a systemic approach.

Assumptions shape the way we define social problems, which in turn shapes the way we implement policy and practice. If we see problems in a compartmentalised way, we will operate in a compartmentalised way. In order to bring about a change a systemic approach is needed that is not limited to bureaucratic organisations but instead uses a broader range of arenas.

The way people think about key concepts such as development, education, employment, transport, health and citizenship impacts on the way in which policy to address social and environmental issues is developed and implemented. Research, policy and development based on the notion of controlling variables in the real world (such as socio-cultural, political, economic, technological/scientific and environmental variables) in order to engineer changes is based on an attempt to control or limit causal connections. This approach to development assumed that by imposing changes on people positive changes could be achieved and that narrowly defined and specialised projects could successfully address specific effects in response to specific causes. The reality is that the complex multifaceted world is not equivalent to a laboratory where factors can be addressed in isolation. Also people need to participate at all stages of a development process because development is only meaningful if it is defined to be so by all the participants (particularly those who are in the process of being developed!). Top down approaches on so-called less developed peoples by more developed 'experts' have been unsuccessful.

Thinking needs to be systemic and health and development solutions need to be achieved through working in partnership across organisations, across subject areas or knowledge disciplines and across the sectors of health, education, business and crime prevention. Social problems are as much a product of the way we think and conceptualise issues as they are the result of social structures. Apparently discrete problem areas are in fact linked locally, nationally and internationally, as conceptualised in Diagram 8. If this is understood then a co-ordinated approach can be developed based on partnerships with a real sense of direction. The systemic links that can build or erode social capital can have a greater likelihood of being addressed at the local level (see chapter 7).

## *3.6 The Nature of the Identified Complex Social Issues*

We need to address the following challenges: (i) compartmentalised ontology rather than systemic ontology that see connections across issues. We need to avoid concentrating on an effect of a problem, such as the high rate of alcohol associated mortality and morbidity and seeing it in isolation from the societal system that contributes to it or shapes it. We need to avoid narrow responses such as crime control, rather than considering the wider context of prevention as a priority and (ii) narrow notions of research epistemology rather than scientific humility that depend on being open to criticism. The probability of 'truth' is based on failed attempts to falsify assumptions, not on narrow dogmatism. If the holistic picture can help to demonstrate the absurdity of compartmentalised social policy solutions, then I have achieved part of my objective.

# 4

# *Missionary, Mercenary, Misfit? Boundary Work and the Policy Research Process*

> But now I am in Jung's land ... I became fascinated by the language ... because of its underlying poetry ... It occurs in Jung's own writings ... if we don't know what we are as individuals, then how can our choices tell an observer anything about our 'individual' values?
>
> *Churchman 1982: 44*

## 4.1 *Being Part of One's Subject Matter and the Implications for Praxis*

Where we position ourselves affects our social, emotional well-being and either places us within the taken for granted or beyond the cultivated boundaries and in the wilderness. All the layers of identity as female, blonde, Australian South African with social justice leanings led to my constructing and being constructed in particular ways. From my perspective it appeared that in a predominantly patriarchal 'Territorian' society (where a hot dry country requires tough responses, mateship and a cold beer after 'footie'[1]), it seemed that playing the game was what was required. This is, to quote a local, essentially a company town of professionals and business people who make a living out of service delivery and research. To differ from the norm appeared to excite interest and suspicion, fueled by the interlinked dynamics of a small town. 'Keep a low profile, play a close hand' was the essence of the advice given, 'because if you are associated too closely with that group you will be ostracized by the other group'.

[1] Australian colloquialism for football.

In her article 'Our responsibilities as systems thinkers', Romm (2001) quotes Banathy (1998) 'what we know about the world becomes projected onto the world'. She adds 'A recognition of this means that we begin to operate consciously in 'constructivist' mode, taking some responsibility for the way in which we generate constructions (and thereby bring forth possibilities for seeing and acting) through the knowing process. Knowing can become an act of intervention in the social network.' She discusses the way boundary judgements (as per Midgley 1996) can define who is a legitimate knowledge maker and who is not. She stresses that the issue of accepting that there are many versions of a story is important in co-creation of shared realities. For a researcher however there is a need for self-reflection on why we think the way we do. Critical examination, according to Romm (who quotes Gregory 2000: 497 in Romm 2001: 9)[2] 'should take account of emotional, intuitive and other psychological features of actors participating in the dialogue.' Romm goes on to say: 'She admits that it is beyond the scope of her paper to fully explore what this taking into account might involve. But she indicates that the motivation of people to be prepared to reflect on their emotional attachment to specific concerns, can be derived from their appreciation that others too are required likewise to undertake such self-reflection as part of the process of dialogue (2000: 500). She hints that it is on these terms that we can hold people accountable in processes of human dialogue (1996: 2000)'(Romm 2001).

The political and the personal are played out through the research process. Churchman (1979a,b, 1982) asked: how can our designs encompass our human frailties and strengths or 'the enemies within', namely 'politics, religion, morality and aesthetics?' In his work he draws attention to the paradox that without these we are lost and with these values we act in ways that make us less than human and less than compassionate and considerate. He asks: what do we do when our designs are based on logic that fails us? Induction can enable us to leap from the known to the unknown (Banathy 1996); it can lead to creativity and 'the pit' (to cite Flood and Romm 1996). By drawing on Churchman I began to realise that deduction could be flawed, based on mistaken premises. Further, Churchman argues that faulty equipment and flawed readings are based on human senses (sight, hearing, touch and smell). He stresses that being tired or emotional can cloud judgement, but suppression/denial of feelings and ignoring

[2] Romm, N. (2001) Our responsibilities as Systemic Thinkers 45th International conference of international Society for the Systems Sciences, Asilomar.

intuition can 'cut off' useful insights, to use his phrase. He stresses that recognising a perfect system incorporates the notion of goodness; that elusive quality that is by definition part of the perfect monad (Churchman 1971:41), we can be more accountable to others.

Interactive design for the future (as per Ackoff in Banathy 1996, 2000 and Churchman 1979a,b) requires exploring issues and ensuring interactive processes that preserve the sacred and envisages new approaches (built on a sense of what is most important to the human condition and the environment). It co-creates (in the sense used by Reason in new paradigm research and 2002)[3] a shared vision, and then sets in place a process for implementing the ideas. It is based on the belief not only in the creative potential of human beings, but in the hope that they will re-shape ideas and structures. Even though our life chances are limited by ideas and structures, the process of thought and dialogue can help us to 'leap beyond' the limitations of our constructions and to re-formulate choices (Banathy 1996).

The very notion of who we are and what our values are as a citizen is implicit in the process of drawing up the research parameters. The struggle in terms of ontology (the nature of reality) and epistemology (questions about the nature of knowledge) shape the research process, the content and the outcome. Through the use of broad approaches to looking at communities systemically and through shaping the research methods to include qualitative and quantitative data gathered (not only in terms of the narrow parameters of questions) but in terms of the issues that emerge beyond the parameters, we can move closer to becoming accountable to one another.

In other words the picture that describes the reality beyond the rigid research frame is swept in from this perspective. It is the detail that is not captured by binary oppositional or simplistic categories that is important when we strive to be accountable to others and ourselves. It is the questions that emerge through conversation that give an insight into what people think about relationships, values and the research. Research is a political process from conceptualisation to evaluation. This is a story of attempting to operate in terms of a participatory and systemic approach rooted in values of ecohumanism. The operative word is 'attempt'. A Pandora's box of issues emerge that stretch beyond any limited frames of reference.

[3] *Justice, Sustainability and Participation* (2002) Concepts and Transformations 7(1): 7–29.

The issue of power (associated with different worldviews) is perennial and not one that can be set aside. The 'might–right' (Flood and Romm 1996[4]) issues that one observes and in which one participates need to be surfaced, whilst mindful of the strategic implications for the stakeholders, in order to assess the way that power and values impact on data that are presented as fact. To sum up my attempt at critical thinking draws on Habermas (who is concerned with developing truth and rightness through rational discourse) and anti-foundationalist approaches that strive to preserve the jump lead of diversity in the way knowledge is constructed. In this context, the difficulty of developing meaningful research is forefronted. The work of Churchman (1979) is also important in so far as he stresses that the process of 'unfolding' and the process of 'sweeping in' create a dialectic of self-examination and examination of the context in the broadest manner possible and that 'to decide', 'to create boundaries' is to 'cut off' (1979: 85) possibilities.

The possibility of systemic analysis of might–right issues by planners who do not look within, to assess the quality of the thinking associated with the self is problematic. If construction is recognised as relevant to knowledge creation, then the personal context of the individual mind is relevant to the process of co-creation. The paradoxical interplay between the construct of self and other is relevant. Churchman (1979) discusses the need to look inward and outward which resonates with Jungian and indigenous ideas of self-knowledge, in order to know others. This is vital for all social policy makers, not only those who focus on health and well-being in a narrow sense.

Globalisation also contains a central paradox. Whilst on the one hand it leads to the increasing control of markets, there is a simultaneous increase in the liberative potential of social movements to address individual, social, political and economic issues as a result of global networks. I describe the impact of globalisation on management styles in public sector research and development and the way that liberative research and development can be controlled. Chapter 7 addresses social movements for change and how social movements at the international level for social justice[5] can play a role in bringing about change, provided space can be found within the public and private sector and provided that the ideas are not colonised and contrived to mean something entirely different, whilst using the rhetoric.

[4] Flood, R. and Romm, N. (1996) *Diversity Management*. Wiley, London.

[5] For example through Indigenous movements for land and health and through the World Health Organisations' Healthy Cities and International Council for Local Environmental Initiatives Agenda 21.

The case study provides an opportunity to discuss methodology and the context of research management and its implications for policy and practice at the local level that has resonance for other researchers nationally and internationally. The case study gives layered narratives and analysis, but also attempts to unfold the text and context of the research findings and process. The study addresses my role as it intersected with the subject matter. In this research process I acted as a catalyst and was constructed in different ways by stakeholders with different sets of assumptions and values. I was seen as 'Trojan horse' (illustrating the hope that I could achieve some change from within the system, as perceived by the labeller). I was also seen as a 'blow in' (one who arrives quickly and disappears equally so). Simultaneously some saw me as an ' ideologue' (an outsider with ideas of equality, who is essentially out of touch or perhaps without an understanding of the status quo). Fortunately some also saw me as a sincere facilitator whose ideas were in harmony with many who wished to consider their options. These labels will be unpacked and analysed.

The use of dialogical vignettes enables me to explore ideas through a co-creative process that *unfolds* ideas and *sweeps in* relevant concerns (in the sense used by Ulrich 2001). A version of retroductive logic is used as a tool for analysis of the data in chapters 5–9. It is adapted from a Marxian critique, from CW Mills (1975) for an emphasis on history and biography and from Churchman for taking into account 'religion, aesthetics, politics and ethics' (1979). Whilst Habermas (1984), Ulrich (1983, 2001), Flood and Romm (1996) are valuable for learning about critical questioning and triple loop learning. Foucault and Gordon (1980) are essential for understanding constructivism and its relationship with power (without taking it too far down the track of post-modernism, because of implications for human rights and environmental justice). Zimmerman (1994) and Beck (1992) are relevant for 'sweeping in' (as per Ulrich 2001) the environmental issues. In this case study it means unfurling the coils of a system that leads to disadvantage. 'Objectivity, subjectivity and intersubjectivity' (as per Zhu 2000) are thus considered important parts of the ontological whole and that if we address what Jackson (2000, drawing on Habermas 1984) characterises as 'technical, strategic and communicative knowledge' then we need to accommodate space for all three levels of knowing. Also we need to accept that constructs may be rational to some and irrational to others, or subjective to some and objective to others. We will only find out the contexts through talking with one another in an open and respectful way.

Recursiveness (adapted from Giddens 1991) is the concept used to address the vital ingredient of citizens (including this writer) being able to make a difference by virtue of the way they construct and reconstruct their thinking in response to their lived experiences. These two tools (retroduction and recursive thinking) were useful to me in understanding issues. Minus this jargon the essence is looking at oneself and exploring the paradoxes imbedded in stories and the way structures are constructed. These are important for building local governance capacity. PAR is such a useful process to enable people to (i) make the connections across their own lives and the lived context (ii) to work with the boundaries of sectors and knowledge areas to bring about changes for social and environmental justice. Citizenship in non-Indigenous terms is about the cash work nexus (Pixely 1993) and about paying one's way in society.

The notion of non-Indigenous citizenship has been explored by Rowse (1993, 1998a,b) as a parallel concept when he discussed dependency and exclusion from citizenship rights and responsibility. A deeper question is not whether citizenship in non-Indigenous terms is the same, but whether there is in fact a homogenous response, which does not appear to be the case, given the diverse narratives that I will share with you in later chapters.

Given the diversity of responses to self-determination it is likely that having multiple options and multiple forms of representation may be important to some, but undoubtedly the nation within a nation is the overarching/central response, because of its political and emancipatory function at this stage in Indigenous history.

But nevertheless the issue faced by many of the most marginalised Indigenous residents living in a housing association on town camps remains:

(1) Should we fulfil our immediate obligations to our family first and share whatever resources we have at the time with them (as would be appropriate according to traditions, where all things are shared with extended family who come to town for a range of reasons).
(2) Do we save our resources so that we can fulfil our civic responsibility to non-Indigenous forms of rights and responsibility?
(3) Who decides what questions are useful? Who decides who should answer them?

Family affinal (relationship) and agnatic (blood ties) play a role, as does language and culture. Many Indigenous people form many interest groups by virtue of language and culture. Criss-crossing family ties and status positions make the matter complex. Resistance in some ways and participation in some ways to non-Indigenous forms of rights and responsibility is helpful for self-determination. What kind of self-determination can be explored?

Participants raised only some of these questions and which ones are useful questions or not is open to question. Some of the questions were explored by Indigenous people in a governance workshop. Some further questions were raised and although I facilitated some of the discussion, I was not privy to much of it. Being non-Indigenous precluded my involvement. Sometimes co-creation requires stepping back and respecting that others have the right to their own space and their own time to consider their options and the implications of multiple narratives for Indigenous models of governance and obligation.

To sum up, drawing on Jackson and Flood (1991) Critical systemic praxis (CSP) is characterised by: complementarism, co-creation, emancipation, critical reflection, 'systemic sweeping in' (see Ulrich 1983) and commitment to a revised form of the enlightenment approach and humanism. It is mindful of the contributions of idealism and materialism. The skills that are needed are: (i) participatory design and decision-making using tools for policy development, such as triple loop learning (a diversity management tool), (ii) an ability to think critically based on theoretical and methodological literacy, reading available statistical data (for example ABS and Australian Institute of Health and Welfare statistics) and (iii) ability to apply qualitative and quantitative research methods PAR processes to needs expressed through service usage (Bradshaw on definitions of need in Kettner et al 1988) to establish 'normative needs, perceived needs and expressed needs of current service usage'. (iv) communication skills, counselling, advocacy and negotiation, networking and lobbying skills.

Although most of this section was written prior to reading Ulrich (2001), I would like to acknowledge the sense of resonance I felt when I read his article on competence. I support his argument (drawing on Singer and Churchman) that competence in research is based on 'sweeping in' areas that are considered outside the boundary. In this research the methodology is used in order to 'test' my narrative.

> While the tradition of systems theory has long been rather weak with respect to the philosophical foundation and critique of applied systems thinking, the tradition of critical social theory has been similarly weak with respect to the practicability of its critical ideas. If we are to learn how we can practice the pragmatic maxim in a critical way, we need a way out of this impossible alternative of practicability versus critical defensibility. The way out that I propose with Critical Heuristics is what I call the process of systematic boundary critique (Ulrich 2001: 12).

## 4.2 *Entering the Field and Reflection on the Approach: Time, Space and Working the Hyphen*

> "Critical subjectivity means that we do not suppress our primary subjective experience that we accept that our knowing is from a perspective; it also means that we are aware of that perspective and of its bias, and we articulate it in our communications. Critical subjectivity involves a self reflexive attention to the ground on which one is standing and thus is very close to what Bateson (1972) describes as Learning III)" (Reason 1994: 327).

I was told (by a newcomer) that some locals greet newcomers jokingly as they fix their gaze. They initiate conversations with the sly comment: 'there is a saying in Alice that everyone can be classified as a missionary, a misfit or a mercenary'. It is a very useful poser, one I have attempted to address, because understanding requires that the lens be pointed at both the self and the other. My ideas about social justice for all (and particularly young people) and the need for systemic approaches to development have made some people think that I have operated with missionary zeal. Certainly, as a sociologist/social anthropologist employed to undertake research for future planning I am a little unusual, in that I use methodology to straddle sectors and disciplines. Also I use participatory action research and tools for communicating across paradigms to enhance meaningful communication and participation. Then I have tried to come up with social policy ideas that address all interest groups, in order to enhance the life chances of those in the greatest need. And yes, I have worked with government, non-government and the private sector in a way that is different, in so far as I work with terms of reference rather than within frameworks. I worked for a salary, but I have framed the research task to address the concerns of the marginalised without compromising the larger picture. Research can be used to empower or disempower participants. This is well understood by people who have been the subject of social research in Central Australia. 'Putting the last first' to use Chambers' (1983) term, has been my creed without a romanticised sense of present or past reality. Research (designed in terms of a cultural model) has been 'a mask for non action' to quote Robert Chambers (1983). This makes undertaking research a challenge amongst people who have been regarded as 'a honey pot' for researchers doing thesis work that leads to little change.

Ethnographic accounts that attempt to address the interrelated facets of reality and frame an account of them (in order to shape social policy) to first attempt to present pictures of themselves, so that their assumptions and values can be seen for what they are. And so that others may be able to read their research findings with a greater sense of understanding.

As a 'blow in' (a newcomer to the Territory who may move on at any time) with an obviously non-Australian accent in their opinion, I was considered to have strange insights into the local situation, because I was able to draw parallels from elsewhere, whilst having equally strange areas of ignorance. I was as much observed as observer. As a newcomer in a small town the gossip network positioned and framed me the researcher and me the person. Questions and comments such as these indicate the range of opinion over the period of 3 years and the growing friendships, whilst the interpolation of my comments give a sense of the way I have interpreted them:

> Can you cope with the heat (locals pride themselves on being 'able to take the heat' and being tough minded); 'Are you from overseas or down south?'(Down south is the ultimate insult, because 'the bleeding hearts come from down south'). 'It depends on how long you are going to be here' (so much time and energy is wasted on newcomers who buzz off just when they are getting helpful). Well what would you know ... you are new? We will help you to speak to the right people. (If you are going to address these issues then you had better get the story right) Don't trust anyone in X (in a small town, that is highly politicized, the gossip networks are active and everyone in town is in the web). You will be shot if you say that. Ethical research may mean that you are unpopular with some people. They will crucify you! Good on you keep at it. Anything you need, I can try to help. I know someone who had to go to Tasmania! It is more than my jobs worth (to tell you that). Its like the Middle Ages ... if you are too different ... . The research is too big, how can you cover so much? Why do this research; we know what the problems are. I will introduce you ... You are a Trojan horse ... good on yer ... you are a bloody academic though (maybe you can make some contribution). Be careful. What are you trying to do? How will you get any change? You are as busy as a lizard drinking! (This is a compliment). Yes, but your teeth are very blunt, aren't they? You will find that we are very powerful (this was not a compliment made by a political player who was invited to a meeting on the WHO healthy cities approach. This was not what was wanted on his patch!). Yes it is nice to be part of the same network (after three years. "We of the Never Never" stick together (a phone call on Christmas day from an Indigenous friend). I wondered whether to be involved. Some Indigenous groups don't work with others. We need to address that, but I need to address that. That is why I am going to help. It is hard to do the right thing ...

The questions and comments indicate the range of opinion over the period of four years and the building of rapport. Documenting the comments is in line with anthropological tradition of reflecting on the process of research. It isn't intended as an exercise in narcissism! But it does serve to bring me into the research frame and to demonstrate that understanding the affiliations of who says what and why is a starting point for co-creating meaning through working with different interest groups. If a sociometric diagram[6] of the local leaders and interest groups can be mapped out and if in addition the comments made to one can be analysed in terms of their assumptions and values, their attitude to you and to the research, can be mapped out, then the task of creating links and alliances can be addressed systemically. This monograph is largely about development methodology per se and addressing my values, perceptions and emotions as they play a vital part in this attempt at understanding the complexity of a development initiative. By vignette I mean a picture that telescopes reality and in so doing typifies an issue or a perception. In all instances initials only are used and are not necessarily meant to signify any particular people. The conversations and situations are real in so far as they are based on a distillation of sharing in similar contexts. The first question I asked as a participant observer and systemic researcher is where do I stand? It is easier to attempt answers through dialogue and reflection.

Narrative dialogue creates the connection within oneself and across self and other. It is a useful tool to ensure that the self (so often carefully removed from the frame of the research, in order to preserve one's own privacy and position of power and authority as 'the expert' studying 'the other') is part of the story. Ethnographic accounts that attempt to address the interrelated facets of reality and frame an account of them, in order to shape social policy need to first attempt to present pictures of self, so that our assumptions and values can be seen for what they are. And so that others may be able to read our research findings with greater sense of understanding.

The personal is indeed political and despite the hackneyed usage of the feminist phrase, it remains a truism. Nevertheless I am mindful that 'writing and representation, regardless of style, always involves treachery because of the creative impulse associated with both' *(Carter in Cvetkovich and Kellner 1997)*. That is why the dialogue of working the hyphen (Fine in Denzin and Lincoln 1994) across self, other and one's environment is so important without trying to pretend an understanding one can at least explore the tensions involved in the process of understanding. In this work, my voice is the loudest and the most dominant. Perhaps the very reason I have written it is not only to give voice to marginalised others, but also to give voice to marginalised self. Perhaps sharing a sense of marginalisation can give my voice some authority and authenticity.

Values, perceptions and emotions of the stakeholders (in any community) play a vital part of systemic research. I try to show the implications of values for our positions on social justice and to avoid slipping

[6] See McIntyre (1996) Tools for ethical thinking and caring, where I explain how to use sociometric diagrams.

into zealotry (based on one paradigm) or cynicism (based on nihilism associated with the extremes of post-modernism) when I strive to co-create share outcomes. This can be very difficult at times but acknowledging values and emotions is essential for research that is honest, in the sense used by Harding (1992) in that it strives to address the layers of meaning that taken together could give a metaconstruct of social problems. The process strives to avoid bounding meaning constructs in terms of a single set of values.

This approach does not fit any particular category of research in terms of methodology, theory or discrete topic. It is critical and systemic in approach and as such employs multiple methods to work across rather than within boundaries. Detailed information on health, education and employment and the many examples are used to emphasise the fluidity of issues and how intervention or non-intervention in one area impacts on the rest of the system. I also stress that my being part of the system as a change agent had a significant impact on my own life in a physical, mental and spiritual sense. My role and motivation for co-creating meaning is not outside the boundaries of this research. I reflect on an experience of undertaking a study of the life chances of citizens in Central Australia and how this impacted on my own life.

The purpose of this study is as much about an attempt to achieve self-knowledge and self-healing as an academic, citizen and as a private person as it is to understand the dynamics of a place. Perhaps it is also to give myself (and the other marginalised voices by virtue of age, gender, culture, values and particular views on ontology and epistemology) a front seat and a hearing. And so in this work the voices of the inarticulate are articulated (admittedly through a co-created series of conversations). In this town the majority of people work for government or government funded programs and residents who understand the dynamics of the place have called it 'a company town'. One of my first challenges was to assess the comments for content and for an indication of the way in which I was viewed. Also I situated the people in terms of their social, political and power standing. It became clear that those with most power were least enthusiastic about the breadth and scope of the research.

## *4.3 Location and Dislocation: the Space for Writing, Individuation, Recollection and Reflection*

A wide canvas appeared to be risky, because it provided a different perspective. In order to apply CSP, I need to explore the emotional content of the research as well, because it gives an indication of the full story.

Heuristic, composite conversations about the research from discourses at the domestic, the local public and the local private and the wider international context give quite different layers of meaning to the material. *First,* I include vignettes of conversations located in public workspace with a visiting colleague who saw himself (quite rightly it seemed to me), as a leader in social policy reform in Alice Springs. *Second,* I recall vignettes of meetings with people living in Alice Springs. *Third,* I recall a conversation with a long standing colleague who knew my work from an early stage of my career in South Africa and thus could give a perspective on my path to CSP. The conversation takes place in a conference space at an international conference. *Fourth,* I introduce a vignette of a recalled conversation with two friends at my home in Adelaide. This is the most personal and it gives an insight into emotional dimensions that are woven into my work, albeit less visible in the finished product.

My locational and positional space moved from an urban centre to a regional area, that paradoxically was a centre for service provision to the surrounding desert. The desert provided a space for thinking, an alternative point of view and a shift to another level of thinking about boundaries.

No one goes to Alice without reason. I went for personal as much as for professional reasons, to leave behind memories and to create memories, to heal myself by being of service to dignify others. I think that just about sums it up. As a participant observer, I tried to analyse what I saw, heard and felt. I lived the dynamics of the town I describe.

Essential for analysis were my sense of sight, my hearing, touch, emotions and the will to think conceptually and reflexively. Attempting to co-create meaning was a key challenge. At times the cynicism and zealotry—my own and that of others—made my task of co-creation more than difficult and the temptation to lapse into cynicism and zealotry became almost overwhelming. The struggle was expressed with familiar muscular and mental tension! This struggle is reflected in some of the dialogical vignettes. These are a means to explore recollections and to learn from the experience.

As a researcher stereotyped as global researcher/knowledge worker I found that mining the minefields of meaning is instructive, because it helps to locate me the researcher and the research per se. As an outsider, although people with whom I worked were interested in me as a person, they saw that I had a wider life beyond their place. The phone calls I received after I left Alice and the notes meant much to me and enabled me to continue to work as a facilitator.

The initial questions posed to me about the rationale for research were very useful and so I have attempted to answer them, because transcultural understanding requires that the lens be pointed as much as at the self and as at the other. I am an embodied social actor with locatable cues. I have a particular layered accent and vocabulary, product of the many places where I have lived and my background.

### 4.3.1 *Vignette: A Conversation About Identity (Political and Personal) as Expressed Through our Role as Academics, Teachers and Researchers*

The research data are based on what I observed, smelled, felt, thought and sensed as a result of interacting with others who co-created all my experiences:

Hi Janet, at your desk I see, what are you up to?

Hi G, you make me sound like I am plotting something!

Well mate in this town it doesn't take much to be labelled a bit of a stirrer, even if you seemed an unlikely stirrer at first. The white tribe are very keen to look to their own and even though I am only a middle of the road academic you would be surprised how I have been labelled! Talk of red under the bed!

I don't really understand how anyone can talk such nonsense … by the way what did your students think of the seminar on my research? I replied a little stiffly, trying to sense his intention.

Oh well once they got beyond the big words and you started to talk from your heart about the politics of research then it became worthwhile. Indigenous people have heard the word 'participation' so many times; the point is 'on whose terms' and 'to what effect'. What will happen to the research once you have done it?

I can only do it as honestly as I can and then try to convince decision-makers that the findings are relevant for everyone's benefit. What is interesting is the extent to which people are concerned about what I will say and whether I will consent to being 'reined in to quote a concerned party'.

I guess we can only do our best … you look really pale …

Yes, I am OK, just stressed, a bit absurd really isn't it … I feel like a puppet trying to wriggle against the pull of authority, who can't quite see beyond their blinkers[7] and are also not sure which way to pull me, only that pulling regularly constitutes management. It isn't comfortable … . Particularly as I am always trying to go beyond my own blinkers. Do you know the work of Bateson on ecology of mind? He cautions that we often operate at a level one (within a single paradigm) rather than a level two or even three where we operate within narrow confines that can be atheoretical and without any conscious understanding of methodology, purely bound within a framework of theory or methodology or preferably able to see and locate frameworks and consider their ontological and epistemological implications.

That is all quite interesting I am sure, but is what you are doing really worth it … I mean there is more to life and getting people off sides isn't smart … it can affect they way you are seen as a future employee. Remember this is a company town. Most people here (just about) are employees in a business that needs government support or at least does not want to get off sides, or more likely works for a government department. If you fall out of line here, you leave town. Most of the people here are mortgaged to the hilt. They play the

[7] Refers to horse blinkers used to ensure that they would not be distracted by other traffic on the road.

tune they are paid to play. This is a service town, service to government and service to the surrounding desert.

Sure, I understand what you are saying, but I cannot do research in a way that I believe is not only unethical, but that it does not really try to represent the other. The work after all is to represent the other isn't it, so as to provide policy directives to address the identified issues? I take my work very seriously. I try to represent the views of stakeholders within the social, political and economic context. I believe that research is about representing the truth of different stakeholders and my own truth. Of course the stories are linked and they shape one another. But telling the whole story can give a better insight into the research content and the way in which it was gathered. It is not that what I am saying is in any way not in the interests of the state, I believe that it is, it is certainly about achieving access and achieving a sensible approach to rights and responsibilities through better understanding of the axial themes.

That could be interesting said G, cautiously in a way you could do an ethnographic study of your time here, all the details.

Yes but not ethnography about me the researcher doing pristine data collection about the locals. It needs to be more than that.

Good on yer … that could be an interesting read … a step away from the colonial approach of learning about the other.

Yes that's right, it will be about my experience of living here and how I was perceived, but also about the research content, which is important. It is just that if I am to do critical systemic research then it needs to include a study of how the research context changes as a result of the presence of the researcher and also that what she thinks and feels contributes to that reality.

G looked vague and said, Oh well sounds very academic to me but I wish you luck with your project. I could sense that he was trying to look supportive, but he had his doubts, after all I was an unknown quantity.

Yes, the story needs to be more than just a distillation of all the literature I have read, all the data I have collected, more than a montage and a reflection on the material. I want to work the hyphen as Fine[8] expressed it; but we need to add the environment. Some of the Indigenous people I have spoken with understand the last part very well. I want to portray the experience of being a researcher here and how the other (made up of many interest groups) has perceived me. The dynamics of research on policy and policy on research are very important and the way we operate can change things … sometimes we may be told that we are wrong and not be given any credit for the ideas, but eventually they are taken on board and normalised. Sometimes this leads to real benefits and sometimes the words are just used, without any real content. But even so sometimes it leads to a few changes.

### 4.3.2 *Vignette: Who Make a Difference?*

I walked through the large domed construction that housed the stands for the Agricultural Show and found the stall for Reconciliation, set up as a Commonwealth Government initiative to facilitate reconciliation between Indigenous and non-Indigenous Australians. I had agreed to answer questions for a few hours and to hand out colourful

[8] Fine in Denzin and Lincoln (1994) *The Qualitative HandBook*. Sage, London.

posters of a smiling Kathy Freeman and stickers of 'Walking Together'. The time flew by. Some people were glad to receive policy pamphlets, few were anything but polite. At worst they shook their heads or avoided my eye.

There you are said M, well done!

What do you mean I asked? M always tended to be more enthusiastic about my contributions than I thought was warranted ... although this enthusiasm was always appreciated.

You won an award ...

What for I asked? As I didn't enter a prize vegetable?

No but your project received a commendation at that stand over there. She pointed at a stand at the end of the hallway.

Really, I exclaimed ... are you serious?

I am she said ... you can now regard yourself as officially sanctioned in a positive way!

For a newcomer to a company town, it isn't too bad is it?

She walked off satisfied that she had handed me a bouquet. She had in a way and the feeling of receiving a gift of belonging warmed me.

So what do you mean by marginal lives and what is wrong with current policies? asked a colleague in the public sector after reading some of the material and talking with me about it.

I see the implications of marginalisation, for all the reasons discussed, the lack of empathy for people who are seen as dole bludgers by those who are working (very often for longer hours than they ever have before). Here in Alice it leads to a particular sort of contempt. Alice is a particular case. Although there are more human service professionals here than elsewhere the social health outcomes are so much worse than elsewhere. That is because the approach to social welfare policy is too compartmentalised, despite the rhetoric to the contrary and because intervention is in terms of survival needs, not in terms of meeting the holistic needs of people to build social capital. That is what worries me, I am sorry if I sound like I am lecturing.

But what do you mean? So much money is spent on health and education and it is not anyone's fault that people are unemployed here. How can you employ people who cannot read and write in English? How can you employ people who do not want to fit into the market system?

Perhaps we need to take each part of what you have said and look at it carefully. Certainly this is the opinion many share and it is opinions like these that shape social policy. First of all being deserving or undeserving was the basis for deciding you should receive welfare in terms of the social wage when it was first introduced in Britain and it is certainly the basis for residual welfare (qualified welfare) today. The social wage is different from having access to a healthy setting that gives access to health, education, recreational services so that people feel that they can participate in the social fabric of the town and not be isolated in their homes or town camps. Just because we can't see poverty, it doesn't mean it is not there.

Oh I think there is plenty of poverty and it is because people waste the social wage on alcohol, cigarettes and gambling.

Yes it is wastage and I dislike wastage as much as you do. My particular dislike is to see people destroying opportunities through spending money unwisely, but so many people do waste money from all walks of life, don't they? Just think of how much money organisations spend on fireworks for public entertainment in Australia.

So why talk about fireworks?

Well affluent spenders in the public and private sectors like to celebrate life with fireworks, here in Alice and just consider the display for the Olympic games in Sydney. What an extravagant use of resources, for immediate pleasure. Emotion and pleasure are factors in decisions. I think that is also why people use recreational drugs, for immediate pleasure, but there is also another side. People use drugs to make the unbearable in the lives more bearable. That is the difference. Drugs provide an escape from reality. And gamblers are also pursuing a way out of the norm, the everyday.

So what has that to do with the deserving and undeserving poor? Why do you think the welfare state isn't doing enough for people that waste so much?

I find waste abhorrent too actually. Waste of money that could be spent on food for children and cleaning materials for the home so that basic hygiene can be maintained also bugs me, if I am honest. But the point is that people have had different lifestyles and the use of water for cleaning was not common in desert communities in the past. Also one is measured by what one gives away, not by what one saves for later. If you have resources in an egalitarian society, you give them away. It is the norm to be generous. But the point is that greed and waste is part of capitalist society and the market economy. When you think about it, it is western capitalism that has taught Indigenous societies everywhere about waste on a grand scale: pollution of the air and water supply and the dumping of toxic chemicals. Also the construction of want and desire through advertising that renders objects desirable and in fashion or as undesirable because they are cues to ones status as being poor.

That argument doesn't really address the point though does it? People can't expect to be given the social wage by the hard work (a lot of it unpaid overtime) of taxpayers and to expect to go on wasting it. It makes me sick, to see the way people get away with things. I drive to work in the morning and think how nice it would be to be lying in bed listening to the news and drinking a second cup of tea. Dole bludgers have a standard of living this country that is the envy of people in the rest of the world.

You are right that people who are working are spending longer hours and I also resent the lack of spare time. It would be much better if work were more evenly divided. Long hours are bad for our health. But also bear in mind that even taxpayers can lose their jobs. Then they would be pleased to know that there was a safety net for them. We tend to forget that don't we? But to get back to wastage, I do agree with you about wastage, but not that everyone lives in luxury on the dole. The social wage is fine provided you can access the full amount, are not already in debt and are able to budget and provided you are numerate and literate and are not sick or dealing with life's contingencies or requirements of Indigenous family members who expect to stay with you. And provided you do not have to fix broken household equipment or repair damage to the house caused by domestic violence and provided you are not suffering from depression so you just don't care anymore ... you see what I mean, it is not so much the social wage per se that I am worried about, but the systemic nature of problems, that is not brought into the policy picture.

But the fact that people are uneducated is their own fault, we have a low student to teacher ratio here and there is no excuse for parents not sending their children to school. It is free after all and in fact there are free buses too for some!

Yes the school is free and when the buses arrive early the children who are hungry and tired because they have not slept or eaten because of the drunken fighting and arguing are not ready to go to school. Can you see what I mean? There is a need to consider issues systemically, not in isolation. Children look at their parents and emulate them. Parents look at the wider society, compare their circumstances with despair and resort to (or if you like medicate their disease with) alcohol to numb their hopelessness. I am not saying this is the right choice, but the world over this pattern repeats itself as a culture of despair faced by dislocated, marginalised people.

### *4.3.3 How do We Know and Who Cares?*

Hello there you are said N.

Hello, I replied seeing N make her way through the crowd that milled outside the auditorium in the conference centre.

I am glad we found each other.

Mmm we did ... you seem the same since last time we met said N.

You too ... not much change I said looking at my friend whom I had seen sporadically but with whom I had kept in touch ever since the days I had met her at the University of Cape Town. I had been drawn to her as much for her ideas as they way she interacted with people .... A constructivist who listened to the ideas of other people with respect, hoping to co-create meaning in conversation.

'You also look the same too' I said '... the large bag containing snacks and items for every eventuality ... just like your approach to making room for all points of view'.

We laughed, 'so how was the experience?', asked N warmly, her eyes twinkling.

'Well you must have some idea from the emails', I said. The emails were important, changes in my life and moving around from place to place really challenges boundaries. It shakes the very sense of who you are. Then in the new place you have to retain the core of who you are and remake yourself each time. I have been thinking a lot about identity these past few months. I think it was the question who am I that really drove me to write this manuscript the way I have. I was so relieved when Karen Scott Hoy gave me her thesis to read and there was the key that helped to unlock the approach to working the hyphen. I had so many nested documents linking levels of analysis that just did not seem to cover all the aspects that I wanted to cover. The use of dialogue allows for reflection in different contexts and this helps to deepen the analysis in a flexible way. It allows for slippage and context.

I started with the research report on social policy '... you know the two part report that was a compromise and even then I felt as if there was too much filleting and not of my choosing'.

N laughed 'ever the comedian Janet' ... I am glad you don't allow yourself to take even that too seriously.

'Laughter and pain are very close friends ...'

I know that said N, smiling.

'Research can be very difficult in a context where the policy emphasis is on individual responsibility for social issues, rather than both social and individual responsibility. This research reflects that tension'.

'What happened to the guts of your experience'? asked N.

'Well the issues are addressed in what I am writing now about trying to draw all the strands together. I think bureaucratic compartmentalisation is a painful experience. It is indeed like have a steel cage around one'.

'Yes Weber would agree with you', said N.

'Sociologists like Weber have said it before. But Foucault has covered the pain the boundaries give to the thinker and to ordinary citizens .... But it remains such an important point. ... I really wanted to opt out, but I could not. Not only was there a lot at stake for me as a researcher committed to doing the best work I could, based on a sense that the issues really needed to be looked at differently and that research and practice needed to reflect, or at least try to reflect complexity. I had re-located, rented out my house and my dear friend, now my husband followed me to Alice Springs and commuted each fortnight from his place of work in south-western Queensland. The logistics involved in the commitment to the project meant that I simply had to stick out the daily challenges. Because I am sincerely interested in the task I was able to reflect on the difficulties and that kept me going, but the editing and reining in process tested my thinking and wellbeing to my limits. I tried so hard to co-create meaning, but the powerful do indeed shape the boundaries of knowledge and it can require so much energy to continue to co-create and not to lapse into the anger of zealotry or the neutrality of cynicism. I think that the emotions attached to different philosophical outlooks are so important'.

Come on let's get some lunch said N.

There is a Chinese restaurant nearby I replied. I ate a nice, inexpensive lunch there yesterday.

A smiling woman who beamed at us and seated us at the small table in the restaurant. She bowed and turned away. On her back were the words 'Jesus loves you'. I noticed that each of the waitresses wore a similar T-shirt. It seemed a traditional family affair and each participant in the business conformed. Asian cultural and family values seemed to merge comfortably with their version of Christianity.

So you were talking about thinking and the fact that when you are limited in what you are allowed to do, that it is painful, because it means resisting the desire to give up the co-creation process.

Yes, I replied, thinking carefully, trying to draw together the strands of thought and emotion. In the end there was some co-creation and a shared memorandum was drawn up between organisations to address some of the issues that had caused difficulty. Also there was a real attempt to develop governance skills and to create more opportunities for self-determination. I have often felt rather limited by the need to conform and to play the academic role. Written work not only needs to mirror the complexity of what is really happening, we also need to apply our thinking in the real world.

You see N, right from the outset of living the life of a sociologist/social anthropologist; I was working the boundaries and trying to sweep in wider considerations than any one discipline focuses on. The very decision that sociology and social anthropology has much in common, was a little different back in the early eighties. I can remember how amused people were at Uni that I tried to do research on mental health in Guguletu where the migrants lived as a result of the Group Areas Act and the now infamous pass laws controlling migration from 'homelands'. I was working on how people coped with those structures ... trying to show how intwaso in the 1980s was a product of poverty and apartheid even if this traditional concept covering disease was around in the past, the causes and processes of healing were very different.

At that point their choice of steamed vegetables and noodles arrived.

Are you going to have the sweet and sour pork? they were asked.

No, we are vegetarian said N, their host smiled and moved away silently.

Yes said N, but what has this to do with what you were saying before ... it seems quite a leap in another direction.

Not really I think when I was doing that research I began to really understand the complexity of social issues and that the methods and knowledge disciplines were inadequate. Really to get to grips with the topic of intwaso, I needed to do systemic thinking incorporating socio-economic, political and economic variables. I needed to use ethnography but also official epidemiological statistics, clinic and hospital statistics to get a sense of official versus actual reality of ill health. ... Also what really worries me is that even back then I understood that TB symptoms were the trigger, the last straw in the ability of people to cope with poverty and the rigors of apartheid that led to intwaso, but that was just one set of triggers. There were a web of factors and of course TB by the early 1990s was beginning to be understood as the disease that was suffered by those who had been infected with the AIDS virus and it was realised that as more and more young people had TB and intwaso that there was a wider issue at stake. I realised even then that I wanted to work in such a way that multiple variables could be borne in mind … Of course I was limited by what I could do alone … I needed to work in a team with other people … with other skills … I have a friend called G. L. … who I would dearly liked to have been able to draw into the research for her math ability and her ability to see that number, value and meaning were linked. She radiated intelligence and was able to apply her understanding to research design and problem solving that went beyond any disciplinary boundary, a rare intellect … I can remember trying to get up a cross-disciplinary thesis and trying to explain that what I wanted to study spanned disciplines to challenge categories in a divided society. My age, culture and gender in a deeply racist and sexist context did not help, I felt silenced, just like so many others at that time. I was seen as blonde, not enlightened! That gave me just some insight into what it feels like to be powerless by virtue of age or race or level of ability and thus to be less visible and less audible to those in power.

You sound angry, said N, with her usual perspicacity and taking a sip of green tea.

Yes you are right … and I have not come to the point … you see there is more … even if we do work along the lines suggested by the WHO Healthy cities/settings report of working across disciplines and sectors it is still not enough.

What we really need to do is to think even more deeply about the boundaries that I was talking about before. Our emotions, values and assumptions shape thinking and policy. I tried to put these ideas together in an email. I can't remember if I sent it to you.

No I haven't seen that said N, but let me see it now.

I pulled out a heavy manila folder from my conference bag.

Well you also seem to carry around quite a lot, said N laughing.

Yes, I do tend to carry around papers. It is part of the process of trying to make meaning out of chaos! This email is about the need to research how theoretical and methodological literacy can help us to be more compassionate and how it could ensure sustainability of biodiversity (including of course human diversity) on this planet, if we could see the connections across things …

What session are you going to this afternoon? asked N.

I think the one on paradigm dialogue that you are running … that's tomorrow … said N … let's go on.

Let's sit outside and talk for a bit there.

Here is the paper,[9]

[9] The proposal is to undertake research in the area of systems thinking and its relevance to social justice in particular. The work of Ludwig Von Bertalanffy is of interest in this regard. Although systems theory is able to deal with only some phenomena at the moment there is a vital need to address the area of compassion/empathy for the other and the way social and environmental contexts are linked with styles of thinking assumptions and values. The area has been the preserve of compartmentalised disciplines such as the law and human/environmental rights, philosophy/ethics, religion and psychiatry/psychology and recently areas such as conflict resolution or peace studies within sociology and international studies. But a systemic approach to the interrelated area of social and environmental justice could introduce another approach and build on the UN and WHO charters that stress the need to work across sectors and disciplines. Human beings have become increasingly literate in a technical sense, but we appear to remain less literate about our ability to understand one another and our place within the universe. The number of wars and human-orchestrated disasters

Thanks said N smiling as she quickly absorbed the contents.

'What makes you think it is so important to change the way we think?' ... asked N, adopting her 'ex-duco'[10] approach, to lead out further reflection.

'Unless we do we just go on wasting our resources ... thought has the power to change the world. Thought is very powerful and thinkers have a great responsibility.' I pulled a wry face at my own pomposity, but meant it nevertheless.

N laughed ... that is so ...

I am such a small cog, but even so I would like to think and discuss things more openly ... really that's the nub of it.

'Yes, but boundaries can also protect can't they?' mused N.

'Yes' I replied, 'we have to be able to redraw boundaries though to incorporate more ... also if we take our thinking further physicists know that the 'off on' or 'one zero' or 'yes no' or 'right wrong' state is only one level of understanding. There are other levels where continuity is the truth ... So that is why I love the concept of paradigm dialogue and why I am here at the conference ...'

'Yes', said N, 'I do understand. Of course you must read more of the work of Ludwig Von Bertaanffy he had already said all that at a time when you were just beginning to struggle with those issues as a new researhcer'.

'Yes, he was fortunate in his broad understanding of the world and his mental capacity!'

'Hang on let me look at that program ... the session is on this afternoon ... they changed it again ... . Thank goodness we noticed ... let's go; we will make it in time ...'

*Continued*

demonstrates the problem. Hubris appears to remain a relevant theme for this millennium. (I was relieved that a celebrity physicist such as Steven Hawking has recently stressed this concern.) Recently the International Society of the Systems Sciences has focused on the way in which systems science can serve humanity. The theme of the 45th ISSS Annual Meeting and Conference is service to humanity. I was encouraged to think that perhaps I should write up my research on compassion, styles of thinking and the meaning of social and environmental justice. The rationale for the work is to create space for a much neglected but vital area for the future of human and biological diversity in general. A review of styles of thinking and their implications for understanding/framing social and environmental justice. Analysis of the implications of compartmentalised versus systemic problem-solving and knowledge management. The approach involves literary research and reference to a body of transcultural development research data and case studies. After one indepth study has been conducted a further more complex study is based on interviews with social and environmental development thinkers in the public, private and volunteer sectors in so-called developed and less developed nations. Australia, South Africa, Canada, India and Indonesia could be useful. These nations have diverse populations and Indigenous populations who have experienced colonisation and the full gamut of development thinking. The goal of the research would be to examine themes in order to discover regularities. These would be the basis for theory generation. The approach is called empirico-intuitive, to use Bertalanffy's terms. Essentially it is exploratory research that will identify themes and suggest interrelationships based on patterns.

[10] Latin for lead out. It is the root word for educate.

### 4.3.4 *Vignette: So How do I Feel About the Research Process?*

This is another vignette of a conversation at home between friends about me the researched and about me the researcher. It is useful as an indicator of how the research affected my thinking and practice and how it affected me emotionally as well. It cuts to the heart of what my work means to me and how my work and me are one and the same.

Hello Janet, how are you? asked B.

'Oh rushed as usual', I replied.

'You need to take time, life is not a dress rehearsal'.

Her words triggered a sudden, vivid memory of my thinking of the meaning of life when I attended the public funeral for Charles Perkins. There must have been at least four thousand people there. Quite an achievement because the venue had changed earlier that day from an outdoor occasion at the Telegraph Station (previously the site of the Homestead for so-called half-caste children) where Perkins had spent some of his childhood. Only once people started to arrive at the memorial service did it start to rain and we were all told to pass on the message that it would be held at the Show Grounds. For me the new venue was also significant because the very same hall at the show grounds had been the location of the booth we staffed as members of Australians for Reconciliation. Perkins made a specific request not to have representation from any politician in favour of mandatory sentencing, an issue that had concerned many others and me with an interest in the life chances of young people and their families. We travelled to the venue and more and more people gathered. All sides of politics were represented. Members of his family gave speeches as the rain drummed heavily on the tin roof. Then when the speakers from his family stood up to talk the rain eased for a bit. A roll of thunder provided a dramatic moment that could not have been better scored in a choreography of the event. When we walked out about two hours later a perfect rainbow provided yet another aspect and reminded me of the ochre song. Ochre is collected for ritual and for decoration: 'Put colours in the bags and make the netted bags all the colours of the rainbow.' (Strehlow 1997: 94).

I looked up and saw Jenny walking slowly up the steep, slate pathway. She bent to make her way through the bushes. Her post polio syndrome and diabetes made her visit to me an act of love. Did you enjoy your time in Alice? Jenny asked me soon after she had taken her customary seat on the veranda and she had taken her first sip of tea. She needed it. As a diabetic every few hours she measures her blood. Small, every day acts are an effort for her, but they are also a ritual, carried with care for herself and care for others. I thought of the difficulties that diabetic sufferers faced in the Alice where 'life on a machine' was the ordeal of many with Type 2 diabetes that had been untreated. Jenny is honest, because life has required her to be courageous. The answer I give her must be honest, not the usual 'it was challenging but worthwhile', or the flippant one 'the political heat rivalled the temperature.' The bitterness of experience tasted like gall. It was familiar, but self-reflection is quite different from self-pity. I replied carefully, controlling my voice: 'I was told the joke in town is that any newcomer can be classified as a missionary, a mercenary or a misfit.' I remembered the title of the book 'with head and heart and hand' (Kelly and Sewell 1988), which I had used as a reference for students, my heart and head are engaged in research. And I was particularly receptive to what was really a rhetorical question.

'Self-reflection is not a luxury, I realise it is essential for honest research and intervention that tries to represent and address wholes, not parts but to open spaces for participation in civil governance. Research cannot occur without looking inward and outward, to ensure that bias is surfaced. It is also about identity and it is about how we see the world.

My way of seeing is a result of gender, geography, class/culture position and personal history. The way I am and the way I have been portrayed as a sociologist/social anthropologist is part of that context.

B smiled at me; 'we are drawing all the threads together.'

Perhaps we could discuss the questions I asked my two friends.

Sure they said, they are quite pithy said B.

Also very confronting, said J. Are you a missionary? People have said things to me that make me really wonder about how I see life and my place in it. I have always thought that my lifework has been as a co-creator of meaning, rather than as a missionary. I try to work with people to create solutions that they want to achieve and I am satisfied with the role. I think that some see me as 'rude' or 'opinionated', often this is because I do not listen enough. Sometimes it is because I interrupt. This is sometimes because I can't wait to disagree. This makes co-creation so much harder. But I am learning patience and a sense of self-worth, that does not depend on proving another wrong, has sustained me through the worst patches of my life after re-locating to Australia.

'Was it really not your choice?' asked B. 'I thought that you were quite keen on exploring new places and being open to new experiences'.

I thrive on new and different experiences, but sometimes I felt torn from context, like a word ripped from a sentence.

Are you mercenary? I was brought up along the lines of the Protestant work ethic. In South Africa if you did not work you starved, or depended on your family. I was lucky. I was born white into a professional family. I am an only child. I went to university and unlike my father, who worked when he was studying architecture at the same university, I was supported by my family whose forebears had benefited from colonialism, even though some had been religious refugees. Then I worked as a tutor and received merit scholarships. The fact that I was white meant that I was privileged. I was aware of this from an early age and I felt the injustice and the guilt, but I also understood how precarious the privilege was and at what cost in terms of injustice.

Are you a misfit? The irregular pattern, event and person is attractive to me. I can remember Adelaide Dlamini, the Indigenous healer who studied me as I studied indigenous healing as part of my MA thesis said 'Janet you are called by your ancestors to heal yourself and to heal others' She constructed my role in terms of her framework of understanding the world, but for me it was a significant moment.

It is so important to look inward and not only outwards as a researcher, if one wishes to be accountable in a professional and personal sense. Perhaps I carry the guilt of being a white South African as if I was personally responsible but, we were all part of the problem, part of the colonial heritage.

Leaving one's history and one's family and one's sense of place isn't easy. Nor is divorce and the sense of failure. I find that generosity to self and others provides a healing synergy that gives me energy, which I can use in my work! Distrust of self and other saps energy and without trust involvement with people and place is limited.

The process of turning the lens on oneself is necessary, if we are to aim for some kind of objectivity (Churchman 1971). As an Australian citizen, I take with me the eyes and the experience of a white South African woman. I see the world through the filters of my experience and I am received, portrayed and understood in the eyes of some others in terms of their (contradictory) stereotypes of what a South African citizen and Australian citizen constitutes: 'a privileged and spoiled person', 'a refugee, like the Zimbabweans', 'a South African will understand (read conservative)' versus 'an ideologue who brings too much emotional baggage from South Africa'. In all instances the stereotypes tell me something about myself, the way others perceive me and about the social context in which I am operating. Also, reflection on perceptions of the way I was seen has helped me to ensure that my identity is strongly integrated in order to respond sensibly to the way in which I have been stereotyped in the quick short hand of the categories that are available or within the frame of reference of the labeler. The outsider/the other are stereotyped according to the way in which the labeler knows the world. The labels tell me about the society from where I came and the society where I now live. I soon learned that if one identifies too much with 'the other' and one is marginal, one becomes more marginal. Where you live, learn and engage with others shapes your identity. The dissonance and dislocation of the researcher who can be constructed as a mobile knowledge worker or as a worker who must respond to global markets and the fractured life associated with mobility. In this monograph I discuss issues beyond the margins of a defined research project. Discussion includes the way that the research process can be commodified through attempts to reshape boundaries, even when the agreed upon parameters had been previously accepted. Decisions were made on the basis of what served the status quo best. Objectivity was defined differently depending on positions and the likely ramifications for future career prospects. I can do no better than to quote Hardin (1992: 581–582)[11]

> What does it mean to 'start thought from marginal lives?' 'Marginal lives' are determinate, objective locations in the social structure. Such locations are not just accidentally outside the centre of power and prestige, but necessarily so. It is the material and symbolic existence of such oppositional margins that keep the center in place: ... 'Matrix theory' which focuses on the systemic social relations between such macrostructuring forces as the class, gender, and race systems, provides an empirically and theoretically more adequate account of these social structures than do the earlier class theories, gender theories and race theories that did not prioritize the way class, gender, and race construct and maintain each other. The thought that develops from such a starting point emerges from democratic dialogue-the sort characteristic of coalitions-between various marginal communities and, also, the dominant ones. ... These accounts are not fundamentally about marginal lives, instead they start off research from them; they are about the rest of the local and international social order. The point of identifying these problems is not to generate ethnosciences, but sciences-systemic causal accounts of how the natural and social orders are organized such that the everyday lives of marginalised peoples end up in the conditions they do ... .

[11] After the Neutrality Ideal: Science, Politics, and 'strong Objectivity'.

The silencing (existentialist nausea in the Satrean sense) helped me to empathise with others who were without a voice. One of the defining moments of my life was working with an Indigenous healer in South Africa who taught me the philosophy of ubuntu and holism and demonstrated its healing power when she worked with those who were most marginalised by apartheid. The calling traditionally known as intwaso was interpreted to be a way of saying that a person could no longer cope with life. In many instances the symptoms were those of TB (also the first symptom you are most likely to suffer if you are HIV positive) a poverty related disease that continues to be rife amongst the poorest members of the population. By setting up a holistic approach to healing providing emotional and physical support within a network of supporters, those who had limited access to public support were able to access support from a network beyond their family. To use the phrase of Moss Kanter (1989) 'pooling, linking and allying' resources, most importantly food in this case, enabled the sufferers to co-create a caring community within the ghetto-like townships of apartheid South Africa. The symbolism used by Indigenous healers was redolent of nature, despite being in the city. Ancestors appeared in dreams as powerful animals and plants gathered from the bush and the wilderness, beyond the boundaries of their city life, helped to establish the links across self, other and the environment and to re-establish a sense of harmony. Music, drumming and rhythmic dance are also important elements to re-establish a sense of control over time and space.[12] Achieving ubuntu through being in harmony with self, other and the environment is a challenge, both pragmatic and rational for all systems thinkers. It is this image that inspired the idea of setting up a community of practice that could resonate with shared ideas and could lead to some practical benefits.

No doubt I see the world in a particular way as a result of my life's experiences. I cannot forget the feeling of horror of seeing the ash remains of a 'necklace murder' during the civil war in South Africa, an act of group brutality, a systemic product of centuries, nor the many other significant incidents that contributed to my understanding of what power and separateness actually meant to ordinary people. From young homeless children numbing their pain by sniffing glue, sold to them at a premium by some adults, whilst others stole their blankets as a means of discouraging them from being out on the streets, the only space they had, to the so-called mentally ill adults who sought solace in the voices of their ancestors in a bid to co-create meaning out of chaos.

[12] See McIntyre-Mills 2000 for more details on healing, energy and communication.

Being a participant observer means that as embodied actors we experience the whole. I have tried to re-work all the boundaries, revealing my layers of personal values and assumptions. Perhaps that is why when I see people struggling to find a voice and feeling shamed in public places. It resonates with that memory of another place and another time where reconciliation was only achieved through a civil war. Not such an absurd comparison in rational terms, because in Australia, although our democracy is rooted in respect through multiculturalism and many citizens express support for reconciliation, tolerance for the other has been undermined in our attitudes towards refugees in a bid to 'maintain border protection'. Multiculturalism is an achievement and needs to be preserved. Further, the active citizenship of Indigenous and non-Indigenous Australians needs to be maintained, lest we forget what has been achieved and how far the state has traveled since 1961 when Australians voted that Indigenous people be allowed to vote as full citizens. All democracy needs to be constantly co-created, to prevent lapsing into cynicism and zealotry evident towards the so-called 'boat people'. This monograph attempts to contribute to this process of examining assumptions and values.

Perhaps I have no right to a voice from the point of view of cynics; I am merely an academic with dual citizenship, which could be constructed as a convenience. A cynic could argue that 'bleeding hearts from Down South' blow in as regularly as they 'blow out' of the Northern Territory. All these debates have already been played out. The statistics I quote may be more acceptable than the qualitative data that are indeed filtered by values and assumptions. I argue that values and assumptions are part of the construction of what constitutes personal lived knowledge. As a researcher my knowledge is filtered by my own experience and the same is true for all the participants in the study. The best attempt at accountability or a kind of objectivity is to acknowledge or surface these biases.

This section raises the issues (albeit tangentially in relation to my own experience) of the new conservatism and addresses the themes of globalisation, managerialism and the control of social participation, research and policy. I argue that compartmentalisation of thinking along with the colonisation of ideas and marginalisation can be regarded as vehicles for maintaining the status quo. Outsider status confers possibilities and restrictions. On the one hand it confers stranger value, because people cannot locate one immediately, but they quickly search for cues and locate one in categories that make sense to them and that they are familiar with according to their frame of reference in social, cultural, political and economic terms. The political and the personal are expressed in mobility, outsider status and the tenuous nature of contractual employment. Professionals are employed on contract and their activities and contributions are shaped by the contractual conditions. In a context of multiple meanings managing diversity is essential to address relational tensions.

*What is the point?*
*What practical use is it?*
*Does it represent an Indigenous standpoint?*
*Does it represent the business people?*
*How do you deal with secret knowledge that you can never know?*
*You are non Indigenous.*
*You are blow in*
*You are from down south*
*You are a bleeding heart*
*You are South African,*
*You are a Zimbabwean*
*You are female, you are white, you are a white Australian, you are a Trojan Horse, it is like the middle ages, it does not pay to be different,*
*You are very pale/blonde. Can you take the heat?*
*Can a manuscript ever represent all the voices equally?*
*How do you decide when the whole story has been told?*
*You can never address all those subjects or those issues at once*
*In trying to tell the whole story (when it is impossible) is this not just another way of developing a grand narrative or big lie?*
*People just say what they think you want to hear.*
*Whose story is this?*
*You can only heal your own thinking and your own people's thinking.*
*Good on yer*
*You are as busy as a lizard drinking*
*You will need a lifetime to do this work*
*I am not sure of the responses to these questions and comments, but they reveal a lot about myself and the others and the context*
*The story that follows. is just a segment of the whole, although it tries very hard to be systemic and to do justice to the stories of everyone, the writing can only represent some of reality.*
*Song, dance and textures cannot be represented, only hinted at, nor is the heat the warmth of emotions, the cold of contrasting opinions, or the rhythms of the seasons.*
*This is only an attempt to make some meaning out of my own experience and existence as it connected with other people in a particular place at a specific time.*
*It is just a case study of everything that seemed to matter to the people who talked with me.*
*It is also an attempt to record the shared meanings and the differences of opinion. The narratives attempt to speak for them selves, but they are mediated by the way I experienced them and by the way I heard them.*
*How does a white women who experienced some of the worst aspects of colonial history see the world?*
*To what extent is she post colonial or perhaps just neo colonial?*
*To what extent has she traveled from her place to another space?*
*So what? This is a legitimate question. but justice is perhaps best served through looking at ourselves as well as others whilst thinking and doing.*

# 5

# *A Landscape of Multiple Cultures and Interest Groups: A Panning Shot[1] of Place*

## *5.1. Cultures as Maps of Meaning*

I have been as open as I can about my motivation, which seems appropriate and accountable, as I shift the lens from self to other. Following Janzen (1978) I will give a panning shot of place that blurs spaces and boundaries prior to critically honing in on specific systemic issues. Dialogue across self and other and environment sharpens the focus of this overview as I shift the lens to give a panning shot of place. The dialogical vignettes give a sense of the range of issues faced by residents (Indigenous and non-Indigenous) living in town. The purpose of the vignettes is to make space for people to construct their own maps of reality and to show how their own maps relate to other constructs. People see the same reality in many different ways. The vignettes and narratives give a sense of the constructions and how they relate to one another in a particular place at a particular time. The perspectives diverge and converge. Areas of convergence and divergence are explored in later chapters as interpolations across the landscape of perceptions.

It is necessary to consider multisemic realities or multiple maps of reality when undertaking research in diverse cultural contexts. The maps of people are presented first and then attempts are made to find interpolations across maps that will create shared lines of reasoning and shared understanding (if not shared narratives), because they occupy different parts of one map in a shared landscape.

[1] The terminology of panning and focusing shots was used by Janzen 1978 in his work 'The quest for therapy in lower Zaire'.

Finnane[2] poses the question 'image or illusion?' In an article she begins to analyse the content of the tourist signs and boards which tend to gloss over the detail of Indigenous history. The Telegraph Station, site of the Bungalow is not explained within the context of removal of so-called half-caste or quarter caste children.[3] Arrernte history is not told from an Arrernte perspective. Sacred Sites Legislation can be invoked to protect areas considered to be part of Indigenous identity and well-being. This is also an aspect of political self-determination as Indigenous citizens. Mead (1996) analyses the context in which stakeholders involved in the Hind Marsh Bridge litigation involving a bridge constructed for development purposes in a contested sacred area by the Indigenous caretakers. In the analysis of the Royal Commission, commercial interests stressed that the Indigenous women who contested the construction were lying about the spiritual significance of the area. The commission found in the interests of commerce. A second legal challenge considered other material and found in the interests of the Indigenous women. Maps of meaning are based on ontological and epistemological considerations about the nature of reality and how we know what we know. Mead (1996) explains that Fergie, a social anthropologist acting for the women stressed that contradictions in testimony were a result of different women having different levels of knowledge about women's business. Secrecy is considered akin to sacredness and only some are custodians to the most secret stories. The contested area has intrinsic natural beauty and the area is a sanctuary for birds.Whether or not stories are constructed as myths or fact, depends not only on a range of factors, social, cultural, political and economic, but whether or not open dialogue is able to co-create meaning.

Post Hindmarsh there is a realisation that spiritual significance is a matter of definition in cultural terms. The systems of meaning or ontology or worldview need to be seen as having an integrity of its own, not merely a cynical response to developers.[4]

Finding ways of sharing definitions and a sense of space and place is the challenge.

> Despite demands to extinguish Indigenous rights and to establish the sort of discrimination rendered illegal by the Commonwealth Racial Discrimination Act 1975, there has been some progress towards new relationships and processes in the wake of recognition of native title (Howitt 1998: 29).

[2] Alice Springs News vol. 6 Issue no 16 May 19th 1999 page 15.

[3] See Caughlan, F. (1991) for a history of self-determination in Alice Springs.

[4] McWilliam, A. (1998) Negotiating desecration: Sacred sites damage and due compensation in the Northern Territory Australian Aboriginal Studies 1998 No 1 pages 2–10.

The challenge is to ensure that progress continues. Howitt refers to the geography of land use and land rights and the attempts at reconciliation as geopolitics for just, equitable and sustainable outcomes from regional development activities. ... (Howitt 1998).

Aboriginal people draw

spiritual sustenance from nature and the land. It was this spiritual sustenance that gave them their real strength and the power for such long endurance. They celebrated the land and their closeness to it, even oneness with it, through various ceremonies. More importantly, ceremonies bound them together on the deepest religious level. Here was their strength. Materially, they deliberately had little, apart from the undiluted treasures of nature: spiritually they possessed much. Some ceremonies heightened the knowledge of and respect and reverence for significant religious places: at the same time, bringing the people of the present into contact with their past and their dreamtime. Some ceremonies were for promoting the increase of knowledge and responsibility in relation to 'the Law'. Some were for initiation and for the discipline that went with it. Some were about promoting the survival and increase of the species, including humans. Some ceremonies were for reconciliation among groups and for coping with sorrow or grief on the death of members. ... All these religious practices had their roots in the land and 'country'.

The people were part of the land and the land was part of them ... (Miriam-Rose Ungunmerr Nauiyu Community 1993)[5]

Indigenous people move from place to place for practical reasons to access a range of services and for cultural reasons, such as 'for sorry business' associated with death and grieving for relatives and in order to perform rituals. People are said to come to town and then 'get lost', because they spend their money on alcohol to ease the pain of sorry business. But movement is more than merely cultural or social, it is also political. Social movements for self-determination and movements across space are part of an initiative to achieve greater self-determination. Coulehan (1997: 6) describes the process as follows:

From the 1960s and with increased momentum in the 1970s, Aboriginal people in the Northern Territory were taking steps to assert their right to choose where and how they lived. Aborigines began to 'walk off' pastoral properties and to leave mission and government settlements to return to country and establish small dispersed and more autonomous communities. By the 1970s, the exodus of Aboriginal people from the settlement sites to which they had been attracted or compelled by the agencies of mission and government intervention and mining and pastoral industries, had assumed the momentum of an Aboriginal social movement. The return of Aboriginal people to lands of social, cultural and economic significance has come to be called the 'outstation' or 'homelands' movement. This social movement was an Aboriginal initiative that preceded the advent of Aboriginal Land Rights legislation, which dates

[5] Traditional Aboriginal Medicines in the NT of Australia by Aboriginal Communities of the NT.

> from the introduction of the federal Aboriginal Land Rights (Northern Territory) Act of 1976. ... While the outstation movement demonstrated Aboriginal affiliations to country and autonomy in patterns of movement and settlement, it ought not to be interpreted as an Aboriginal retreat from change nor as an intention to remain isolated in 'country' Aboriginal groups have always insisted on their need for modern ... services ... in their remote homeland centres and outstations. Contemporary Aboriginal patterns of mobility and settlement reflect the desire of Aboriginal people to be more self-determining, whether in 'country' or in urban centres such as Alice Springs.

Maps of reality amongst Indigenous people are not without differences and certainly the maps of reality between Indigenous, non-indigenous people and the government organisations is contested. The role of the Strehlow Centre is regarded as a museum, a research institute, an extension of the gallery of art and artefacts and also regarded with suspicion as a vehicle for the support of land claims. Amongst Indigenous people the centre and the work of Strehlow is held with varying levels of regard, due to differences in opinion and the non-return of religious artefacts. Indigenous language groups form different interest groups. Meaning maps differ at many levels and there is a need to work on co-creation of Indigenous meaning as well as Indigenous and non-indigenous meaning maps.

History, language, religion, politics and the environment are swept into maps of reality and thus cannot be framed out by rigid approaches to research categories (see Ulrich 2001).[6] This systemic approach strives to hold in mind multiple variables. Ignoring 'just one variable' can make all the difference: misunderstanding, misrepresentation, ignoring cultural nuances, token considerations, forgetting the importance of social dynamics, social distance, power and their political/historical context could undermine the viability of a project.

In the context of cultural maps we need to discuss self-determination. It is defined very differently by Indigenous people, but the challenge for Indigenous leadership, according to the Kalkaringi statement is to develop a nation within a nation. In terms of the Tangentyere Council protocol[7] the residents on town camps are not transients who will move from the camps. They are seen as permanent residents with a right to the land. There is debate about whether moving off the town camps is the solution. For some Indigenous people the solution is to move away, for others it is to remain

[6] Ulrich, W. (2001) The quest for competence in systemic research and practice. Systems Research and Behavioural Science 18: 3–28.

[7] The Tangentyere Protocols developed by Tangentyere Council in collaboration with Central Australian Division of Primary Health Care (CADPHC) Centre for Remote Health (A joint centre of Flinders University and NTU).

on the camps and enable them to function more effectively. According to Vadeveloo (personal communication 2002), public housing is available for people who wish to leave the camps. For others the camps provide a way of life that protects aspects of a culture that they wish to retain (Tangentyere Council Protocol 2000). Tension in definitions have been expressed in political terms. According to an Indigenous Country Liberal candidate (viewed with some cynicism by the Indigenous electorate).

> residents of town camps must realise that we live in a user pays society. They will be left behind if they do not realise this. My sisters live on the town camps. They would like to leave. What is needed is another option for upwardly residents so that they can have their own private properties. The land is valuable, why can't it be used like other people use land? Why can't people get small mortgages or borrow money against their land?

Another difficulty is that the town camp residents (represented by the Four Corners Council) are not seen to be the rightful owners by local Arrernte people. Those who come into town (for instance Walpiri and Pitantjara) are not seen as the original owners of the land. The argument in the past (which has been resolved to some extent as a result of the call to unite) is that permission needs to be given to visit other people's land. On the one hand, is the idea that Indigenous town camps are a special case where culture will frame options. On the other, that culture is a resource for living and that a range of options should be available.

The user pays option couched in extreme language of economic rationalism and being 'left behind', makes town camp residents and their advocates fearful. A third way, namely learning about rights and responsibilities and enabling people to determine their own futures (on town camps or other options) is desirable.

Diversity of needs is a reality because although overall Indigenous people form a marginalised group (that can be seen in terms of class/culture proxies), the reality is that Indigenous people have life chances that span the full spectrum of options, from owners of business, to professionals, skilled workers, unskilled workers and the unemployed. It is true to say that the majority of Indigenous people are unemployed or on CDEP programs. They are six times more likely to be unemployed (Totham 2002, personal communication) and the educational outcomes remain poor, although there are more students in year 12. Demographically the number of Indigenous young people are growing, but the number of unskilled and semi skilled jobs is shrinking.

The Four Corners Council was set up originally to provide governance and to manage the different Indigenous interest groups. The challenge is that Arrernte people feel outnumbered by the other 'tribes'. As far as the Arrernte are concerned, they own the land and traditionally the people from other areas were supposed to visit, do their business and then return. The return to country program is considered worthwhile by all stakeholders as a way to facilitate the travel of those who wish to return home. An Arrernte spokesperson said:

> We must find a way to work together, but the people on town camps were not supposed to be here for a long time. They were supposed to return home. When my mother visited other people's lands she had to wait to be invited to use their water supply. Now we still have to apply for a license, but anyone can come onto Arrernte Land.

The mission for the town camps is to declare that the people are not transients that the land they are living on is theirs and that they have a right to the land (Coughlan 1991, Tangentyere Protocols 2000). The town camp residents are thus viewed less enthusiastically by some, who see themselves as the traditional owners of the land. They are also seen as temporary by non-Indigenous people. Others, who do not subscribe to the nation within a nation agenda, see the solution in terms of education that will enable people to move off camps and to live independently in town.

Self-determination for residents of town camps and for some Arrernte residents is thus not the same. The challenge according to the Kalkaringi statement is to find ways to overcome differences and to unite as one voice in order to establish a shared Indigenous national identity. The other challenge is that the attempts to develop local independence could be seen to undermine the national Indigenous agenda. The Kalkaringi Statement has a centralising mission.

Seeing the issue of Aboriginal marginalisation as a class issue is also problematic, if it does not give attention to the identity and meaning of Aboriginality, but nevertheless class is also a means of uniting marginal people, rather than focusing on cultural differences.

The views of people are shaped also by gender (male, female) and age (the young, the middle aged and the elderly), by social class and status and by virtue of length of residence. In particular migrant's maps can be fractured and building a sense of identity is essential. Also the mapping by the disabled is quite different. For some with physical difficulties they see the environment as fraught with areas of uneven surfaces, gutter edgings that are dangerous and heavy doors that are difficult to open. For those with mental illness just making cash withdrawal or plucking up courage to ask a librarian for help can be challenging. As a spokesperson for the disabled said:

> It is a citizen's right to go shopping, not to feel isolated and to use public transport, not just a special bus. The world is fraught with access difficulties

> that have to be negotiated. It becomes increasingly difficult when the disabled person is less empowered as a result of language or feels powerless because they have had limited opportunities for employment and have little confidence.

Young people who are obviously not consuming services are affected by limitations of their use of public space (White 1997). When young people are visible to the police they can be asked to move along because they are perceived to be loitering or because they are considered to be disruptive or 'bad for business'. Both young people and service providers stressed negative contact with police who tend to profile Indigenous and young people for attention in public places. But an elderly informant at the Migrant Resource Centre also stressed the way in which she had experienced problems when using seating space as a non-consumer in a shopping centre. Targeting non-consumers in public places was common, because they are seen to be potentially 'disruptive to business' or 'taking up public spaces'. In a hot climate access to air-conditioned or shaded public places is a particular concern.

## 5.2 *The Service Centre for the Remote Region*

Alice Springs, despite its remoteness from other urban centres is itself a service centre to the remote communities within a 600-km radius or more. The political economy of remote Indigenous communities and Alice Springs are closely linked. Indigenous income from mining plays a role in keeping the remote local economy alive, despite the fluctuating commodity prices. The most important mines include the Tanami gold mine, the Granite gold mine as well as vermiculite and energy mining by Santos. Most of the mines are on Indigenous land, but the royalties have not made a positive impact on the well-being of communities. The systemic reasons are numerous and stem as much from a misunderstanding of development assumptions and processes on the part of a number of organisations as they do from the history of colonisation that Australia shares with many other nations. The life chances of residents in remote communities remain well below those of other Australians living in urban centres. The rising cost of living associated with the GST and the privatisation of telephone and postal services will impact on quality of life (nutrition, food security, ability to communicate with the rest of Australia are aspects that need to be considered). The living conditions in remote communities are of particular relevance in so far as Alice Springs is a regional service centre for the surrounding areas.

According to Agnote[8]

> the pastoral district encompasses an area extending north-south from approximately 21 degrees latitude to the South Australian border and east-west from the Queensland to the West Australian borders ... properties vary in size from 8,000 to 11,000 square kilometres with the majority (65%) between 2,000 to 5,000 ... Water and the siting of boreholes is vital to the industry that is effected by droughts in one out of every four years (op cit.). Transport, markets and commodity prices are the other key determinants of the fortunes of the farmers.[9]

### 5.2.1 *Pastoral Voices and Themes: Vignettes From the Bush: 'The Price of Beef'*

#### 5.2.1.1 Falling Commodity Prices

At the time of the research a consultant, was undertaking a study to add value to the Cattle Growers product. 'We are working on adding value to the beef business through canning, dehydration and providing balanced meals through the Central Land Council for remote communities through a local abattoir and through prepared meals at food outlets. The challenge is to sell the less than prime beef through other ways' (Monica Bradley 1999, personal comunication).

[8] Department of Primary Industry and Fisheries No 269 G2 July 1996.

[9] There are 239 pastoral leases according to a 1993 study comprising 86 separate stations including 10 Aboriginal organisations or land-trust properties. At the time of the research the global economy for primary industries was poor and the collapse of commodity prices led to many re-evaluating their future, it was expressed as follows, by a consultant: 'The pastoral industry is suffering from the collapse of the Indonesian economy and the consequent depressed export beef prices. Aboriginal cattle stations are thus also badly effected and need to diversify.' The Rural Enterprise Unit responded by striving to develop bush resources through a research project (oils, flowers and food). Like other primary industries mining and particularly gold mining was also affected by fluctuating commodity prices. But within a year the fall of the Internet economy raised the value of commodities. Also the contamination of beef in the United Kingdom and Europe changed the outlook for locals after the fieldwork had been completed. These qualitative snapshots give an insight into the spectrum of viewpoints at that time. The data are presented as in thematic themes and give a window onto the lives of residents. The following themes were raised in a series of interviews focusing on the challenges some pastoralists face on a daily basis.

### 5.2.1.2 Generational Pressures to Survive on the Land

Bradley explained that her research found that the (non-Indigenous) cattle growing male is now third generation on the land. They have faced the difficulty of reducing profits and not being able to provide for all the sons. The self-esteem of cattle growers is linked with maintaining the land for descendents to inherit and a sense of failure because in many cases it is clear that the next generation will not be able to continue.

> Each generation finds it harder to meet the needs of all the family. This generation faces handing on asset worth much less than the one that they received from their parents. They are trying to consider cattle as just one option and exploring possible ways to use the Internet to market other services ranging from tourism to station-sitting one another's stations to allow for time out, to marketing beef differently. According to Bradley, the women are more prepared to listen to other stories and are open to suggestions whereas the men in the family have competed with one another all their lives. Competition with one another will not help to address the shrinking commodity prices.

S.F., a pastoralist and her family live on a station. They are 'the fourth generation, but this one is all girls which makes it easier. I believe that the industry will pick up, we hope so, but now we have to be experts at understanding the market, and it's not like the past … . We still have clean beef; this must mean something'. The contaminated beef crisis in the United Kingdom and Europe led to a rise in the economic possibilities of beef. But at the time of the research the situation looked bleak for cattle farmers, as illustrated by the following:

> Beef fetches the same as it did 20 years ago. Commodity prices are very low currently, but the cost of living has risen. Pastoralists are attempting to diversify with the help from advisers.[10] Some of the options are (i) Providing tourist station visits, which are popular, nothing flash, they just live the life we do for a while, which most really enjoy. We take a lot of things for granted out here we could set tourism farm stays and advertise through virtual farms on the Internet. (ii) A local abattoir is needed; this is something government needs to support. We can't set up a business, although it was suggested that 16 of us should, just too expensive. We would appreciate some assistance. Perhaps government could support the development of an abattoir … surely skins could be processed locally … . Bradley suggested cooperation and pooling resources to develop markets in town and even develop virtual farm stays on the Internet. The three pronged approach is to reduce costs, diversify income and develop electronic infrastructure.

[10] Monica Bradley (consultant) and David Kennedy (Department of Primary Industries).

### 5.2.1.3 Access to Services and Communication

The Flying Doctor Service and Distance Education are the two main services.

> For health service delivery 'the mantle of safety idea' is good, but in reality there are only two local planes and if they are not available other planes have to be called upon from further afield. It also depends on who is on the radio, one person is excellent, she knows how to assess cases, some are not as good. Really bad cases are left whilst people with less serious cases are attended to first ... . We waited for four hours for someone who was unconscious to be airlifted; they had to take him to Queensland ...
>
> As far as education is concerned our choices are School of the Air, which is good, X school, which we believe has changed and no longer meets our needs, we will be sending our child interstate as have others ... In the past we employed a governess, but we cannot compete with schoolteacher salaries anymore. At best we can offer an opportunity to a young graduate straight out of university, who wants a new experience, before settling down to a career in a city classroom. The govvie (governess) option has not worked too well in the past, girls only stay for a few months, because most get positions in town halfway through the year. The government does not help ... . Also the cost of coming to town to provide face- to- face teaching at the School of the Air, we had to fund our own childcare for the younger kids ... The Parents and Friends Committee of School of the Air was helpful though ...

One of the informants explained that she had to come to town for a few years for the sake of her girls who have speech disabilities. 'They need specialised speech therapy and they are improving'. She lived with her family in town, whilst the children received therapy. The children studied through School of the Air and distance learning exacerbated their speech impediment, in her opinion.

Pastoralists are concerned about their isolation and the possibility of air postal services to the remote areas being cut and that the postal services that they need for business and recreation could be privatised.

> 'We have to maintain our own airstrip ... . We don't seem to have too many problems only the airstrip here is not so good, it's a bit bull dusty. I get the pharmacy to mail out medicines through the RDF and I go to Fink Mission for immunisations. We have only had one evacuation.' Coming to town for services is difficult and expensive. The other problem is transport and accommodation when going to the hospital for tests ... unless the specialist notifies you of ways to access funding you get none ... this is not exactly fair ... . It can cost 800 dollars in transport costs, let alone the wear and tear on the vehicle ... . We are lucky with accommodation, never had to do the motel bit, R ...'s Mum lives in town and she looks after the kids. I have sat in the library when I have been waiting in town a few times ... didn't use the library mob before because I thought I would have to pay for the freighting mob for books ... then I found our jackaroos (farm workers) made use of the library and our 'govvie' (governess). I will look into it I think.

According to Bradley:

> Isolation needs to be breached through suitable technology: the use of faxes in the region is hopeless because the power is not available 24 hours a day. We have had problems with our phones and we are still waiting for Telstra (a company) to fix them. We can get incoming calls, but we can't phone out … . We have a computer and we need to hook it up to the Internet. We would like to get information that way … it would be good to develop a site with local information …
>
> Perhaps government would like to collaborate in setting up an Internet site with us to promote business opportunities … this would be a good local industry. Government could provide information to beef growers. The time spent in a truck could be best spent listening to educational and recreational audiotapes. The CD approach is still too advanced.
>
> Learning, recreation and business is at stake without communication. Out here if something breaks down it is not fixed quickly.

Informants stressed the difficulties, which they have with Telstra and said that if they did not have a particular contact in Alice Springs they could be without a line for weeks.

> This is a small community, we have 7 adults and 5 children here … The problem is that sometimes you do not pick it up (the lack of a line out) for days, it may be a problem only with in or out coming calls. This is of course not good for business. The issue has been addressed many times … When the phones are down we resort to wireless.

Another informant said:

> The future is in the Internet, we have a computer and are linked through the School of the Air … they have been excellent in helping people access the Internet. We don't download some information like pictures because it takes too long and it is expensive. This is a real problem, which still needs to be addressed … . Telstra (a company) does not provide a reliable service, for many years, the lines went down on a Friday. Not all stations have wireless, nor did they at one stage, only a bush radio. When they had some serious injuries on their station, they had to rely on a mail plane to radio on their behalf; many hours later they received help.

### *5.2.2 Urban Voices, Places and Themes*

The following table of data from the Australian Bureau of Statistics across residential areas gives the impression of a fairly homogenous community, but the data averages conceal the diversity of life styles. The town camps are scattered across some of these areas and also outside these areas, also the median weekly income does not give a sense of the range in incomes in some pockets of households within town camps (Table 5.1).

*TABLE 5.1*

| | Total population | Aboriginal population | Overseas population | Population aged 65 years and over | Population 14 years and under | Median age | Median weekly income |
|---|---|---|---|---|---|---|---|
| Alice Springs | 27,092 | 3,792 | 4,056 | 1,446 | 6,464 | 30 | $469 |
| Area of golf course estate and Eagle Court | 1,890 | 23 | 424 | 68 | 367 | 33 | $593 |
| Area inclusive of Eastside, Sadadeen and Hidden Valley | 4,937 | 772 | 716 | 218 | 1,290 | 29 | $477 |
| Area bounded by Will Terrace, Railway line, Todd River, Heavitree Gap | 3,252 | 549 | 461 | 212 | 349 | 32 | $405 |
| Area of Larapinta west of Bradshaw Drive | 4,925 | 674 | 934 | 177 | 1,291 | 29 | $476 |
| Area of Gillen and Bradshaw | 3,341 | 447 | 495 | 114 | 1,086 | 27 | $475 |
| Area outside of Gap | 3,016 | 484 | 329 | 351 | 626 | 36 | $370 |
| Area bounded by Larapinta Drive, Stuart Highway, Lovegrove drive to town boundary | 5,339 | 804 | 644 | 304 | 1,367 | 30 | $435 |

### *5.2.3 Vignettes From the Town: 'What Matters to Us'*

Participants had very different perceptions of Alice. Some celebrated their love of Alice through a range of art forms, others wished they were somewhere else. 'The hot climate and the lack of a coastline is a problem miss the sea air ... just being able to eat an ice cream and watch the sea ... . Some water sports here would be good ...'.

Some of the non-Indigenous and Indigenous residents said that they went to Alice because they were 'drawn to the spirit of Alice Springs'. 'Came here because of the landscape and mountains – art and study'. Others said they came for career reasons, but missed the convenience and way of life of big cities. 'It's not as good as Canberra, but it is improving ... '

Family responsibilities were highlighted by many as a reason for being in Alice: 'Here because ex partner can share child rearing and because it's easy to be a single mum here. But employment (although plenty of jobs) is still limited, the scope is wider in the cities.'

#### 5.2.3.1 Area 1

The quality of life in this suburb is high (the amount of disposable income is indicated by irrigated gardens filled with non-Indigenous plants, dual car ownership, although some of the units and share houses showed signs of their occupants being 'en route' or 'just making do' with little in the way of furniture.

The informants are diverse in cultural background and diverse in the level of economic security, as some were casual workers doing multiple jobs. They include American Space Base families who are located in Alice for about 3 years. Informants expressed varying degrees of pleasure at being in Alice Springs, from women in their late forties used to moving from one place to another and regarding the move as a challenge to be met with efficiency, to young wives in their early 20s without extended family and obviously finding Alice Springs a place very different from 'home'. Their 'outsider' experience was very different from that of a wife living in a base house who was born in Alice Springs who felt a strong sense of place as she had lived in Alice all her life.

Migrants of other nationalities felt less confident. For instance a Filipino wife and Japanese wife, both with very limited English ability were isolated to some extent and had little sense of wider issues beyond their home-life. The respondent born in the Philippines relied on the Migrant Resource Centre for assistance and friendship. She did not make use of many other public facilities and had limited understanding of civic life. Another female and younger respondent born in Japan had just had a baby and was involved in a range of mothering groups. She also had support from her husband's family who live in town and felt that she played an active role in the life of the town. Furthermore she had set up a baby sitting service from home. When asked about issues in town, she expressed concern about the number of single mothers at her child care group.

She said that something should be done to ensure 'children have fathers'. She seemed to think it was a matter for government intervention and had little concept of civic affairs or levels of government.

An Australian born woman spent most of her time alone as her husband worked on a cattle station. She had recently arrived from Queensland and gave birth a week after her arrival. She appeared to be coping with some difficulty without private transport, as she had to take taxis to town to do her shopping and sometimes had to wait in a public place in the sun with a crying baby.

The middle-class professionals with businesses from interstate used the golf course regularly and were aware of many of the public services and stressed the sense to which they felt part of the fabric of town life. Members of Neighbourhood Watch were helpful and offered to assist with the research in the interests of public safety (and property).

A recently fired construction worker in his late forties was very distressed and angry about the lack of accountability by people in town to health and safety in the building industry. He believed he had been unjustly fired because he had raised these issues.

A 55-year-old married male, who was retired, spent his time maintaining his properties in Alice Springs and said that he was not interested in anything else ... .

A male recently moved from New Zealand with his wife and two young children runs a clothing business. He worked from home. His pregnant wife looked after two young children and helped with the clothing business. She relied on the Migrant Resource Centre to network within the Asian community.

A young couple with two children under 5 years of age made the decision to leave Adelaide and the Northern suburbs to give their children a chance to 'grow up in a decent environment'. Neither had a full time permanent job, but they said, 'through shift work and casual work they would be able to move off welfare. They said they chose to rent a house on the golf course because although it meant that they paid high rent, it also meant that their children would not grow up within an environment of despondency like Elizabeth in the Northern Suburbs in Adelaide, South Australia.' They explained that they felt that this move was for the best. 'Those areas down south are stigmatised and filled with young people at risk.'

*5.2.3.1.a Themes Raised: Citizenship Rights and Responsibility.* Overall, there was an assertive sense in which the women and men were aware of their importance to the town. Some were particularly negative about: (i) litter, (ii) alcoholism being allowed to flourish. 'The cultural models being used don't work ... but this is a problem in US as well on the reserves ... .' (iii) maintenance of pavements and cycle tracks.

Australian born, tertiary educated women were aware of their rights and some had experience lobbying for a number of things such as public amenities, recycling and social issues. The sense of their rights versus the rights of other citizens was expressed in terms of: 'we the rate payers wish to see our money spent on infrastructure and services not on the issues covered by the police and welfare' to 'we the rate payers want to see government address social concerns because public alcoholism, unemployment and littering are very bad for the town'.

A strong sense of class-consciousness was expressed in terms of their being earners of incomes and providers of services and that they had rights which should be met. Citizenship rights appeared to be very closely connected with a sense of being rate payers and tax payers.

For some with a public conscience social issues were a source of concern, but only one person articulated human rights strongly.

5.2.3.2 Area 2

The area also comprises a diverse group of people, ranging from those who live in modern new units to those who are renting or living in subsidised accommodation. Sections are older and tree-lined and other sections are newer without trees. Overall the residents have a slightly lower level of income than at the golf course.

'Just making ends meet' was the way an Aboriginal mother, grandmother and great-grandmother (in a separate residence) explained their situation. They had returned to Adelaide from South Australia. Their concerns were survival and they relied on social security as well as part-time work. They had a new-born baby in the house and said that child-care services cost so much money that their budget and the cost of living restricted the range of their activities. They were also unaware of services: for instance they did not use the public library. The mother summed up her concerns as follows: 'You don't have to be afraid of men any more—I am afraid of children ... its because of the lack of things for them to do, that is why there is crime ... . Looking at her young grandson she said: 'I fear for these young children in the future' ...

Whilst in Sadadeen, the police (both female) drove past to the women's safe house at the bottom of the cul de sac. Down the road lived a single mum suffering from epilepsy who was also isolated because she could not drive a car. The issue of public transport was important, as she had to rely on her boyfriend for support. She lived with him in austere surroundings. She also had little sense of services available to her and her child.

A single woman and her son lived in a unit. They had emigrated from Alaska to find a better life, a better climate, a job and a future. She said that she could make a positive impact as a bank teller in Alice Springs through providing better banking service. She stressed the need for service that is respectful of cultural difference and said that she had plenty of experience in appropriate service delivery where she used to live.

*5.2.3.2.a Retirees.* 'The pension is good enough.' *said a* married woman and nationalised migrant from the Philippines. She expressed her pleasure at living in Alice Springs. 'I have all the amenities' she and her Australian-born husband considered themselves fortunate. The informant had no family in Australia, but after working for 15 years in Alice she was happy to retire and make use of the library whilst her husband bowled. She compared her life in Australia with Manila and the rural parts of Philippines very favourably and said that she was satisfied because the town was small and she had had the benefits of a good education (partly in Australia) and had enjoyed the courses at Centralian College. 'We own our home, we own our car and the pension is good enough!'

'Cost of living is too high.' An elderly man with severe arthritis said that it prevented him from doing his volunteer work. He was retired and on a pension, on which he said he found it very difficult to make ends meet. He focused on the limitations his illness placed on him and said that now they were in Alice they would have to stay, as they could not afford to move. He did not like the way the town had grown in size, he thought that it made it a very different place from the way it had been 25 years ago. He said that he wished he could watch the sea and smell the sea. He would prefer to be in Port Augusta. ... In his opinion he had had the best years in Alice. 'Used to do a lot for a club, used to fix furniture and so on and drive the bus, had to cut back because of me health.' He told how he had received a traffic fine, which he found difficult to pay and said, 'if it were not for my daughter I would have gone to jail ... he laughed.' and shrugged his shoulders and said 'it wouldn't bother me!'

*5.2.3.2.b Recycling.* He expressed very strong views on recycling and said that bottles and tins could be used to make edges in the garden rather than dumped. He also collected tin and was paid for the tin, which enabled him 'to buy quite a lot of meat from the butcher'. Another resident stressed that in his opinion recycling would only be addressed when there was a way to use the material in Alice Springs. This requires innovative manufacturing, using the materials for building and construction locally for local job creation.

*5.2.3.2.c Mandatory Sentencing.* 'Alice is OK for older people, it's the young people I feel sorry for', *said a* retired policeman living alone in a unit. He explained that he was divorced and 'had brought up his daughter'. His teenage daughter was visiting at the time. He spent much time telling me how he had spent the years of his retirement de-institutionalising himself from the force.

> When my daughter faced personal fines that would have landed her in jail (not for one but for a few offences not served simultaneously), my interest in mandatory sentencing increased. Young people do not get a fair deal and some feel that the NT is not the place to be. Good kids who make a mistake and whose parent can't afford to pay some of their traffic fines can be sent to jail; never mind what happens inside, they will meet the wrong element and in a small place like this, they never get a chance to keep away from them ... you could walk down The Mall and see them ...

As a volunteer at the Tweweretye Club, he felt strongly that Aboriginal people were not given a fair deal.

> They need separate clubs. This is a well-managed club, which is a safe place for families to be together. The license would help it to survive. He felt that Aboriginal people were discriminated against. Why should Aboriginal people spend their money at other places? Why not spend it where it can do some good? They only have X to spend; having this liquor outlet is not going to make them spend more. Why is it OK to have liquor licenses given to other clubs, just because they are non-Indigenous?

His daughter joined in the discussion saying that after going to this club and seeing how it brought family together she 'changed her mind that all Aboriginal people just have a drinking problem'.

Another very important social issue is the way in which taxi drivers charge Indigenous people different fares and in some cases are perceived to be potential or actual rapists of young women ... . She spoke passionately about the rights of Aboriginal people to some dignity and respect in Alice Springs ...

*5.2.3.2.d Retired Widows, Living Alone.* The following cases exemplify varying degrees of social integration.

> Glad to keep to myself. I have lived on a cattle station for many years and don't wish to make use of any of the clubs ... the people who go to the Senior Citizens club ... just gossip about one another ... very boring ... I just visit my

> neighbours ... I don't use the library because I am well catered for, have all my own books, I take a taxi to town when I wish to shop and I don't like the cinema; it is too loud ...

Her major concerns were not expressed

> Can't say what I think of the problems in town, we can't say what we think anymore can we?

In terms of food hygiene she said mildly: 'I never buy grapes ... or things that aren't packaged ... the sight of the Gins[11] running their hands over the grapes ...' she shuddered ... 'I am 80 and not young anymore' said an elderly informant suffering from a skin disease on her arms and obviously losing short-term memory. She lived alone in her own house and she had no support from social services. Her shoes were broken and her clothes unwashed. She had no family in Alice Springs and minimal social contacts. She visited the shops and walked 20 minutes across Ross Park each day to town and returned to buy her groceries. She stressed that she never used public transport because she preferred to walk and was proud of her fitness. But she also gave the impression of one waiting to die because she felt there was very little meaning in her life. She spoke at length about her husband's death and that she had no children.

'Look on the bright side, I can manage' was the life philosophy of an 83-year-old woman with 5 dogs and strong ties in the community. She also stressed her fitness level and explained that she walked to town. She said that she managed well on the pension because she owned her home and had the support of her two nephews. At that time she was an active member of the church and did voluntary work: 'I help others and they help me. It doesn't help to moan, even though I am on the pension I manage.' During the interview her nephew visited with an electrician to install an outside light in her home that she had owned for the past 40 years.

#### 5.2.3.3 Area 3

Themes raised

'Good place to raise families', said a resident, but a bit isolated and not a long-term destination. Others stressed that the good life could be lived easily by all age groups in Alice Springs.

*5.2.3.3.a Environmental Health.* A male resident stressed the need to pay attention to men's health. He had set up accommodation for gay and lesbian people and played a role coordinating a conference on men's health

[11] Colloquial, derogatory term for female Indigenous women.

covering issues such as:

> Parenting, sexuality and cultural differences and commonality in growing up, good practices in men's health, masculinity and resilience, alliances between women and men's health ... Men and women need to rediscover the magic in their lives ... we need to think broadly not narrowly to find options and solutions in life ... .

*5.2.3.3.b Planning Concerns.* Residents expressed an assertive sense of human services being poorly delivered because of the lack of commitment from short-term professionals.

> In Alice Springs the problem of about 72 licenses and alcohol associated violence needs to be addressed. The root concerns are social political and economic and cannot be addressed by thinking narrowly in terms of departmental responses or in terms of bean-counter responses which only calculate what the cost will be. We need holistic healing in Alice.

Another resident stressed that the public toilets were a problem and that having to pay was the only way to keep some clean, although tourists could find this inconvenient. There was no mention of any other section of the community that may find it difficult to pay.

#### 5.2.3.4 Area 4

The area also comprises a diverse range of residents.

An elderly Japanese women who had recently joined her daughter, lived on a street named after a pioneer Chinese market gardener. She complained about the bottles and grass in a public space between her property and a block of flats. She said that she was fairly isolated but that she preferred to rely on her daughter. She conceded that the Red Cross had been very good when she had hurt her leg earlier in the year.

An Aboriginal student at Batchelor College showed me her artwork and confided that although she was delighted to be in Alice Springs (because she wanted to be an art teacher); she missed her family and her daughter. She said that she tried to live a balanced life in Alice but that it was difficult for her, as she was homesick.

#### 5.2.3.5 Area 5

An Aboriginal woman with young children spent much of her time at home. Most of the services for young children and youth activities are not too far away ... She used the park opposite her house regularly. It is well maintained in her opinion, although it is clear that it was regularly used as a spot for drinking alcohol as the very small fragments of glass, bottletops and tops of cans indicated. The Queen of the Desert Hotel is nearby and has 'a takeaway' outlet on Gap Road. New trees have been planted and the play equipment, adventure tunnel, climbing ramp and swings were well used.

This road with two mature trees and a splendid view of the Gap is lined with Housing Trust Accommodation and serviced by Arrernte Council in so far as welfare services are concerned. At least one of the residents also worked for Arrernte Council.

On Gap Road there are other Housing Trust houses, only one of which had a very high fence in order to 'keep out the drinkers', although the informant said that he used 'to give it a bash in his day!' This 70 year-old widower lived with his son (in his 30s). His wife had died only a year previously and he was still grieving. He explained that he tended to stay at home, because his legs (which had blood clots) made it difficult for him to walk any distance. Having the hospital nearby in the Gap Road made him feel secure. He said that he had to give up his car because he could no longer see well and his lack of mobility made it difficult for him to walk to the bus stop. It used to be closer to his gate but had been moved recently. He said that he could no longer take his dog for a walk:

> I just open the gate and she takes herself for a walk when she wants to go out ... she has good sense not to go in the road. Just the young son lives with me ... he and I manage together ... he cooks well and I don't need any help. I can fry bacon and eggs for myself if the son is out. I am retired now ... I have plenty of time. I have family all over Adelaide, which is where I come from, and also a son in New Zealand ... .

He explained that he liked to stand at the high gate and look at the passing parade, seeing life pass ... it's a break from TV ...

A devout Catholic rugby player and volunteer referee and coach in his 50s, lived in a unit. He spends his time promoting rugby, refereeing games. He was concerned about finding a job despite having had a serious accident. He expressed strong views about the need for the public sector to sort out its image: 'Cut out the squabbling and crap and realise people are elected to serve – there are too many egos.' He and his wife had just returned from grocery shopping, the phone rang to ask about a football match. He demonstrated great community involvement and a clear sense of coping well despite his work-related injury and the recent loss of a family member. He emphasised that the public sector should realise that they should

> not only serve the white collared workers in town, but also the rest ... It is their duty to think beyond rates and to care for citizens ... The Council should stand

> up to Territory Government and tell them what they think of the Alice in Ten Years Plan ... we don't want a railway line though town either!

*5.2.3.5a Themes Raised.*

> 'Violence and a lack of adequate street lighting in some parts'.
>
> "When fights occur" the police won't come if you call ... Injured people have come to the house requesting an ambulance ...'
>
> 'There is too much grog and too many drinkers in the park'.
>
> 'The traffic is too fast along Gap Road, maniac driving, prangs, screeches people getting killed and poor lighting contributes'.
>
> 'Rubbish in yards is a breeding ground for cockroaches.'
>
> 'Not enough bus stops.'
>
> 'Re-name the streets to reflect Indigenous history and reconciliation.'
>
> 'Make more opportunities for talented young people.'
>
> 'High cost of living.'
>
> 'As a resident of housing commission I and many others see $245 a week rent is way too high.'

### 5.2.3.6 Area 6

The life chances of residents vary widely from those living in a caravan park on a permanent basis to those on privately owned small holdings where the key issues were access to basic infrastructure and services. An informant at a caravan park told how she spent most of her time tending, her tiny 'fairyland garden' at the front of her van. Another informant on a small holding enthused about the wonderful lifestyle for her children and how they enjoyed the views around their property.

A young man of Russian descent said that he liked living outside Alice because he could develop his own lifestyle and that he had more freedom than in a big city.

### 5.2.3.7 Area 7

A diverse range of residents experience a range of life chances. A single woman aged in early 30s living in an affluent complex of townhouses kindly agreed to be interviewed, despite being in her pyjamas, as she was a night shift worker and about to get some sleep. She stressed that she believed in the value of hard work and that people should be responsible for their own lives. She was conservative and anti-reconciliative but very pro-job creative for all young people. A young male living in a shared house with three males and a female said: 'Alice has a lot to offer', whilst a young Aboriginal woman with a child expressed strong anti-government views. She had moved recently from Port Augusta to find work and talked at length about her concerns for the future.

An elderly Aboriginal woman who had just caught a taxi from town was carrying armloads of groceries. She was concerned about the lack of activities for young people in the area, the cost of public transport and the number of mice in their house. She also stressed that she was disappointed with the way in which the 'Old Cemetery' was managed. She had requested a map to locate her family's graves but she was unable to match the markers to the map because the markers could not be found.

### *5.2.4 Vignettes From the Town Camps and a Case Study of One Family's Struggle at Mpwetyerre*

'People get weak from talking about alcohol. We must be strong.' An interview with a leading family is included as a case study because it demonstrates quite poignantly the cluster of issues which one family faced on a daily basis. My first contact with the informants was at the request of the housing association coordinator to act as witness at a Liquor Commission hearing. The family members wished to make a case for establishing a dry community because of the social problems associated with alcohol.

The town camp is next to a number of liquor outlets and opposite the Todd River, which is used as a space for public drinking. Because of the 2-km law, which stipulates that people should not drink liquor within a 2-km radius of the town, many of the people drinking in the Todd River opposite enter the camp. The difficulties of raising children in this environment and preventing violence is detailed in the following case studies and in the discussion on family violence and environmental health.

The women stressed that it was so hard 'to keep strong' when every day they are worn down by the same issue, namely alcohol availability which people within their own family and the town camp could not handle well. Even worse are the drinkers who visit (as family members) and the strangers who come in from the river to escape the 2-km law. The damage is physical, mental and property-related. They have to deal with the rubbish dumped each day and with the mental and physical hurts within their community. When the effects of property damage spill over into the wider community they could be charged under mandatory sentencing and the cycle of hurt deepens. Violence is modeled within the community and children learn about violence and its effects at an early age. 'Sorry business' and 'grieving' are part of life in town camps. The damage is caused by the perceived greed of business owners. This town camp is also named after the liquor outlet next door. The challenges faced by members of

Housing Associations located in Town Camps are to maintain a standard of living within a context of alcohol abuse and associated violence.[12]

L.A. and D.A. define their family as being both nuclear (within the boundaries of the camp) and extended family, including local people and those on the border of South and Western Australia. D.A. explained using the term foster (from the language of government welfare) that she looks after three younger children for her other sisters. So although they speak the language of nuclear families when they talk of the need to limit strangers from their town camp, they continue to have a strong sense of extended family responsibility. In this family the impact of alcohol has resulted in the lives of two sons (K and C) being badly effected. K and C completed year 8 at Yirrara College but were in trouble with the police. Whilst inebriated they committed a property offence. They were charged for breakage and were represented by a lawyer from Legal Aid, after running away from Yirrara College. The family was very concerned and believed that the majority of the problems could be prevented if they did not have a service station, take aways, a tavern and a grocery shop selling alcohol in close proximity to this camp.

Nevertheless the seniors try to set an example for the younger generation; for instance a senior male and president of the camps demonstrated responsibility as follows:

> This is my grandfather's area, but others tell me it isn't. I have been to a lot of cities … .
>
> I have put in information on the take aways.
>
> I have worked for X (an Indigenous Organisation) for many years.
>
> He stressed that he had overcome difficulties to me and to an audience of his wife, children and other members of the extended family, who showed varying degrees of respect. The younger women looked down and ceased to contribute to the conversation, but his wife showed annoyance because she considered that he had interrupted our open conversation with the women in the family.

[12] Attempts have been made recently to declare a camp a dry area with the support of the Alice Springs police. The erecting of high fences is inadequate to keep the community safe from others including extended family. The 2-kilometre law does not assist them, because people get arrested if they drink in a public place and so because their camp is convenient to the Todd River, a favourite drinking area, they are the next port of call for the drinkers. Another problem is that taxi drivers bring alcohol in and hand it to people through the fence. They have already used the Warden Scheme for 2 years. Witnesses from Abbott's Camp and the Housing Association administered by Tangentyere Council stressed the 4 corners Council proposal that Two-way law be implemented, 'White fella and black fella law' should work together to combat the problem of alcohol abuse.

> I did not go to college—like this mob-pointing to two young grandchildren in their late teens. They must use this place they have, it is a block of land … My mother was tough—she beat me up. I learned what I know in the bush. The young people need to stand up, to be strong.

The two young women lowered their heads, as they were the ones the speaker referred to. This speech really nettled L.A. his wife who said that she had already been speaking about the important concerns with me. The major issues highlighted in the conversation with the women are drinking, visitors, rubbish, breakages and fighting. These are similar to all the other camps, namely 'too many visitors bring coolabahs and sit down for a long time here … they make rubbish and upset this mob … we don't want them to humbug us … People get weak from talking about this problem. We must be strong, even though I have a heart problem. There was a sense that people or the system had failed them. … For instance, school was found to be unsatisfactory for L.A.'s youngest child aged 9 because his school money was always stolen and he was afraid of the fighting. There was also a sense that IAD had failed them because they were closeby and her daughter C aged 17 who had attended Yirrara College was still waiting to hear from them … there was little sense that they could do anything to change things, even within the boundaries of the camp, because visitors and passing drinkers fleeing the 2-km drinking law could come in and disrupt the fragile peace.

D.A. and her sister L.A. try to maintain a sense of community through sweeping up and picking up, through painting the rocks on the drive-way, through painting the walls with their dreaming stories and encouraging people to join in, but they get tired … . No one helps us … they say they will—they do not come back … .

### *5.2.5 Themes From Town Camps*

Selected town camps were included in the study. I was formally introduced at Housing Association meetings after which I conducted the conversations. At other times I joined the toy librarian and sat under a tree with mothers and children talking about issues they considered to be important.

#### *The key issues*

A detailed analysis of the data follows in later chapters. Briefly, people on low incomes pay more for basic services, for instance the cost of electricity is higher than for the standard rate for residents of town camps who make use of a metered card system. Also the lack of private transport in most instances means that reliance on taxis is usual and can be an expensive addition to a grocery shopping trip or a trip to access a service. Residents on town camps are diverse in terms of their aspirations. Some wish to hold on to the old ways. Some wish to move to a less communal way of life, in order to escape the negative aspects of camp life.

Many Housing Association Committees face debt as a result of a high level of damage and poor maintenance, there is difficulty in achieving financial independence. The decision made by the NT government and ATSIC to make town camp residents more self-sufficient has been implemented via incentives such as The Indigenous Housing Authority of Northern Territory. At the time of the research they funded $1,700 per house if residents pay a minimum of $30 in rent and all visitors pay a portion of their salary through Centrelink deductions. The 'house boss' is responsible for registering visitors. People who move to camps now have to pay rent for being at the camps. The possibilities for domestic violence need to be considered. In some instances there appears to be little understanding of the implications of so-called 'strong rent policy' for their monthly budget. To pay rent appears quite reasonable, but the rent has to be collected not only from the locals but also from visiting family from remote areas. If they break property, then the local residents are required to pay. Much of the damage is associated with drinking and the 2-km law that encourages people who engage in public drinking (because there are not an adequate number of serviced drinking clubs for Indigenous people) to take refuge in the town camps. The number of strangers and visitors who drink and contribute to damage is out of the control of some urban families trying to gain a foothold in the capitalist economy by educating their children and trying to get a 'real job' (the term used by town residents to describe nonCDEP work) with some long-term prospects.

Through Tangentyere Council

> A space was starting to open for indigenous people to articulate the distinctive forms of their urbanity and modernity. (Rowse 1998: 203)
>
> The work of Tangentyere Council since 1978—developing custom-designed houses and town camps—has made it possible to write in more positive terms about cultural differences in ways of living in a town such as Alice Springs.

Paying rent is considered a step in the direction of fostering a sense of responsibility. The housing committee membership is used for creating a sense of self-direction and control. The notion of women's citizenship rights and responsibilities is particularly difficult. Women bear the responsibility in many instances of being the house boss, but have little ability to address the never ending problem of uninvited visitors needing accommodation and often contributing to property damage. At the time of the research the housing association was addressing the management of residents who are houseguests through the notion of rent deductions. The implications for households in terms of providing food and basic necessities have to be considered in relation to budgeting, the automatic deductions from their social wage and the pressure on non-government organisations to provide basic necessities to people without access to cash or credit. But the question needs to be asked what are the systemic implications of the user pays policy initiatives?[13] A resident of Akngwertnarre (Morris Soak) said that he thought the number of men and women who are 'house bosses' were about equal at that time and that both men and women (husband and wife) could be house bosses together. According to the Housing Association Coordinator, this could be an issue in some cases where the woman happens to be a house boss and a male does the damage. But according to this informant, the Housing Association would ensure that 'everyone chips in not just the women. They could also take the money out of the social security cheques of the person who damaged the property.'

The so-called *'trashing of camps'* is a male activity (particularly younger men) associated with drinking binges often linked to large football events that attract family from out of town. The women living on a town camp explained that on one night everything could be trashed and that they could not control the situation. The notion of suddenly becoming responsible for paying for maintenance (although in theory it may appear quite fair that all people have responsibilities not only to their families, but also as citizens) had practical difficulties, namely the building of increased indebtedness. A full month's social security funding can 'be booked' so that the following month people have to go to 'St Vinnies' (St Vincent De Paul's) or 'The Salvos' (The Salvation Army) for help. This practice occurs frequently according to informants.

[13] According to Rowse (1998: 185) 'The housing association has attempted to create an opportunity for people to be citizens with rights and responsibilities without being clones of Western culture. The town camps provide an opportunity for better housing and services, but those who do not comply with the rules or who have difficulties because of cultural differences choose to live in the informal camps in the Todd River. Tangentyere can be regarded as a specific service to residents of town camps, not to other aboriginal people. Each camp has a housing association, which is grouped under Tangentyere Council. Most camps now have brick houses, community facilities and CDEP programs (op cit.) It provides integrated service delivery including housing, education, usual council duties, such as refuse etc., transport to and from home base within camps to school, welfare services and some health services for renal patients as well as primary health education and employment training, as well as a youth education program with a special emphasis on self-esteem.'

Unemployment and changing gender roles are dual challenges faced by Indigenous men and women.

The role of women in leadership and decision-making on town camps needs to be considered in all development interventions. The changes in gender role need to be re-negotiated in a changing world and opportunities for women to achieve leadership and self-determination need to be considered in development initiatives.[14] The requests by older women for a women's group (or camp) 'to be set up' is important to balance the power of men within the camp and also possibly as a means to prevent younger women becoming too focused on nuclear family life within their own homes. Nevertheless local women need support so that they can develop leadership skills to address meeting the health, education and employment needs of their families. The strong rent policy could be of particular concern, many house bosses are women and are likely to carry more responsibility for paying for damages, not the perpetrators who are most frequently younger males.

The identified themes are analysed and discussed in more detail in the following chapters. At the time of the research the residents of town camps were being initiated into the capitalist system through user pays programs. Their narratives highlight the following barriers: depletion of social wage through alcohol use, poor health in terms of physical, mental and social well-being, low levels of literacy and numeracy, payment for rent and damage by visiting family and strangers, funding cuts and the user pays policy, known as 'Strong Rent Policy'. In some instances there appears to be little understanding of its implications for their monthly

[14] According to Rowse (1991: 80): 'Bell herself is inconsistent on such issues as to whether women's camps (jilimi in Central Australia) are an asset or a diversion in women's political projects, or whether women have more or less economic autonomy in an economy of cash welfare benefits. She has been accused of representing the interests of Aboriginal women through non-Aboriginal ideology of separatist feminism (Huggins 1992: 418), in particular, she has had to defend strenuously both the ethical and the ethnographic basis of her comments on rape and violence in Aboriginal men's behaviour towards women (Bell 1991). Bell's persistent (and courageous) concern for Aboriginal women as victims of Aboriginal men shows the many difficulties (ethical, empirical and conceptual) of combining respect for traditional custom with advocacy of women's human rights ... .' Rowse (1991: 80–81) emphasises that 'the understanding of Aboriginal social order developed to include an understanding of male control through ritual dominance and thus dominance over the way' material surpluses of women's labour, and disposition of young people's sexuality and fertility achieved through violence, needs to be balanced by an understanding of women's separate power base, albeit subordinate.

budget. One of the most difficult problems faced by residents of town camps is the perception that visitors 'trash the camps' and that uninvited people 'squat' on the perimeters of camps.

The notion of who has a right to be in camps is fraught because drinking and violence is considered by many in camps to be largely caused by visitors or outsiders. Control of outsiders is perceived as vital to promote functional camps. A camp can be rendered dysfunctional in a week as a result of outsiders visiting, according to the Coordinator of the Housing Association and the residents at Inarlenge, Nyewente, Ilyerperenye and Ankwertnarre. Another related challenge is creating private, safe spaces within public town camps. The younger generation appeared to want more individual control over personal space. This is very difficult to achieve on a town camp, where many things are regarded as common property. Breaking the cycle of violence and alcohol needs to be multifaceted: creating employment opportunities is probably the most effective way for people to break out of the destructive cycle. The cost of electricity is higher than the standard rate for residents of town camps who make use of metered card system. Lack of private transport in most instances means that reliance on taxis is usual and can be an expensive addition to a grocery shopping trip or a trip to access service. A lack of trust and limited life chances makes everyday a struggle for survival. Gender roles are also problematic and need to be re-negotiated to suit changing circumstances. The difficulties faced by unemployed males who feel powerless to control their lives and who express their masculinity through football and through drinking is a result of many structural and historical challenges. The difficulties faced by women who live in a situation of domestic violence and need opportunities for leadership and their own self determination are yet another issue which needs to be considered. Informants stressed that they felt that attempts to control their lives 'were blocked'. Whether self-determination be declaring a dry area, attempts to control visitors and strangers or getting a real job, the informants felt that they were always struggling to communicate with people who did not really understand their story.

#### 5.2.5.1 Town Camp 1

*5.2.5.1.a Housing and Infrastructure and Maintenance.* A fluctuating population of between 80 to 140 people live at the camp in 14 houses. The housing ranges from formal cement and brick houses to those living in less formal corrugated iron structures. It was said that there are only three public toilets. At the Housing Association meeting it was explained that a new house would be built and that the previous committee (which had been set up to give information on the housing design) were invited to meet the following Tuesday to formalise the design. One of the members of that committee was in Adelaide and so another person was asked to stand in so that the building process could be completed in the next few months.

The Housing Association Coordinator proposed on behalf of the housing association that they consider a strong rent policy, because they needed to build up their bank account to ensure that building and repairs could occur. It was stressed that the houses on the camp are their own houses, not Tangentyere's houses and that they needed to be responsible for mending them and looking after them. The house boss was requested to ensure that they were signed on for rent payment and that each person 'stopping for more that 2 weeks be encouraged to sign on for 12–14% of their income'. She asked if someone would move and second the motion. It appeared that more discussion was required. NS said that she realised that those who were at the meeting were usually the people who paid their rent.

A list of details of maintenance was put forward and these included broken windows and doors and a ceiling that needed to be mended because electrical cables were exposed and dangerous. The latter would be dealt with immediately because of the risk factor.

A washing machine was requested, which could not be provided as the camp was in debt to the extent of about 36,000 dollars. After the meeting the women complained that they needed to use the machine to wash blankets. It was explained that Tangentyere Council could make it possible for them to wash the blankets at the shared machine at Council. They would have to wait until their funds had built up in the account before they could replace their machine. Someone added that they needed tools. The response was that Tangentyere Council had bought many tools and that they kept being lost or borrowed. The coordinator suggested that they 'should buy their own'.

Another complained that they had fixed a tap themselves, and this was praised and it was said that this was the right thing to do ... The residents said:

> No 13 house is dangerous, no 7 needs painting ... if you give us paint we will do it ... The response was: there is no money for paint. ... I wish there were ... . I feel terrible ... if you chuck in rent and those stopping with you then by next year everything will be better.

If they contribute to housing them IHANT (The Indigenous Housing Association of the Northern Territory) provides funding to assist with maintenance.

> If the damage is malicious damage, not just wear and tear then the person who damages the house needs to pay or the house boss, not Tangentyere Council.

> If you do not pay rent then you do not live here. It was said later that the problem is that family come to visit, particularly for a football match and they trash the place, throw toilet seats out, break doors and windows.

The women shook their heads and said it made them tired. The implication being that it was a 'men's problem' and particularly male visitors. I asked if they could go and visit these relatives for a few weeks if their money ran out, after fixing their damage. They laughed and said it would not be possible for them to do this.

The comment was made by elderly women:

> We need to have a women's camp here, I like to sleep out at night ... it is healthy and sometimes it gives you company and a change ... I get lonely because all the young women stop in their houses with their husbands. The young women should do more to keep the place healthy ... they should not just sit, they should clean the toilets and rake the rubbish ... too much sitting indoors and watching television and drinking too, even the grandmothers ask their grandchildren to get a beer ... .

At the Housing Association meeting the Vice President (E.E) and the president (E.W.), I.D and A, both elderly women played a vocal role and said that they would like to set up a women's group.

*5.2.5.1.b Environmental Health*
*Alcohol*

The issue of alcohol was raised as the cause of all the problems and they said that they wished a specific aboriginal organisation would help them set up a women's committee to manage the problem ... we should have a bush camp first and give the women support.

I asked them more about a specific area and how they felt about the proposal about declaring it a dry area ... whereas in the meeting they had been very derogatory about camp, because it is 'just one family ... like white people' ... . Afterwards when I asked them to tell me more about what they thought of the idea at camp they said that perhaps it was a good idea ... but they thought it would be very hard to maintain the rule ...

*Cost of living*

The cost of living was stressed as making it very difficult to find money for everything each month ... the cost of electricity can be about $300–400 a quarter and when this has to be paid it is a problem. Many have had their power cut off as a result. The cost of food and the quality of food from the nearby supermarket was commented on. Once the rent has been taken and the money taken for food vouchers there is nothing left to buy from other places.

We met J and her cousin sitting outside her house at about 9 O'clock absorbing the morning sun. J greeted V and I and when V said she was looking well, she explained that she had attended workshops on self-esteem and had done painting. One of the mothers aged in her early twenties said that on the whole the mob at this camp knew how to drink responsibly.

Whilst the young children played, puppies played with them, the children kept saying 'down, Missy and gently cuffing them away when they licked their faces'. I counted about five well-cared for dogs and about four cats, which were also obviously pets. They fight said C, so do we mob ... its all family here ...'

The day was slowly beginning, a group of elderly women sat outside a corrugated structure and boiled tea in a billy can, a group of men a few houses away sat together eating and talking and Rene arrived with a younger family member and sat down on a piece of canvass and watched whilst the younger woman played solitaire. Rene smoked as did V and C whilst they sat and watched the children.

*Hygiene*

Don't even get detergent from association for cleaning up ... .[15]

*Access to health services*

One of the women specifically asked that a message be taken to a specific aboriginal organisation to ask why they did not come to the camp anymore.

> Dr C Dr W and Dr M good, they came here, now those doctors sit in their flash office to make money ... they are frightened to come here ... they should come. here ... the old people don't want to go to appointments and we can't make them ... we call an ambulance and if they won't go, then the ambulance workers won't take them ... it costs 70 dollars to call an ambulance ... this is too expensive ...

When the story was shared with an officer at aboriginal organisation, he explained it was difficult to get health service delivery onto camps, because the housing association did not want service delivery,

[15] Tangentyere provides vouchers for purchasing domestic cleaning items.

they wanted the health workers to be trained and to deliver (preventative) services within the camp. It was also suggested that they wanted a separate health service through though the housing association, but it was doubtful that another service would be set up ... The officer said that ATSIC had also become involved in the issue and it was about family politics. The issue is that it is now difficult to get a Congress vehicle onto camps because protocol has changed and or there is a communication breakdown on this point between Tangentyere and Congress President.

*5.2.5.1.c Transport.* The only person who owned a car said that she tells people 'it is off the road, not working, which is true some of the time, because it is borrowed to go around buying grog'. 'Taxis are expensive and the (public) bus does not stop here'.

*5.2.5.1.d Communication.* The lack of a telephone for a month because it was broken was raised as an issue, 'because children get sick at night and we have to walk to a nearby town camp. One person on the camp has a phone because of a heart disease, but this is not generally shared. They confided A doctor organised the phone, because he could go (die) quickly.

The camp received a bill for 7,000 dollars from Telstra for a damaged phone ... this has been renegotiated to about 1,000 dollars.

*5.2.5.1.e Recreation for Children and Young People.* A toy librarian from a central toy library provides a service to town camps as well as to women in suburbs whose children are in need of toys. She receives names of women from Congress and other agencies. The children at this play session numbered about 11 and ranged from 8 or 9 years to about 2 years. Some had chosen 'to wag school' because they liked the visit by the toy librarian and the activities which they undertake each week under her supervision. The activities take place under a large Pepper Tree on a sale cloth. On colder days she will set up at Tangentyere's Child Centre (not far from the Gap Youth Centre, run by Congress) on Gap Road. This service is specifically for Town campers.

Children present said that they usually went to Gillen or Bradshaw Primary School. There seemed to be no sense of concern that the children had chosen an alternative to school for the day, because they were seen to be learning.

The camp leader is known as 'Auntie' and everyone deferred to her. The people living at Truckers are all her family according to my guide, V from Tangentyere Council. V had spent some years studying in Melbourne and had also worked in Perth and said that she loved Alice because of the hills, not like Perth, which is flat. Discussions of the landscape were the usual start of our conversations with town camp residents.

Research from the point of view of Aboriginal people has to be undertaken according to their time frame and pace and according to the extent that rapport is established. There seemed to be little difficulty in getting the children to show me themselves in photographs or to show me pictures they had drawn which they described as 'deadly' a phrase used by all the children, particularly the older children at the Gap Youth Centre. 'Deadly' means 'colourful, attractive, cool!'

A small group of elderly males sat down with 'Auntie D' not far from the children and the toy librarian (ST). Dogs joined in the activities of sticking cut outs of material onto the banners. According to ST this is to teach hand eye coordination skills. The sticking activity, that is sticking cloth cut outs onto banners was carried out by children with varying degrees of ability in determining that the glue was different from the paint! Many children put glue on both sides of the material and had difficulty detaching it from their fingers and the dogs. The approach to teaching was gentle suggestion with a great deal of praise and laughter.

The children liked this activity and concentrated for about 15 minutes, then played ball before returning to colour in. On the second visit the children aged from 12 years to 2 years played with letters in different colours which were attached to the board. It was clear that the 12-year-old who usually attended a primary school was too big to fit onto the small plastic seat attached to the board. He said that he 'was crook' and that was why he did not go to school that day. Two girls aged two and three sat at the board and found it difficult to understand the logic of the letters because the letters were different colours, it diverted their attention from the shape. They had just learned their colours and so were grouping letters according to colour. The game would be better designed if the colour letters were uniform and if the small and large letters were not introduced at once. It almost appeared to do more harm than good. One little girl expressed her frustration by pulling all the letters of the board and shuffling them with exasperation. The design of early learning tools needs to be carefully considered.

The women requested a load of river sand so very young children could play safely without getting too dirty. 'We shout at them all day to keep clean … . We need Panadol (a painkiller) at night … much laughter.'

*5.2.5.1.f Education.* Young people go to the local primary school but they 'wag (avoid school) more than they go'. said older women … younger women said that part of the problem was because the Tangentyere bus did not collect the children for two months. The child got behind in his schoolwork ... but he is doing well. I asked how it was possible that the bus just did not come ... They explained that they thought the bus driver was doing something else ...

Yirrara College of the Fink River Mission is popular for older children, because the children board there and learn consistently because they do not get distracted and because they do not have to be transported to school ... less likelihood of wagging. It was stressed that they did not choose to send their children to Yipirinya School because they wanted them to learn matters other than culture ...

*5.2.5.1.g The Needs of Older Children.* Those aged 10 to 14 have very little to do except for football and school. The older ones tend to go out drinking ... . They also watch TV. They should be taken by bus to the Youth Centre near Anzac Hill. When asked if they ever used the Gap Youth Centre (run by Congress) they said 'no that is not for us'.

*5.2.5.1.h The Needs of Older People.* Older people have problems associated with mobility, incontinence and the fear of hospitals. The needs of elderly people who are physically frail are not being adequately met. They said it was 'hard to care for those who soiled their blankets ... we have no washing machines, they are all broken, also the toilets are difficult for them ... too high'. Mental frailty was also raised as a problem, one elderly man who came to ask the women a question was discussed afterwards as 'one who was losing his mind and they thought he could be a worry, even violent'.

*5.2.5.1.i Definitions of Self-Determination.* Self-determination is the goal but it is defined and conceptualised very differently by different age groups, people with different levels of education and depending on whether the informant is male or female. Responsibilities are difficult to accept, not merely because of the frequently quoted welfare mentality or the high cost of living (both of which are aspects of the situation), but because the notion of rights and responsibility in terms of Western notions of citizenship, do not necessarily have the same meaning for all Indigenous citizens with a culture of sharing and a culture of community and family rights and family responsibility. The annoyance expressed by the women of Trucking Yards about the way in which Abbot's camp functioned illustrates the point that this group of people lives in a more traditional manner. Family are acknowledged and welcomed even if 'they trash the place and cost them money'. The notion of rent being paid by visitors could be a dramatic turning point in the notion of family obligation to allow people to 'stop with you, simply because they are family'.

It was clear that on this camp women of different age groups had different priorities some were lonely because of the change in the way in which the camp was organised in terms of nuclear family houses. The women stop with their husbands, not with the other women ... I.D. wanted more communal meeting around tables outside the houses on camps. She wanted river creek sand to be placed in the communal spots and the CDEP program to be used to get the men to make tables and chairs. It is clear that the CDEP program is seen as a way not merely to keep people employed and active, but also as a way of ensuring that tasks are performed around the camp.

Some young mothers wished to ensure that their children received the best chance at education and succeeded at school. Some women have drinking problems, one in particular was patronised a little by the others asking her to tell me about how she was progressing as a result of CARPU's program. She said she was better but looked very uneasy about the way the conversation was going and left the group soon after.

5.2.5.2 Town Camp 2

The camp is an unserviced camp on a dirt road. It comprises approximately 30–40 people living in 9 sheds and informal structures and is headed by M and P who are brothers and sisters. M decided to 'sit down' there because it is her grandmother's country. It is at the edge of what is known as X Station. M used to work for the Education Department; she has retired and looks after her grandchildren who moved with their mother to be with her when she lost her husband and the father of her grandchildren in a car accident. This was mentioned quietly to me when no one else was around.

*5.2.5.2.a Housing and Infrastructure and Maintenance.* M led much of the discussion along with her brother at the meeting of the Housing Association called by Tangentyere Council. I was asked to introduce myself and explain what the research was about. I stressed that I was attempting to find out the needs of all the residents in Alice Springs. Then the meeting began and they listed their needs for their mob, which comprises one family group of Arrernte people.

The Housing Association Coordintor explained that they had 60,000 dollars and needed to decide how to spend it. They listed ablution blocks, a new road and 5 sheds for housing. It was said that the ablution blocks themselves could cost almost 50,000 because *'the material has to come from Adelaide.'* I asked why good recycled material could not be used, not just tin. This seemed to be regarded favourably by the camp members. They later discussed how they would like a truck to help them obtain the material. Residents have building skills within the camp, for example L.H. explained to me that he worked for Arrernte Council and had been trained at Centralian College as a builder. He brought his certificates out to show me. 'If they would give us materials I could build and could show others ...'

'We may as well build our own humpies ... we ask and nothing happens, said M'.

> We can't fix the pipe, because we do not have tools ... I tried ... We pay 20 dollars rent for these sheds, they have no light ... There are only 4 taps for running water.

The existing shower block has two showers, one of which has a missing shower rose, the drain is blocked with a wide range of rubbish, chicken bones, kangaroo bones, old material and paper.

They asked for 3 double tin sheds and 2 single ones. It is important for the young boys aged 18 and above to have a place for themselves. They also need another two ablution blocks. Currently there are only two showers for more that 40–50 people. The two pit latrines are moved regularly by Tangentyere Council when they fill. It was suggested by Tangentyere's architect that there are other types of technology which could help to ensure that the latrines did not have to be moved, that is double sided ones which could be allowed to fill and be used alternately as one side decomposes. He was asked to bring plans and explain further, because M did not understand and as the person with expertise in health she felt it her responsibility to investigate further.

*5.2.5.2.b Plants, Livestock and Pets.* 'This is a good place to live' ... M has made it her business to decorate the camp. M showed me with pride her tall Sturt Desert Roses, her climber beans, grapevines, pumpkins and watermelons. She also keeps chickens and turkeys 'but everyday one of the dogs catches one ...' this was borne out by the chicken corpses under the shower blocks and next to the chicken coop. A visiting teacher said, 'they should get the CDEP workers to make the fence stronger'. Once again they expressed the idea that maintenance on camps was CDEP work.

Cats and dogs play a central role in town camp life. The pets looked healthy. 'Biscuit' the cat and 'Cheeky' the dog made their presence felt during the meeting to the amusement of everyone. I admired the hawks that swooped and circled around us.

*5.2.5.3.c Transport.* There is one Toyota here, but it is not always available 'because they drive around in it and when we need it, it is not here ... 'The (public) bus does not service this area.

*5.2.5.3.d Recreation.* M explained that the grandchildren come here and sit on the mattress and read comics. What about books I asked … . 'No one' reads books or uses the library. 'They are too shy' said M.A four-year-old rode a plastic bicycle and other children played with toy vehicles. The toy librarian did not go to this camp. Those aged about 12 or so attached a wheel to a piece of wire to make imaginary vehicles to drive.

Football (Australian Rules) is a great favourite with everyone and two of the boys aged about 11–14 were practising at the time. Two older boys aged about 18–24 said that they also played at a nearby field and that they belonged to a team.

*5.2.5.3.e Education.* Children attend Yipirinya School and the Detour Program and are reliant on the school bus and the Tangentyere bus. Some of the children of school going age were at home because they had colds or because they did not wake up in time for the bus, which arrives at 7.30. As described to me by an Indigenous mother, keen for her child to obtain a better life, the links between not being able to catch the bus in the morning to school because children are too exhausted after sleepless nights caused by alcohol related violence need to be understood as part of the web that leads to non-attendance at school. The non-arrival of buses at times also contributes to this. According to her when a driver decides to have a business on the side, because he thinks no one will notice if a few more children from the town camps do not appear for school, providing an additional ripple effect to the other ripples caused by lack of sleep and motivation.

*Spirituality and well-being*

An elaborately thatched structure for church meetings dominates the camp. It is built on a rise and has a tall wooden cross at the front. A grotto is filled with statues of the Virgin Mary and freshly picked pansies in small vases decorate the grotto.

*Health and well-being*

A priority is lighting for each shed. As Mr H said, I would like my kid to watch TV. The maintenance of the four existing tall stemmed lights was a problem. One had been broken. They provide light from about 6 or 7 o clock until about 4 in the morning.

They also requested solar panel heating and were aware of the CAT models for appropriate technology. They asked that the architect obtain quotes for ventilated latrines.

A request for a road was made to the Housing Association Coordinator. They explained afterwards that if a road was built it would mean, 'work for the young people'.

Children were said to have problems with their stomach (probably gastro-enteritis). Another child had asthma and has 'a puffer' to aid his breathing. Two adults have to use kidney dialysis regularly. Drinking was perceived to 'not really be a problem only on the days when they are paid, then they go to town and they drink in the Todd. They usually walk back before dark'. M's home (the most elaborate shed with a veranda at the front) houses the radio that connects the residents to Tangentyere and enables them to make emergency calls. They said this was not always satisfactory and in an emergency they had had to walk to town for help. M pointed to the caravan opposite her home that she said was used by children at the camp when their parents were drinking too much and they wish to feel more secure.

A woman said that the church services were no longer held regularly because the minister complained about people coming to services drunk … .

Another key problem identified was that the camp was not recognised officially. Many years ago they applied to the Alice Springs Town Council for a license. They did not succeed because it is beyond the town boundary … They have also applied through CLC for a land claim and a politician visited them. Nothing has as yet happened.

### 5.2.5.3 Town Camp 3

Many people moved here because they did not have land elsewhere and have made the camp their home.

Families live in 11 houses, each of which have about ten residents or visitors at any one time. There are at least 110 people in total. The language spoken is Luritja, but according to an informant the problem is that other people come into the camp and cause problems.

This was borne out when a camp warden arrived and said that he would like to get more assistance from the camp president, Mr B.W., to ensure that a tourist operator did not drop off his employees on town camps at random, because they have a disruptive effect on the life of the camp. He has been repeatedly told and he repeatedly ignores their requests.

*5.2.5.3.a Housing, Infrastructure, Costs and Maintenance.* At the Housing Association Meeting they discussed how to spend 50,000 dollars to upgrade the infrastructure. The decision was discussed with the architect from Tangentyere and an initial list was prepared: namely verandas, outhouses, new washing machine (which had been broken for 2 years) and maintenance tasks were also listed.

A tap gushed water just behind the place where the meeting was held. Maintenance is an ongoing problem as the cost of maintenance exceeds the income from rents. The list of requirements outstripped the extent of the funding and so the list would be prioritised at a forthcoming meeting.

The Housing Association Officer asked if the members of the meeting would support a strong policy on rent collection from the house boss and all those living in the house. The rent paid by the house boss is 30 dollars and all those living in the house (excluding visitors who are classified as people staying for two weeks) would be required to register at Tangentyere and 12% of their income would be deducted by Centrelink. The issue of budgeting and paying for food could become an issue.

The washing machine had been taken away to be fixed but had not been returned because it was beyond repair. Until such time as the funds had been built up in the maintenance kitty, the machine in the public ablution section could not be replaced.

*5.2.5.3.b Environmental Health.* Camp dogs were raised as an ongoing issue by the Housing Association Coordinator: 'Do you have any cheeky dogs you don't want? If so the dog catcher will collect them.' The decision for camp dogs was placed in the hands of the community to decide.

Participants gossiped along the sidelines at the meeting to me that elderly people have difficulty in camp houses, because they are not adapted to meet their needs, toilets are the wrong height and steps are difficult for those in wheelchairs.

The needs of those on dialysis were also raised. They said three people needed dialysis at their camp .... Leaving home for dialysis treatment is seen as a problem because people are afraid.[16]

The problem of household hygiene and health of residents was highlighted by J who showed me her home and trusted me to say what I thought about the leaking shower which she said made everyone (who slept on the floor) wet. The visit to her house underlined that in central Australian culture the sun and wind remain a source of health, it dries out and sanitises. In a house, with the windows shuttered against the sunlight and an electric light switched on permanently, the dirt accumulates and does not blow away and the house is not sanitised by the sun. The cost of electricity for lighting and the television (on permanently) was not understood.

She explained that cleaning the floor was difficult because she did not always have a broom, it was borrowed by other people and not

[16] According to the Tangentyere protocol (2000) 20 people are on dialysis and 40 have diabetes.

always returned quickly, people sleep on the floor and cover the floor, so it is difficult to wake them up and clean, also she said that the floor was not right … . The broom lay on the floor and a rake had been returned. She said she would begin when she had cleaning materials … .

This house was identified as being in the worst condition and was by no means typical of all the households.

There is considerable variance in living standards across the residents in town camps. The amount of cleaning seemed overwhelming and it was obvious that the house had become uncomfortable to live in. The sink was covered in rags and teabags and also pieces of wire scrubber. Puppies played on mattresses on the floor and piles of rags. A young child whom J was fostering was sick and J is being encouraged to learn about how to keep herself strong in her home. According to the Housing Association Coordinator, J is of the opinion that the child will always be slim or thin because it is in her dreaming (snake dreaming). Old wine and new wine skins make for interpretations which although old are constructed in a new cultural context and in this case may be viewed (in part) as a rationalisation for the illness.

Much of the community development that is done can lead to unintended consequences such as: a housing structure, which does not promote healthy living because the notion of constant maintenance is not culturally familiar, nor is the use and application of cleaning agents. In a culture where everything is shared just keeping track of basic implements can be a challenge. Also sharing a structure with many visitors who make the floor their personal space makes things very difficult.

An officer said that the floor could be cleaned and that she could purchase the equipment that afternoon with a Kmart voucher, if she came to Tangentyere to collect it. She was then taken to Congress to collect some tablets.

The environmental health officer said that the house would be sprayed with jets of water.

People were invited by Tangentyere to join an environmental health officer-training program. Making the connection between learning off site and implementing it within the environment of family and cultural expectations is a gap that which needs to be constantly bridged by the graduates.

*5.2.5.4.b Recreation.* A key source of interest for all in the community is football and this is one of the key connections between camp life and wider town life. Very few facilities are regarded as accessible, because people do not want to be shamed. This involves being asked if one can pay or asking people to unfold their money or simply an attitude of suspicion and lack of respect for their feelings.

### 5.2.5.4 Town Camps 4 and 5

*5.2.5.4.a Housing, Infrastructure and Maintenance.* A meeting was cancelled on account of sorry business. The camp is located next to Old Timers and comprises seven brick houses and a communal ablution and shower facility. En route to the meeting the Housing Association manager stopped to turn off a tap and to discuss the tap with a resident. The meeting was postponed until the following day because so few of the residents were prepared to attend the meeting. This was largely because there was little interest. The dynamics of the camp had changed as a result of a number of visitors in the camp and many appeared to be suffering the after affects of alcohol.

Those who attended the meeting were motivated and keen to proceed. They tried to get a quorum by going to ask people to come to the meeting. Whilst we waited they talked about camp life. It was explained that one of the houses owned by an artist, (who was in Melbourne for an Art Exhibition) was away.

She normally played an important role at meetings. Visitors were occupying her house in her absence.

The discussion centred on the problem of house repairs. Emergency problems involving electricity were followed up by N.S. The problem of electrical cables being exposed by damage to structures and damaged by mice was discussed. Rat poison cannot be used because children and dogs could eat the poison 'They will think they are lollies'. Cats were discussed as being a useful alternative. A comparatively new stove was no longer working because the wiring had been eaten despite diligent attempts to maintain the house. M took charge of pointing out where she thought there was faulty wiring. N.S. apologised for entering houses and said: I am not looking at your stuff, just the electricity.

More than 10 dogs scuffled at the meeting space and it was decided that because there still wasn't a quorum that we would return the following day at the same time.

*5.2.5.4.c The Housing Association Meeting.* The camp president J.P. was effectively blamed for not being at home to organise the meeting because she was in Melbourne at an art exhibition. It was moved by a senior woman that another president be appointed, a male and relative of hers. This was seconded and approved. At the meeting the list of priority areas identified for the architect were: (i) Replacement and securing of windows and doors, (ii) Security screens on homes, building fences around individual houses. (iii) The houses were said to be too small; for instance 'I need a bigger room for my grandchildren'.

The cyclical repairs listed were: taps, wiring of houses because of a mice plague, replacement of washing machine, maintenance of drains and sewerage systems.

The committee moved to accept the strong rent policy. But after the meeting it was clear that the full impact of the policy was not really understood. The perception was that Tangentyere should continue to help them. They said that it was a shame that they had to pay for everything, even the transport for 'sorry business'. This was regarded, as so beyond the moral pale that it did not require any explanation.

The need to show hospitality to family is not only part of their cultural fabric but is essential for those coming to obtain medical assistance and for family visiting those who are in trouble with the law. They said that it was not possible for them to deal with the problems of unwelcome visitors and that for Wardens to tell them to get rid of the people was unrealistic. A catch 22 situation existed which rendered many of the house and camp improvements vulnerable to constant damage.

*5.2.5.4.d Telephones and Transport.* Communication via telephone is also a problem; they are unable to call for taxis. The problem is that public telephones are constantly vandalised. The option of situating a phone within someone's house would not necessarily solve the problem because they explained that people would be 'humbugged' constantly by those wishing to phone. Transport to town, using the public bus would be appreciated, because although Tangentyere provides a service it is only on Monday, Wednesday and Friday. At that time they did not use the public bus, because they did not feel welcome.

*5.2.5.4.e Education.* Yipirinya school is said to be the best option and the children are serviced well by the Yipirinya School Bus. Yirrara is the school of choice for senior students, because they have limited opportunity to get into trouble in town as they board out of town.

#### 5.2.5.5 Town Camp 6

*5.2.5.5.a Recreation, Learning and Leadership.* M.W. explained that she had been an Art teacher some years ago, but was no longer doing this ... she said that she still did a bit of painting herself from time to time ... I choose three colours when I paint to show the berries when they are ready to eat and when they are not ready. I teach the children these things ... Children need access to the Toy librarian once again; she has stopped her visits. They said that after school many of the younger children make use of the public library and that they considered the service to be good.

Young people who are at home for the weekends walk to town and go to the river to drink. They use the Gap Youth Centre occasionally. The Night patrol needed to be more active in their opinion.

*5.2.5.5.b Environmental Health.* The biggest problem facing this camp is the use of the perimeters for camping by unwanted visitors. The Wardens and Night Patrol have removed people but they return with 24 hours. They wish to remove the thatched bowers and the tin sheds so as to dissuade their use by visitors. Another said:

> I am leaving here going to Katherine … there are too many problems here. A third said; 'Contact the wardens and tell them again … we have been telling them, they know but they do nothing' …

They requested that something be done urgently about the unwanted people who kill one another and create a sense of unease.

Another concern is the level of associated drinking and violence, which leads to, 'houses being wrecked'. They have tried to ensure that houses are never left empty for this reason.

*5.2.5.5.c Housing, Infrastructure and Maintenance.* G won the tidy yard award and was solely responsible for planting and keeping her yard clean. A fence around her garden is kept wired closed to keep out unwanted visitors. Her house is nearest to the entrance gate. Her home used to be close to the public phone box. It is now across the road and in a public place. This is considered a better position for it.

*5.2.5.5.d Transport, Telephones and Access to Services.* Accessing services is made difficult by the cost of transport. Taxis cost 5,6,10 dollars … anything they like. The bus would be cheaper. There was enthusiasm for a bus route past Morris Soak and a bus stop at the shops next to the public phone booth. This would assist school children who attend Bradshaw and Alice Springs High. Tangentyere does provide a bus but sometimes they miss it and also having a regular public bus would make trips to town easier.

They said that the use of the ambulance is expensive, about $70, the problem is that some people do not claim the amount from Tangentyere and so they get into debt.

*5.2.5.5.e Access to Services.* Dialysis services from Congress seemed to be regarded as part and parcel of camp life, three of the 10 informants were on dialysis and told me the three days of the week on which they were collected for treatment. Dialysis is a way of life. They explained that Arrernte Council provides services for the elderly and that Arrernte Council also provided CDEP programs for those in the City. Congress provides doctors services and Alukra provides services for mother's with young babies.

Tangentyere Council provides for home maintenance and CDEP programs in the town Camps. They were highly critical of Tangentyere because of the lack of money. 'Why do they drive new cars if there is no money?'

The dirt was identified as a health problem and two of the women present said they were interested in enrolling in Tangentyere's health worker program. An elderly women said she picks up rubbish every day and the task is ongoing.

In general there is a sense that Tangentyere and Aboriginal Congress should do more and that they are 'complacent' in the words of a health worker who was visiting.

*5.2.5.5.f Employment.* None of the residents who spoke to me appeared interested in becoming involved in CDEP programs, because 'the wages are the same as the dole' was their explanation and that most of the time it just involved 'cleaning up'. They said that the other lack of incentive was that they were tired of cleaning up because when they did the visitors camping behind Morris Soak just make a mess again. They said that they did not use the public spaces on the camp, such as the children's play area because it was too dirty … . They talked about what should be done to improve CDEP and to create real incentives to work.

They put their energy into maintaining their own yards instead, because they had more control over their own space, if they could fence it off.

*5.2.5.5.g Recreation.* There was some ambivalence about the use of the Youth Centre because of family differences and political rivalry within the Indigenous community. Also not many of the women are prepared to sit with a visiting recreational officer, nor would they play with the children when she visited. There was no elaboration as to why, but they gave the impression that there was little personal affinity, despite her being popular on another camp.

### *5.2.6 The Homeless and at Risk of Being Homeless*

Critical systemic thinking helped in discussions on homelessness. The discussions began with an attempt to define the scope of homelessness. We included the following definitions of homelessness based on narratives provided by service users and providers of emergency accommodation and temporary accommodation options.

Visitors and strangers sleep in or on the perimeter of town camps. According to ABS 1996 census 73 Indigenous people slept outside on census night. If this figure is considered against the Night Patrol data/police data and Supported Accommodation data it appears to under-estimate the total. According to ABS 1996 census data a total of 84 young people claim to have no usual address on census night. It can be argued that the Aboriginal data are incomplete and that the questions/categories do not address the reality of multiple families sharing households, moving into town to visit kin for a variety of lengths of time, camping out in areas demarcated and non-demarcated by local governments and those who live in town camps have not received official recognition. Youth Accommodation and Support Services caters for about 20 at most on any one night in emergency accommodation. House had only 10 places on any one night at the time of the research. The Domestic Violence Shelter provided for about 20 women and children and boys under 18 on any one night at that time.

The shelter was frequently unable to accommodate all the applicants and they were turned away or if possible provided with funds to go to hotel accommodation. The NGO hostel accommodated about 16 clients on any one night and Bath Street and Aboriginal Hostels also provided options to those who could afford them. Caravan parks provided accommodation at rates higher than the other options. It can be concluded that certainly more than 200 people are homeless per night. Women, children and young people wishing to escape alcohol-associated violence across all the residential areas of Alice Springs, is an issue according to service providers, because of the lack of affordable accommodation for short-term stays.

#### 5.2.6.1 Non Government Welfare Organisation (NGO)

The problem is that people think that there are jobs in Alice Springs, because the unemployment figures are low in comparison with other parts of Australia. Single women and children who are not specifically fleeing domestic violence also lack accommodation. Unlike men they are unable to qualify for the men's hostel which provides accommodation and three meals for $15 a day for 16 men only. The cheapest option for women is the Street accommodation, which is about $30 a day and does not include meals. Street costs $13 per night. The challenge for women is to save enough money for a bond and two weeks rent when they receive $380 a fortnight for the dole. There is very little accommodation available for women, which is affordable. This NGO work closely with an youth NGO (YNGO). 'We have situations such as a young women aged 18 who chose not to go to YNGO because they would have been split up because her boyfriend was aged 24 and could not be accommodated at YNGO'. We have men who say:

> what do we have to do to find somewhere for our wife and kids to stay? Do I have to hit my wife, I will in front of you so that she can find a safe place … .
>
> The options are shared accommodation, a room advertised in the newspaper, a caravan park (close to town), which costs about $48 per week plus site fee of about $25 per night. If families or single people can survive for 3 months then they can qualify for Housing Commission accommodation. The waiting period is the hard time for people to get through.

According to the spokesperson people who use the service do so for the following reasons:

> People spend all they have to get here and then find it difficult when they are knocked back … there are seasonal jobs though … They come here during the height of the tourist season, they think that they will get jobs. … People leave other states and come here in the winter because they think it is warm, they soon discover that it is very cold in the winter. We had to turn five people away on a day … . The churches know of the problem, we work closely with YNGOs.
>
> There are no heroin and methadone programs available for users of heroin. A couple came to Alice Springs from Adelaide with children and wanted to go on a methadone program. They went to every possible agency in town and were turned away. The spokesperson identified one medical practitioner who provides methadone to a few people but did not wish to be identified, because he did not want to have to cope with heroin users en masse. The couple was very agitated and abused every agency in town.
>
> They left us and we do not know where they went … They could be robbing shops … The spokesperson was very concerned about heroin and the lack of detoxification programs available for both heroin users and those who wish to detox from alcohol.

Things are worse now than they were 5 years ago. Then you could walk in off the street, now we put in place barriers to try to minimise the number of people we see each day. We have 15 appointments a day. So far we have seen 13 (this was said at 2 p.m.). The general sales tax (GST) will make things much worse; the cost of food is a problem now. We say to people that if they pay for accommodation privately we will provide them with food … . Also people from St Mary's who have mental health problems make use of the NGOs services.

People aged 15 to 85 years have come through their doors for help. The NGO provides food, clothing, money for medical prescriptions and tries to help with emergency relief. One of the NGOs has less than $2,000 per quarter to provide cash to people for emergency relief.

We are unable to provide accommodation. Those seeking help have been advised to go to the police station or to the hospital for help. Very few wish to make use of government services, because people with family fear the welfare system. On many occasions people have to be turned away including people with mental health problems. At best they can provide blankets for the night and food and advise that people try to find somewhere safe to sleep.

The irony is that people cannot afford accommodation here because of the cost of tourist accommodation, but 50% of the caravan parks and hotels/hostels stand empty. Some sort of arrangement could be made whereby a subsidy is paid to bridge the gap between cost and what people can afford. Access is denied to people without passports in some establishments that wish to maintain a tourism focus.

The challenges are to attempt to assist people to survive in the short term whilst they wait for the three month qualifying time to apply for social security. Rent assistance is not always adequate to prepare people for paying high rents and large bonds (about four week's rent and two weeks rent in advance). Other places to which people are referred include a hostel ($32 a day excluding meals), An Indigeneous hostel ($16 a day), X Caravan park ($190–197 per week), Y Caravan Park ($175 per week).

# 6

# *History, Citizenship, Life Chances and Property: Implications for Governance*

*It is two o'clock in the afternoon,*
*an Indigenous man of frail appearance*
*sits on a wall and decants alcohol from a cardboard cask into a plastic bottle.*
*He hands it to his female companion.*
*In a neighbouring area the visitors who arrive at the house*
*called "didyabringyagrogalong" carry six packs of beer*
*for the afternoon barbecue.*
*I see a staggering woman dressed in a T-shirt inscribed 'I love Paris'.*
*She carries a coolabah, today's cask of wine.*
*Her glazed eyes focus painfully on the dry riverbed*
*and the circle of family and friends who wait for her.*
*Nearby the rows of cars park along the Todd River.*
*The tourists come and go, they point their cameras in curiosity*
*and the public officials point their fingers in exasperation.*
*Overhead the cockatoo shrieks and flies above Mparntwe*
*It is 6 o'clock at night*
*I hear a family argue. The walls are as thin as their patience.*
*They sound weary of life and of balancing their budget.*
*Bitterness is eased with a drink. It is comforting to blame others in public*
*and in private to seek solace with the same wine.*

## 6.1 *Life in the Red Centre: Perceptions on Governance and Lifestyle: A Focusing Shot to Identify Key Issues*

This section addresses citizen's perceptions of quality of life and governance at the time of the research and needs to be read in conjunction with the discussion of the concept social capital in chapter 7. Analysis reveals a sense that governance could be improved through more participation by Indigenous people. But it was stressed that participation could only occur when reconciliation is achieved through health, education and employment opportunities. Informants also demonstrated that they understood the connections across issues and that alcohol is part of a web of cause and effect that could not be addressed merely by means of the control of alcohol availability. The other issues identified are that poor people pay more for basic services. Overall the comments made by respondents and informants express at worst an unwillingness to engage in discussion because of their cynicism about the role of government. This was summed up by an informant as follows 'To be a resident of Alice Springs is almost like being a member of a company town' (Anon personal communication 1999), because most of the business is government business associated with maintaining the human and economic systems. The Territory at the time of the initial research had been subject to the rule of one party for many years. The legal system, human service professionals, the churches and some human rights organisations led the attempts at reforming mandatory sentencing for property crimes at the time of the research, because of the impact the legislation had on the lives of Indigenous citizens.

### *6.1.1 Governance and Quality of Life Issues*

The following vignettes serve as discourses on social well-being.

#### 6.1.1.1 Voices, Perceptions of Interest Groups and My Interpolations

*6.1.1.1.a Non-Indigenous Residents.* The general issues raised by non-Indigenous residents were: 'Isolation, heat and the need to pay high costs for food, services and housing'. Citizenship rights were perceived to be earned by paying rates and taxes and subscribing to the social norms perceived appropriate by the majority. Most conceded that these are easier to follow if one is employed and has access to housing or ablution amenities.

Also the majority of residents considered that there is a wider role for governance at the local level in addressing unemployment and education of young people in particular. Respondents believed local governance should be concerned about a broad range of areas. Education and employment were prioritised as key areas to be addressed in the interests of all the residents of Alice Springs. People felt very strongly about: the need for functional and aesthetically pleasing infrastructure suited to the specific arid conditions.

*6.1.1.1.b Perceptions of Access to Public Recreational Utilities and Public Spaces.* Overall respondents and informants expressed a high degree of satisfaction with sporting facilities. The responses ranged from excellent at best to too expensive and a burden on the taxpayer at worst. The other major barriers to usage were: inability to access due to a lack of transport, opening times being too short in summer, lack of toilet facilities for people with disabilities, lack of shade and water fountains, broken glass and unwelcoming attitudes. The majority of residents in town camps regarded sport, particularly Australian Rules Football to be very important for males in the community. For younger people, membership of teams, practicing and supporting football are a key forms of recreation. Young females tend to be involved in dance activities.

The usage of facilities is shaped by cost, access to transport and the extent to which people feel welcome and able to use a range of services. Sense of welcome is linked with age, level of physical mobility, level of income and level of education and cultural background.

Environmental concerns included a need for shade, better recycling, whilst a concern about the management of the Todd River and its potential for flooding were also stressed. 'The town is on a flood plain. We should never forget that'.

> Listen to some of the old people. The last big flood was in 1988, I wonder what the next one will be like if they do not mine the sand out, it is too silted up ... Buildings should never have been located where they are.

Informants discussed the fact that government departments are fragmented in terms of their functions and that this often suits the departments because they can pass the buck, by saying 'no its not us it's them, or its difficult because we share responsibility ...' Informants stressed the need to think in terms of the links between things, not separately.

The cost of childcare for families without support networks and the lack of spaces in town for changing babies was raised as a concern. Some stressed 'the need to extend arts activities, because Alice Springs is too sports-oriented', 'The red-necked, sexist, racist and bigoted nature of people is obvious in a small place'.

The major concerns raised were public drinking, violence and littering, along with damage to property and the fear of possible crime. Blaming social problems on Indigenous residents was taken for granted by some. An informant working in a human service organisation said that it was a convenient psychological projection to always be able to blame the other. Many stressed that they understood the links between alcohol, litter, crime, self-destructive behaviour and the lack of employment opportunities for young people. But hygiene and crime were key themes linked with blaming the lack of responsibility of 'the other'. The economic interests associated with tourism were stressed frequently. The extent to which blame for all problems is attributed to Aboriginal people is widespread across class and cultural background. According to one informant 'Somehow new comers feel that by being anti-Aboriginal they are more likely to fit in'. It was clear that most residents no matter how new to town understood that many social issues are linked with alcohol availability and alcohol abuse. For example: on concerns about alcohol one of the (whose husband was on contract from America) was visibly upset when we met and women said that she had 'narrowly missed running over an Aboriginal person lying in the street on and that she felt quite concerned on other occasions when people had 'lurched in front of her car'. She reported the first matter to the police and was outraged that they did not seem very concerned. She is annoyed by the threat to her as a motorist and the liability that it would impose on her. She seemed to have great faith in American Zero tolerance policing and said that police are 'too lax here'. She was also amazed at how few children were in school during school hours and said she could not understand why taking children from inadequate parents was a problem. She did not have knowledge of local history.

Distrust is an issue as much for some Indigenous residents as it is for non-Indigenous residents. Others (mainly service providers) however, understand that alcohol is an effect of historical circumstances of colonisation, loss of identity and cultural despair, prevalent in other places with a history (albeit with specific differences) of colonisation such as Canada and the USA. A service provider summed it up as follows:

> Alcohol is a problem but look at the history of alcohol. When did Aboriginal people get the dole, when did they get the right to drink? At the same time! Also if you go into the bars and clubs you will see plenty of problems off the streets. Private drinking is a problem. Aboriginal people drink publicly ... that is the only real difference.

The embodied social actor needs to be acknowledged, that is, human beings are by nature physical and social beings. Hygiene and pollution were themes raised. According to an informant whose family has lived in Alice since settlement and is a descendant of the Arrernte people, Chinese and Polish: it is an unwritten law that they do not use the public bus, because they are unwelcome, because of a perceived hygiene problem.

> But visitors here (and I include the majority of short-term workers) need to understand that for some people water has not been part of daily life and for some who only made a change to western living in the past 40 years the notion of using soap water and deodorant is as foreign to them as it would be for me to sit on the dusty ground and eat a kangaroo leg that has been burnt briefly in the fire. We expect them to do the sorts of things we would find very difficult... Also the lack of washing facilities makes it difficult. Using taxis is an expensive option for people who are on the lowest incomes locally.

Access to or valuing soap and water by some and not others in a very hot climate makes strong distinctions between people. It results in some people being more welcome than others are in shops. Providing public showers was suggested by one resident, a human service professional:

> Body odour is such a personal thing, we don't talk about it, but it makes social relations quite strained ... .

To sum up: the tensions are around issues of safety and cleanliness and the sense of irritation expressed by some residents about the different attitudes of Indigenous residents to dogs, litter and public drinking (resulting in violence and creating a sense of nuisance, for instance loud abuse or public urination). Some residents perceive that this impacts on the image of the town, its business and tourism potential.

*6.1.1.1.c Indigenous Residents.* Indigenous residents were asked about governance, perceived needs, access to information, services, resources and their perceptions on personal and public rights and responsibilities. Their responses focused on immediate concerns about safety, managing the camps, meeting basic needs and difficulties they faced in accessing services. Their responses also highlighted a deeper need, namely to re-focus on Indigenous spirituality and well-being.

Alcohol and 'unwanted visitors bringing in grog' were the major problems highlighted. In their own words the issues are 'The alcohol'

> ... people are dying from grog. The greed of business people. Control alcohol so that lives are not ruined: Need to control the liquor outlets in the X Area. It's destroying young people. Would like them to be closed and they affect the Youth Centre. People getting lost in town. Fear of arrest.

Associated with this was domestic violence, destruction or 'trashing' of facilities and a sense of despair. 'Keeping strong' was said to be very difficult under these circumstances. The next issue was the desire for paid employment in real jobs and hope for the future for their children about whom most women expressed particular concern. Once again their fear of the future well-being of their children was linked with alcohol. Women said they feared the kind of adults their children would become if they were not given a chance in life. Another priority is the extent to which residents spend their 'key money' or social security on expensive public transport, namely taxis. Unlike the rest of the residents who use taxis largely to 'go to the airport' or because they do not wish to drink and drive after an evening out, town camp residents rely on taxis as a means to access a wide range of services. School bus runs to town camps were said to be insufficient because if the one and only bus arrives when children are not ready, they miss school. It was stressed that access to a regular public bus service would make a considerable difference to school attendance and the use of a range of services (health, recreational and shopping). The cost of a shopping trip for groceries is so much higher for a Town Camp resident who has to pay for groceries and then for a mini cab. For example, it costs on average $10 on a trip to town from Abbot's Camp.

The areas which seem to be most contested are the river (hence the derogatory term 'the River people'), The Mall and sidewalks near shops. Despite the contested public space, Indigenous people do feel welcome in some spaces such as public lawns, one of which is the preferred place to visit when elderly people who are patients come into town. Some stressed a sense that public services such as transport are 'for the white mob and being shamed' if they presumed to use them. 'The sense that opportunities for employment are very slim'. Another stressed the need to 'maintain the landscape by not cutting down and bulldozing'. Barrett Drive is known as 'Broken Promise Drive' because the tail of the caterpillar dreaming site was cut off during the development.

According to informants

> 'The Alice in Ten policy suggestion to control and manage the payment of social wage (as a means to control movement) has a long history in management of Indigenous people, rather than self determination based on rights'. Local government is seen by Indigenous people as separate for 'Arrernte locals, the town campers (who are perceived by local Indigenous people to get everything) and a "White Council serving white rate payers". But Indigenous people are the long term residents even if there is movement within the region, they stay within the region'.

The fact that Aboriginal people were wards of the state and could not move freely still effects current perceptions of rights.[1] Many Aboriginal people tend to use separate facilities where possible. Indigenous people tend to use specific services, which are in turn divided for different Aboriginal user groups. According to Bowden (1998, personal communication)

> Tangenyetere serves about 1,500 citizens who have a claim to live in Town Camps as a result of continual residence. Development as far as these residents are concerned would be to ensure that life chances of aboriginal people are addressed. This involves an element of redistribution or creation of specific

[1] In Alice Springs in the past demand for Indigenous labour attracted people from the hinterland and offered them cash rewards, but settler townsfolk also expected that Indigenous people's access to town be restricted, conditional and under surveillance. Indigenous people's desire to enjoy the amenities of town was well developed before settler townsfolk were ready to admit the people of the hinterland ... (Rowse 1998: 208).

> opportunities for Aboriginal people so that people can move to independence.[2] For this to occur there is a need to move beyond only thinking about infrastructure and technical issues. The challenge of sustainable development is to address the fact that by 2020 water resources will be challenged and that the ecosystem is fragile.

A recent census of camps indicates that the number of people appears to have reduced, but it was conceded that the Strong Rent Policy is possibly a reason why people do not list the full extent of the residents in each housing unit. Moving off town camps could be regarded as a successful outcome by families who wish to have a less communal way of life. For some, particularly young people, living on a town camp is only a staging post for something better.

The most frequent issues mentioned by Indigenous respondents was the desire for paid employment in real jobs. An Aboriginal person said:

> I would like to go into a shop and to see one of my nephews or nieces there, dark ones should be given a chance too ... it encourages people to drink when they feel there is no chance for them. The Yahoos (local posh people and

[2] According to Bowden (1994): 'In 1986, 1,064 Aboriginal people lived in town camps, which made up approximately 30% of all Aboriginal persons living in Alice Springs. This made up 5% of the total population at the time this research was undertaken. The town camps (Charles Creek, Hidden Valley, Larapinta Valley and White gate) are made up of mainly Arrernte residents, who are the traditional owners of Alice Springs (Mparntwe). According to Bowden, 'the migration of Warlpiri, Luritja and Pitjantjara people' has led to the original owners being outnumbered. They are now a minority. In terms of all the social indicators, residents of town camps score lower than the other residents of Alice Springs. In terms of employment rates, health, education they score lower and in terms of, incarceration rates, they score higher. According to Bowden 1994: 26–27: 'In 1986 only 46.9% of all working age (5–67) Aborigines in Alice Springs were in the labour force. Of these only 33.9% were employed. This is to be compared with 72.7% participation by non-Aboriginal residents. Fifty three percent (53%) of Aboriginal residents of Alice Springs were outside the labour force-not seeking work' (Nicolades 1989: 12).

The participation rate for Aborigines within the prime working age group (i.e., males aged 25–54) was 53% compared to 79% for the non-Aboriginal group in Alice Springs. Nationally the figure is 91% (1989: 13). The situation on Town camps is much worse. Unemployment in the Town Camps in 1986 was 86% compared to 52% for 'town' Aborigines. Tangentyere Council found when it began the Community Development Employment Program (CDEP) in 1990/1991 that there were large numbers of young Aborigines not registered at the Commonwealth Employment Services (CES). Thus the figures mask the low level of income coming into Town Camp Households.

outsiders) take the jobs for their own kids not for the local kids; they start businesses and look after themselves. When will we go into a foodshop and see a local kid? Why can't there be more government apprenticeships? There is so much discrimination in this place it leads to unemployment and resentment and trouble just smoldering beneath the surface … 'Since the 1970s they have tried to turn a bush town into a metropolis, problem is the brain of people has not kept up … kids have to succeed now to find employment … old Aboriginal people lost their jobs in the 1960s with the pastoral laws on wages. They drink. They could be tourist operators, they could tell the stories, not just 'The Gap' there is a whole story to it … this would be more legitimate … The only industries now are the environment and welfare and building … but that will end soon … oh yes and the Space Place (called Pine Gap) …

*6.1.1.1.d Basic Needs.* Food featured as an important topic in focus groups with Indigenous informants. Indigenous people talked about the cost of food.

The majority of residents stressed the need to address unemployment and education of young people in particular.

There is less time than people think to address current social issues associated with drinking and unemployment. We could play a strong role in addressing youth unemployment and lobby the business community to take some social responsibility in the interest of social justice and the development of the town. Real jobs not merely Community Development Employment Program (CDEP) jobs are needed. Political concerns: How is reconciliation possible in the context of uncaring and unresponsive government? Landrights is about self-determination and recognising that the land is our mother. We get strength from the land. Having a space, keeping separate is the way in which Aboriginal people address injustice. Aboriginal organisations are keen to work on the issue of unemployment of young people.

Informants stress that policy and planning should respond to

the big issue in town—linked with everything else—is alcohol. There are too many liquor outlets. This effects everyone in town. The town should be dry except for pubs and there should be no take always in town. In a pub the drinking is controlled.

Suicide is a call for help … I know that they think it will get their parents attention, some parents don't notice their kids, and they are too busy drinking … I see the changes in Alice, because I went away to Canberra and now I see …

Sustainable, healthy living, that is what we need. Australians waste too much. I research subjects … . I have been a teacher and although I am 65 there are not enough hours in the day to follow my interests in history and culture and promoting sustainable healthy living using the available resources and not wasting materials … Her ideas have found expression through running demonstration project and recovery program for people surviving alcohol abuse. Arrilhjere Demonstration House was conceived by Olive Veverbrants Peltarre. 'The house provides all its own electricity, from a solar electric system, connected to a battery bank and control system. It consists of 6 Neste 120W monocrystalline panels mounted on the roof at 24 degree on the north facing roof. The solar water system is boosted by the combustion heater, used for cooking and space heating ...' I teach community health workers about basic things, not wasting food, not buying processed food which is expensive the need to grow your own shade such as runner beans which can be put on a trellis and eaten. Everything in the garden should have many uses ... . The kitchen should not be a place where children come to get Coke ... .'

And another thing ... water think about it there are more air conditioners today, no water tanks like all the people have in remote communities and the number of tourists is growing ... we used to have an original golf course a dust course, now we have grass. It is watered by bore water ... that will lower the level ... . What about recycling water ... that would be a good idea ... We should do more ... I am not an environmentalist, I am a man of nature, I am not a religious man, I am a sprit man ...

We waste too much, we could lead the way, we have sun, no need for electricity POWA, he groaned. 'We are drying the basin out ... don't care what the engineers say it is common sense ... We have water restrictions but they still water their ovals ... I have seen them in the middle of the day ... .' We should reuse all our material on the dump for building and furnishing. At least now there are lawn sales they used to chuck everything away. Deposits on cans and bottles would be an ... excellent idea, but it is not in the interests of the businesses, too far from markets ... We have talked for years about rubbish; we used to be able to go to the dump. We could reuse machinery, and materials, but the Johnny come lately's never listen to us ... !

*6.1.1.1.e Barriers Within the Indigenous Community.* Urbanised citizens feel different from less urbanised/bush people. Also the four Aboriginal language groups need to be considered, because they have different maps of territory as do different family groups. These add further layers of meaning and complexity.

Gossip across factions (like all political groupings) is fairly standard, because of personal antipathies and philosophical differences. These include competition across Indigenous language groups, cultural and family lines. Nevertheless key players can be identified across organisations with whom it is possible to work.

Divisions were indicated as follows: Arrernte Council was identified with laughter by a group of town camp residents as being 'for people who are half Arrernte and that they did not get enough services from them, it is for those in town, not us!'.[3] Services are mapped in people's minds as

[3] Arrernte Council provides services to Arrente people who are largely town based. Tangentyre Council provides services to Aboriginal people who live in town camps.

being for one family or a particular language group. Competition for health, education and recreational services was evident.

All bureaucracy, including Indigenous organisations were viewed with suspicion, because of perceived corruption and a cynicism about the motivation of those in power.

*6.1.1.1.f Recreational Concerns.* Access to recreational services for young people (swimming pool, sports grounds and community halls) Planning concerns: Shade is needed, not merely politically correct trees but good shady trees for people to sit under! Don't need lawns just shade in civic places. The number of parking areas is excessive and they are unattractive, cutting down trees in the CBD is not a good thing!

*6.1.1.1.g Themes that Illustrate the Divides.* In response to the question 'what are the major issues for you in Alice Springs?' informants stressed the following key issues. Racism is encapsulated in the comments pertaining to social issues, control and reconciliation.

*6.1.1.1.h Rates, Taxes and Rights.* A strong undertone of perceived rights associated with being a ratepayer and or taxpayer was expressed.

The following quotation indicates a perception that Aboriginal people do not have the same rights and responsibilities as other citizens.

> 'I think that we should adopt the Ethiopian model' which was explained to be 'not giving anything to people who could not sustain themselves … i.e. a revised form of social engineering … if you can't help yourself no one should help you … .'
>
> Aboriginal Communities should not be allowed to hold their sporting events in Town.
>
> Tourism will fall apart if we don't do something about the rubbish, drunks street dogs and abuse from Aborigines. White people aren't perfect but we have standards.

This indicates a sense of the frustrations felt about social issues associated with different socio-economic and cultural life chances, outcomes and perceptions.

> Send the problem makers back to where they come from. Make alcohol their problem, not Alice Springs. Let the problem people pay from their own resources.
>
> I have never seen such backward people … they have no pride in themselves …

The 'us them' divide within the community is illustrated by means of the following quotation: 'Get Aboriginal Councils to clean up the river and scrub the mall …' A sense of common rights and responsibilities as citizens needs to be fostered.

Responses that are indicative of an understanding of rights were expressed as follows:

> Respecting the history and role of indigenous people through naming places in Indigenous terms … Give some more work to Aboriginals (sic) too.
>
> 'Government needs a more humanitarian profile' and Australia Day is not a day for celebrations by our Indigenous people.
>
> Relevant and important issues are put in the too hard basket … Development and reconciliation needs to be two-sided and not a one sided responsibility—too much public money spent on outcome neutral activities of this kind.
>
> Developing mutual respect so that Indigenous people are not made to feel 'shamed' as a result of cultural difference and social distance.

Being shamed is feeling that one does not appear well in the eyes of others. A bank teller who had recently immigrated from a remote part of Canada summed it up as follows:

> tellers in the bank just treat Aboriginal people as if they are all the same. A timid women called X knows how to write her name, she is a regular customer there and the other day when there was no pen for her to write her withdrawal slip, she came up to the counter. The other teller told her to go over to the customer service desk if she could not write out her slip. X was 'too timid' to point out that it was not because she could not write, but because there was no pen. I called out … hello X here is a pen for you. She looked grateful that I had helped. It is because people are treated like this that there are problems. I have told my son, that he is wrong when he says now that black people are stupid … he learns this at school, not at home … this is a problem in Alice … . Where I came from in Alaska indigenous people came first, here they come last. It is a big change …

Finding ways of sharing definitions and sharing land use is a central theme. This requires an ability to think in terms of land sustainability and reconciliation. To think in terms of a care taker role and a sustainability role.[4]

> And through building trust in the community. The Strehlow Centre's view is contested at times because the non return of artefacts has effected

[4] Howitt (1998: 29) refers to the geography of land use and land rights and the attempts at reconciliation as 'geopolitics for just, equitable and sustainable outcomes from regional development activities …'. Aboriginal people draw 'spiritual sustenance from nature and the land. It was this spiritual sustenance that gave then their real strength and the power for such long endurance. They celebrated the land and their closeness to it, even oneness with it, through various ceremonies. More importantly, ceremonies bound them together on the deepest religious level. Here was their strength. Materially, they deliberately had little, apart form the undiluted treasures of nature: spiritually they possessed much.

relations as is the NT Land Corporation and NT conservation commission (see Central Land Council Annual Report (1997)).

Themes that illustrate the divides across interests groups in Alice Springs are: 'purity and danger' and policing: 'if I were to lie on the pavement outside a supermarket, what would happen to me?' Hygiene and pollution themes that are indicated by the following comment: 'You have to be careful of the money because the Aboriginal people put it in their mouth, I have seen people eating things and putting them back dried apricots into the supermarket's dried fruit section ... The following is a vignette of separateness shared by a woman at a club':

> My daughter has a friend at school whose parents let her be friendly with an aboriginal child who has family at St Teresa ... she went to stay with her overnight at Santa Teresa and she was not fed, all her belongings were taken, she had to sit in a bath and eat a sandwich when she came home, late ... she had an eye infection and they took her to hospital ... . She said she slept in a place like a chicken coop. They did not realise where the child was going ... they won't let her go again ... much laughter ...

*6.1.1.1.i Contested Public Space in a Divided Town.* People are divided in the way they define their sense of place (Yahoos, Johnny come latelys, People from down South, the tourists, the Americans, the blow-ins, here for a short period of time), socio-economic position (Haves and have-nots), culture and language groups (Indigenous groups with different interests and differences between Indigenous and non-Indigenous Groups). Differences in levels of education and literacy and numeracy, which leads to different perceptions, differences in attitudes and values (the do-gooders, the bleeding hearts) which express themselves through categories some of which are polarised.

'The town is changing and people have very different lives within it'. There are diverse Indigenous maps and diversity across Indigenous

*Continued*

Some ceremonies heightened the knowledge of and respect and reverence for significant religious places: at the same time, bringing the people of the present into contact with their past and their dreamtime. Some ceremonies were for promoting the increase of knowledge and responsibility in relation to 'the Law'. Some were for initiation and for the discipline that went with it. Some were about promoting the survival and increase of the species, including humans. Some ceremonies were for reconciliation among groups and for coping with sorrow or grief on the death of members ... . All these religious practices had their roots in the land and 'country'. The people were part of the land and the land was part of them ...' Most importantly, ceremonies bound them together.

The mapping is in contrast to that of NT Land Corporation and NT Conservation commission. (Howitt 1998: 29.)

and non-Indigenous maps. Culture leads to exclusion as well as inclusion in Alice Springs. Some of the stolen generation stressed this as follows:

> Feel upset about being rejected by own people—know the stories about the country, but was taken away as a child, and now not considered to be a traditional owner by CLC.

He said that this issue had taken over his life and he was fighting for the stolen generation's compensation in the courts.

A non-Indigenous informant stressed that she no longer felt secure:

> There are constraints to the enjoyment of our property due to a sacred site declaration on our backyard.

Some human services are segregated, as are recreational activities. People tend to live separate parallel lives and tend to use separate recreational clubs and spaces. Some young people and elderly people feel equally concerned about using public spaces such as The Mall and shopping centres, in the case of the former, because they fear police and security guards and in the case of the latter because they fear being attacked or robbed. Other areas where people have safety concerns are parks, the Todd River, pubs, clubs, public toilets and pavements. The way in which different interest groups perceive public spaces is a reflection of the different sense of rights and life chances. A sense of alienation for instance is expressed in part through graffiti and the trashing of public spaces. A long standing member of the Alice community who had lobbied for the rights of Indigenous young people stressed that public spaces need to be accessible, not segregated or contained and this is only possible through community (promotive and preventive) policing using specially trained Indigenous staff, rather than a controlling approach.

Triggers for blaming 'the other' are animal control, food hygiene, questions about perceptions of social concerns, litter and maintenance of pavements. When issues such as pavements, litter or listing the areas for local governance, the majority of informants raised the issue of Aboriginal drinking and littering and camp dogs, also a sense that police and other human services are ineffective and using inappropriate models. The opinions were made mostly from the point of control, rather than from the point of view of social justice.

*6.1.1.1.j 'Locals' and 'Blow Ins'.* Those who live long-term in Alice Springs feel that they are often dictated to by short-term professionals who are not here for the long-term. Some residents tend to focus on the local, without seeing the connections with the wider global context. There is a perceived distinction between local interests and a national or global outlook.

> The town has lost its cosiness. People, tourists come to Alice Springs, because they think they are going to discover a town like Alice ... but the place is has been buggered up ... Excuse my language ... but it has with roundabouts ... making it look good for the tourists ... but the tourists want to see Alice the way it was ... . The tourist population is only here for 9 months a year—we live here.

*6.1.1.1.k 'Tourists' and 'Residents'.* Informants said that governance should not focus on the tourists only, such as colourful pavements when basic pavements are not provided in all locations. Shade is seen as a more basic priority for local residents who wish to see that play equipment is useable in the hot summer months before installing new pavements. Some of the 'rate payers' see government paying too much attention to CBD a tourist area and not enough for the long-term residents. Tourism is seen as another problem, by some particularly the lack of ethics amongst the tourist operators. They 'milk the Aboriginal cash cow' for instance some tour operators pay their guides (and very few give aboriginal people real work opportunities), for example $20 and a cask (of wine) for organizing a tour to see a dance.

> 'This is the last bastion of wilderness but we are destroying the people and the environment, drink and development. Tourists don't come here to drink, they don't need the liquor outlets ... why don't we listen to common sense any more, none listens to the old people anymore, we listen to Big Brother in America'. In other words we do not develop our own local models. ... 'Also the art scene ... too many shops selling digiridoos played by non-Aboriginals ... they are not our instruments in central Australia ... The artists are ripped off because their art is copied and whitefellas can't understand that pictures are painted by families, not necessarily individuals ...'

Informants stressed that planning often reflected some interest groups more than others: a source of concern was the issue of getting rid of people from their homes in The Gap Area, Larapinta Drive and South Terrace.

> Now I hear the places are empty except when the tourists come to town ... Flash town houses have been built in The Gap and people moved out of Housing Commission houses. Places are empty and the town only fills up in the tourist season.

*6.1.1.1.l Prostitution and Suicide*

> People don't talk about the prostitution. This involves some of the men in the community ... also white fellas the young girls who get knocked back a few times when they apply for a job, don't care until later they are found dead in the creek. Another thing which they would not advertise here (said after

> the evening interview) whilst I was being walked to my car out of a sense of concern) is the many murders amongst Aboriginal peoples. That would put the tourists off wouldn't it? The elderly Aboriginal informant laughed wryly. 'Be very careful not to get off side ... not everyone thinks the same ... This is Arrernte Country and people come from many places here.

### *6.1.2 Voices, Perceptions of Interest Groups*

Vignettes as discourses on social wellbeing and ability to access services.

In response to the question 'can you make choices in your life?' respondents raised the following themes, which describe their varying life chances in their own words. These range from a sense of being in control of their life to a wide range of circumstances and life chances. The majority of informants who were employed expressed a sense of satisfaction, whilst those with minimal education, looking after young children, or unemployed expressed a sense of despair at worst and fatalism at best in Alice Springs. The high rates of suicide, high rates of alcoholism and high rates of domestic violence (are indicators of people not being in control of their lives). The discussion centred around not having choices in some instances as a result of a disability or caring for a disabled person, work, cultural choices and lifestyle, family choices (whether or not to have a family), about work and about how to manage alcohol and other drugs.

Different life chances by virtue of education and discrimination:

> Have a degree and about to embark on a teaching degree by choice. I have numerous skills in other areas opening up a whole range of choices. Limited by Aboriginality. 'Limited by low literacy levels'
>
> I don't have enough education and have small baby and little sister to look after.
>
> Don't mind job but can't see possibilities for any change jobs harder to get as get older.
>
> Lack of education was the major limiting factor along with racial discrimination in Alice Springs.
>
> Also a sense that the nature of employment is changing which makes it necessary to move for the sake of a career, which means sacrifice of family and friends for some years.
>
> I am willing to work at anything, but what makes me cross is that when I apply for positions they always want experience. But where do you get this unless you are given a chance to prove your talents and knowledge? People forget where they all started from.

#### 6.1.2.1 Property Owners and those Without Property

Those with a regular source of income (ratepayers and taxpayers) considered that their needs were a priority. At one end of the continuum was the attitude that strategies for social issues should not be addressed by local government organisations, because they are perceived to be health and police issues. Others stressed the need for local governance to work on reconciliation with other organisations to bring about 'acceptable community standards'. It was stressed that local governance should be representative of all stakeholders and that it be perceived to be representative. Specific comments highlighted the need for 'more advertised programs for teenagers and for a Youth Drop in centre'. Suggestions made by the public stressed the need for youth services because 'Youth are the future of this town–what we sow now we will reap later'. In this vein residents also stressed the 'Need for a humanitarian and community profile and to develop strategies for social issues and cultural heritage–more promotion and interaction with Aboriginal groups'. The reconciliation theme was echoed in the following statement: 'The adoption of fair, democratic proportional voting system is essential.'

Issues concerning change of the voting system have been raised in the past, but in Alice Springs minority interests do not overlap with geographical areas. Minority interests could be served by maintaining the current system of no wards, but introducing proportional voting.

Citizens stressed the need to be given more opportunity to participate at all levels of government in shaping the laws that affected their daily lives in Alice Springs. This discussion developed in the following chapter on the relevance of social capital and empowerment. In terms of some of the World Health Organisations Healthy City indicators of political well-being, the following need to be considered: (i) the level of participation in decision-making in a community (ii) the extent to which the diversity of the community is represented in key organisations.

Development approaches are needed for enhancing the life chances of young people. The need to work across sectors to create more employment opportunities, educational opportunities and recreational opportunities as a positive approach to crime prevention was stressed.[5] The need for more crime prevention approaches was unanimously stressed. The notion of working across sectors was stressed as vital. For those working with young people in human services alternatives to mandatory sentencing were stressed, not only because of the human rights concerns, but also because of the social and economic costs of incarceration. It was

[5] Not only sampled respondents and informants, but also key representatives of Human Service Organisations and informants across the political spectrum stressed their concerns about crime and also about the life chances of young people.

said that incarceration, should be the last resort for young people. Service providers stressed that: (i) Young people need to have a sense of being citizens who have rights and responsibilities and that (ii) Building a sense of self worth and self-esteem—and hence good mental health is the first step in the socialisation of responsible citizens. Young citizens who contributed as informants stressed that they thought there were better ways to address crime than the use of mandatory sentencing of young people as the trauma of incarceration could make them more distrustful of authority than they were previously.

#### 6.1.2.2 Mothers and Infants

On the whole services were considered to be 'of a high standard, particularly the Infant Health Clinic and the pediatric ward at the Alice Springs Hospital'.

Childcare is a problem, however, according to a child care coordinator. 'People are so transient. Its good for the children, particularly those experiencing the effects of separation or divorce to know that they will see the same faces each day at day care.' Another result of the transient nature of the community is the need for parents to make use of paid care because they have limited family support and limited friendship networks. The service is slightly cheaper, because unlike the others parents provide their own nappies and meals.[6]

Twenty-five of the children are cared in full time care and permanently booked and the rest are occasional places. Only three of the children were of Indigenous descent because parents prefer to use Congress Childcare services, according to the coordinator.

#### 6.1.2.3 Children

Many students at primary school come from single parent homes or homes where there is only one biological parent in a new marriage or partnership. Some of the children live with foster parents. Few children come from homes with biological married parents. Aboriginal children have grandparents and extended family, which is less likely to be the case for non-Aboriginal children. Only 60% of the children's parents have fully paid their school fees, which is $60 per child per year. The school relies on government funding for Aboriginal students.

[6] The cost was $140 for full time weekly care and $35 per day, $25 for half a day and $4.50 per hour.

No discussion of young people's needs in Alice Springs can take place without considering the home environments in which children live.

> ... 'Part of the problem is unemployment ... if people have a great deal of time to drink they will ... this is the starting point of problems' that is If people do not have jobs and do not feel in control of their lives, they tend to drink. 'Domestic violence is commonplace and children who suffer from mental and physical abuse model violence because violence is commonplace in their home environments. The mandatory reporting of violence means that teachers report families to welfare. This process is found to be acceptable by the school, because it supports the rights of young people. But mandatory sentencing of young people undermines these rights because the punishment is not proportional to the crime. I find that in over thirty years of teaching, this is not the worst message to give to young children. You can't teach a sense of decency when the rules do not appear to be fair, can you?'

The choice to spend money from a household budget on alcohol rather than on school fees was criticised by an informant who said: 'Parents can spend $40 on beer, but school fees remain unpaid. The high level of alcohol use leads to disrupted home lives and lack of sleep which effects school attendance. Tangentyere Council collects children at 7.30 am from town camps and in winter many decide its too dark and so stay home.'

#### 6.1.2.4 Young People

The maps of the elderly contrast with those of young people who do not share the same history of loss of land and loss of family and identity as those experienced by the Stolen Generation.

> Young people feel that they have no right to be visible in public spaces or they will be asked to move along. They feel that they are disrespected and under threat. The NT is no place for a young person said a concerned father (an ex policeman). The life experiences and life chances of young people are shaped by high unemployment rates of young Australians, particularly Indigenous young people, mandatory sentencing legislation, one of the highest suicide rates nationally and internationally. Coping with grief and sadness is a central concern for a community which is worse off in terms of all social indicators, for instance: incarceration rates, suicide rates, unemployment rates, morbidity rates and mortality rates. High-risk behaviour is the way in which young people cope with a sense of being powerless. Much community attention has been focused on the results of marginalisation (high rates of alcohol use and public drinking) rather than the causes.

The life chances varied greatly across the young people interviewed. Young people can be described in terms as those who are (i) socially adept who participate fully in community life and are relatively advantaged, (ii) coping, (iii) marginalised and at risk, (iv) survivors who move from marginality to positions in society labelled 'successful'.

Those with stable family backgrounds and attending mainstream school gave very different responses from those with limited family support or those accessing emergency and looking for supported accommodation. The life chances of Indigenous and non-Indigenous young people are also worlds apart in many instances. Need to consider a range of activities suitable for groups with different life experiences.

> Churches have a lot going, outside churches not sure. Need more understanding of their own history. No good just throwing things (services) at them ... have to really listen to young people.
>
> More on youth suicide ... More accommodation for young people'. 'Need more organised trips and excursions'. Not enough fun for young people 'have to have safe places and fun things to do. A-House problems are just about money.
>
> ASYASS is quite helpful, nice people and they listen to young people and help them with payments, need more services like this.
>
> ASYASS has the right idea, they could do more with more infrastructure.

Negative comments stressed need for 'controls, curfews and keeping people off the streets at night'.

*10–14 Years*

> Sporting services are adequate but there is a lack of activities for those who are not keen on sports ... 'Roller blading is not well catered for because pavements in the residential areas are uneven ... Going to the movies' was highlighted as a key activity but it was stressed that there is a great need for 'more to do'. They stressed that public spaces are not for young people 'not pleasant experiences with police—they have a bad attitude—on power trips'.

Negative comments:

> 'Abysmal—very limited, need a focus on young people before they turn 16, because otherwise they drift away from us'. 'High drug use is associated with boredom'. 'Too many roam the street'.

*15–19 Years*

> 'Need for more discos and blue light discos and more clubs without a focus on grog'. 'Too many hang around sniffing or drinking'. 'Pin ball parlours are not the answer'. 'Youth Centres need more funding so they become more active'. 'YMCA seems good'. 'Schools need to support young people more'. 'More supervised places for young people at night'. 'Lack of services for those under 18 who are intoxicated'. 'Mandatory sentencing is a concern—locking people up is not the answer'.

*20–24 Years*

> 'Not much in the way of bands, plays and other arts Only sport and night life TAFE is good—lots of courses can't get bored'. 'Opportunities are available but

> not well known'. 'Not a lot to do, so much is geared to gambling and drinking'. 'More opportunities needed for business minded young people'.

*25–49 Years*

> Most considered the services to be adequate because people make their own entertainment and have numerous family responsibilities that limit the amount of time for recreation.

*6.1.2.4.a Socially Adept who Participate Fully in Community Life and are Members of a Youth Leadership Program.* Overall young people stressed that a happy home life (where parents built up self-esteem, modelled good interpersonal skills) was considered to be the most essential grounding for coping in later life. These young people had access to an education, part-time jobs, extra mural classes (such as music, swimming or dance) and recreational activities.

> 'The world is ours ... opportunities are there to be grasped' is the approach expressed by young people who have stable families and high self-esteem. ... Even if my parents were to die now I would be OK because I have the foundation in my early years. 'We have trust because our families gave us support. A sense of family security is based on a strong, integrated family and some economic security.' 'We are not necessarily brilliant at schoolwork but we feel we can make a contribution.'

They believed that individuals could make a difference to society and that as individuals they had the power to make choices in their lives. All four participated in youth leadership initiatives from the SRC to a wide range of ASTC and other activities. Also a sense that they would never be unemployed: 'Anyone can get work in Alice Springs, unless they are blind, deaf and dumb'. I have two jobs, another said they were able to pick and choose, a third said they had a family friend who owned a business and a fourth said she was paid to do housework at home and $50 for not leaving home and not wearing a uniform was fine with her. I suggested that perhaps the other factor was self-esteem. The group agreed and said that those with low self-esteem and little family support needed a surrogate family or at least 'alternative role models'. For example at Centralian College a female swimmer with a successful national track record had been invited to talk about goal-setting and goal achievements. Goal-setting was uppermost in the minds and taking decisions on the wide range of choices they believed that they had in their lives.

There was little discussion on young people who felt depressed and little real contact with those who suffered from depression, other than professional contact; for example, one said her mother knew more about young people with problems, as she is a professional nurse. If they felt 'down', they said they would 'go out and have some drinks'. They did not specify venues other than 'the Sportsman's'. Other favourite activities were to go to sports training, drama classes or 'to just visit friends'.

They had little sympathy for the way alcohol was abused in town, but said that it was hardly surprising 'because just about every shop in town has a license'.

Without exception they felt that they would leave Alice to further their education, to travel and to work, but that they would feel that they had roots in Alice as they loved the landscape and their families who are likely to remain in Alice Springs.

On recreation, they said that they 'felt as if the town went to sleep after 10 o'clock at night' and that there was a real need for more facilities for young people, 'even if just a coffee bar with notice boards of youth activities'. They mentioned that some of the other venues were inclined 'to hurry you up' or 'ask you to leave the seats in Yeperenye' if you were not purchasing food. Other young people also raised this theme.

*6.1.2.4.b Those who are Coping at High School.* Overall these young people expressed an interest in making a contribution to decision-making, but were not keen to take part in structured committees. They believed that they spent enough time sitting at desks during school hours. They expressed an interest in making contributions using telephones or Internet to contribute to the Youth Advisory Committee. All these young people felt that they needed more respect and more trust in their abilities and the opportunity to enjoy their youth, a time that is short-lived and precious.

The school caters for about 400 students. The focus group was held in a careers class in which there were more than 30 students aged about 15–16 years of age. Two Aboriginal students (one male and one female), one immigrant from Asia, one from America, one from Britain, the rest were born locally.

Whilst the researcher waited to be introduced the participants engaged in some initial mutual sizing up. The attention of the young people became focused once there was a sense that it was assumed that young people had rights, that their opinions would be listened to and that their opinions counted. They said that:

> Citizens are all people who live in a community and who make a contribution. Model citizens make a contribution 'We would like to do as we please, really'. 'No ... We don't have any really we have to wait until we are 18' they said with much wry laughter. Politics means including opportunities for young people to voice their opinions. We should be able to say things in assembly. We should be able to say things informally to Council, not just through belonging to the Advisory Committee. I don't like going to formal meetings ... there are

other ways ... . 'The trouble is the Government doesn't really care, some people do, but most just worry about themselves. It's the job and what they can get which matters to them. One person from government came to listen to us ... that was good'. 'Alice Springs is the easiest place to get a job'. A young person who had previously participated in the ASYASS focus group had said in the previous focus group that it was possible for him to get a job, but it was not easy and comprised washing dishes. Obviously his confidentiality was maintained in that context.

*Health*

Drugs such as marijuana were mentioned as well as 'texter, glue and petrol sniffing'. Alcohol was raised as the most important problem in town, but they said that they though 'it was hard not to be able to drink anywhere in public' until they are 18. They said that in many cases their parents allowed them to drink alcohol at home. One of the girls said that her parents said they preferred her to drink and smoke in front of them than behind their backs. Another (the young boy who was at ASYASS and now back at school and reconciled with his parents) said that his father bought beer and shared it with him. He added that this was fairly usual. An Aboriginal student said that his parents gave him wine to drink with his meal once a week. Another said that it was only OK to drink once or twice a week, not every night. 'Moderation in all things' said another.

Suicide was only mentioned by the participants right at the end of the focus group. By this time the level of interest in the discussion had peaked to the extent that everyone wanted to speak at once. Suicide was said to be a problem, which needed to be addressed in many ways and not only through clinical settings. 'No one wants to go to a place where you are made to feel ...' the young girl pulled a face, because she could not find a word to express her feelings. Stigmatised, the researcher suggested. She agreed. Violence was mentioned as a problem associated with gangs. 'If you leave them alone you are OK. If you are in a group you are OK. You have to be careful where you look if you are on your own. I was beaten just for looking at someone. The worst injuries can land you in hospital. The gangs are associated with places and with activities such as skate boarding.' It was clear that the young person whom I met at ASYASS was associated with the board rider gang.

I summarised what had been said as follows: 'it seems as if public places are contested and you do not feel safe ...' There was agreement. What if you had a public place for young people? 'That would be good' agreed some, there was much talking at the same time and to one another. Once the researcher made it clear that she wanted to hear everyone's 'ideas everyone slowed down ...' There would need to be people to stop fights from occurring'. All agreed.

> 'The place should be in Town, ... I don't know where the Youth Centres are ...' 'They are s**t' said one young guy ... Another wrinkled her nose and said 'they are c**p ... We want a place to just sit and talk ... no one needs to interfere with us .... Even a private business could run it .... It could be sponsored by other youth services ... .'

*Education*

'We want to be respected by our teachers'.

'People should realize that the education at this school is as good as it is at X school.' 'It is stupid to pay for private schools.' Another said that she was concerned about the cost of higher education because she wanted to go to university.

*Recreation and leisure*

'We are bored, bored! There is nothing to do in Alice'. What can you do about that? I asked. 'Well we can do some things. Someone started a place for roller blading ... but there is no where for us to do board riding'. 'We really need a place. When I go to The Mall or to Yeperenye I feel as if the Security Guards are waiting for me', said one male with a big smile on his face. He is regularly seen in Todd Mall in the evening and is very adept on a skateboard to the extent that he can jump over a bench!

We don't feel as if we can go out to public places, we are told 'move along laddie' at best or 'they glare at us. In a super market, in Yeperenye and in the Plaza'. 'We need a place where we can meet without being told that we cannot stay there'. 'Why do adults and older people fear us? How would they like it?' 'We need a coffee shop just for young people. It could be run by someone in their 20s and there could be an older person just to keep things safe. It could have pamphlets and contacts.' I suggested that perhaps they could access services from there and it was agreed this would be OK provided they wanted help. They just wanted to feel that they were safe in a caring place where they could stay until after midnight.

'Transport should also be available.'

Why can't we have an ice rink, it would be great to skate on ice ... it works in Adelaide .... Please, please ... I suggested that it would be more like a private sector project ... 'We would all support it' ... Wouldn't it be a bit expensive, the researcher asked? This was not acknowledged ... 'We need something new here ... Even a wave pool would be good, yes a wave pool and a place to slide in .... 'There is nothing to do here at the weekend, people just go to the movies, or stay at home ...'

*Responsibilities*

'To obey the rules and get an education! To have fun, we have none yet ...' 'We would get involved more with the Council Youth Advisory Committee if we did not have to sit through meetings. Can't we just make contributions through ringing up or through a more informal way? I get sick of sitting down ...'

*6.1.2.4.c Young People Marginalised and at Risk of Homelessness.* According to a YNGO statistics 175 young people sought refuge in 1997 and 254 in 1998. Support (advocacy, financial and general support) was provided on 2035 occasions in 1997 and 3654 occasions in 1998.

These indicated the spectrum of users in terms of cultural and socio-economic background. The only common denominator was youth and the need for assistance (accommodation, friendship and emotional support).

The mission of the human service organisation is to provide an environment where young people feel safe and respected. Accommodation provision and assistance with social security payments and accessing a range of health, legal and educational support are core areas. Their other mission is to advocate on behalf of young people and address social justice concerns. Overall young people felt distrusted and disrespected. They experienced difficulty in completing school and finding work. In most instances they had a variety of relationship problems with parents, peers and authority figures. Many feared living at home, feared institutions and government services, feared the police and security guards in public places and felt little affinity with schoolteachers. Despite these challenges many young people 'graduated' from the NGO by successfully achieving positive changes in their lives as detailed below.

A young adolescent aged 16 confided that he had been to both a private and a public school but did not conform to either. He had been expelled from both schools. His home life was also not supportive at that time. It appeared that he did not stay on at the public school partly because he felt it was inferior to his previous school and because of learning and social difficulties, which resulted in constant fights in the schoolyard. He interpreted these fights as being picked on 'because I am white and bigger than the other guys'. In his opinion most of the fights were racially motivated. He had slept outdoors when he dropped out of school and left home, then he had found accommodation via ASYASS and a job washing dishes. His friend, (also from the same public school) seemed to have similar difficulties at school and had not succeeded in obtaining an apprenticeship as he had hoped because he had also not completed Year 10.

A 16-year-old mother of a 2-year-old was checking to find out 'if her application for accommodation was working out'. From what she said she seemed to be 'having trouble with her partner'. Kim Wright, the Coordinator took photographs of the baby, to validate that she was part of a family who cared about her. A young Aboriginal girl visiting from Aranda House, (which she did regularly) minded the baby whilst the mother smoked a cigarette outside.

A girl under 15 who described herself as having an Aboriginal/ Indian heritage said she had been looking for a job as a cashier without success for the past 6 months. She was feeling very demoralised as she had not passed Year 10 and in her opinion this, together with her background, were barriers to finding a job. She had reached the conclusion that she would need to return to school.

A young Aboriginal ex-petrol sniffer with mental health problems said that he had 'got into a lot of trouble' and that he 'had heard voices'. He had spent time in jail and had been taken to the hospital and put in a locked ward after the police found him in the street. He believed that he was better because the voices had stopped, but he had problems with his neck, which became stiff and rigid at times. He thought that this was linked with worry. Previously he had lived in Finke and was happy there. He said that he had also liked working in a kitchen when he was in jail. He was looking for this sort of work in Alice Springs but had very little success and was very sad because his grandmother had died. He had come to town for sorry business.

*Accessing services*

For young people below the age of 15 there are minimal services available, which do not involve parents. So if young people wish to escape domestic violence it can be difficult to access welfare. 'It is particularly difficult if you are escaping violence and are underage'. To get the Homeless Allowance a parent needs to 'back up your story ...' Sometimes they are 'understaffed and there are lines out of the door ... what can you do then?'

A young mother who had become pregnant in high school explained how she had to negotiate with schools to complete schooling. Some schools were more accommodating than others were.

*Mainstream schooling*

'Schools require you to sit and take notes from a white board for hours and hours'. 'You can't talk and do things; you just have to sit there. I want to move around and I don't like sitting. It was a waste of time for me to stay because I knew I was going to fail because the tasks took so long for me to finish'. 'We were beaten up at school because we are white and we were big and so they thought they would teach us a lesson … . The racism works the other way too, but at this school we got it'.

*Work, training and the dole*

It is difficult to get jobs except for cleaning and washing and other unskilled manual labour. Cashier jobs are very hard to find for those with less than Year 10.

This was raised as an issue by the youth workers because: 'if the training is done by the private sector they only want those students who are most likely to succeed and who will require the least amount of time and effort for the maximum gain. Initially the Australian Youth Advisory Committee supported this initiative because they did not realise the full implications of the policy.'

*Use of public space*

'Security guards in a shopping centre get children to "move along" and you have to learn how to work the system and wait until the next security guard comes on shift. We like to sit in a cool place and smoke cigarettes. I have never seen one adult being told to move along. We know that we must be responsible for our own actions so we try not to make trouble for ourselves. It is the only cool place in town where you can sit and be cool for free. Parks are too hot at this time of the year.'

*Accommodation*

'ASYASS has 20 beds only, X House is closing and St Vincent de Paul gives food vouchers and Salvos can sometimes help'.

*Recreation*

'We need more place to skate'.
A participant from a youth advisory council en route to a university interstate said she was 'tired of handling those young people'. Most of the young people are perceived as middle class who made her feel that 'if you were on the dole there was something wrong with you'. The membership needs 'a total renovation'. …

A young male in his late teens of Aboriginal/Chinese descent described his home in Adelaide as 'comfortable', but that he had left

South Australia because of the lack of employment. He had a few scrapes with the police over non-payment of marijuana but as it was for a back injury (ongoing) pain, he thought this was unfair. He was doing very well at Centralian College and was 'back on his feet'. He was just visiting staff at ASYASS whom he regarded as his alternative family. He was asked to write a song about young people in Alice and offered a guitar by the Coordinator, K.W. This interested him and he spent the afternoon working on a song in the sitting room at ASYASS.

*6.1.2.4.d Indigenous Young People.* Almost 3% of Australia's young people are of Indigenous origin (1996 census). In the Northern Territory, people who reported that they were of Indigenous origin comprised almost one-third (32%) of all 12–25 year olds. Service delivery to young people in Alice Springs is different from other places in Australia according to K.W. In the Central Region the majority of the surrounding population is Aboriginal, whilst at any one time about 15–20% of the population in Alice Springs is Aboriginal. A solution from many Aboriginal organisations is that young people who come to town to explore 'get lost' and need to be repatriated home to 'their country'. It is the perception of carers that the choice to return home or to country has to be made by young people as they have rights as young citizens. The NPY Women's Council has stressed 'the need to keep people within their country', if they wished to remain there, but also liases with organisations, which are not specifically for Aboriginal people, so that a wider range of options are available. Young people are keen to travel and many young people come to Alice Springs from the surrounding community areas and they get into trouble as a result of risk taking behaviour. The CARPA conference hosted by Congress stressed the need to 'track' these young people and keep them from harm once they come to Alice Springs. According to the Central Land Council when young people wish to leave Town Camps they end up homeless because they cannot access welfare and if they decide to remain in town they may end up in correctional systems (Wright 1998b).

At the time of the research the organisation providing accommodation to young Indigenous people at risk faced closure as a result of a critical review of the management of funds and the delivery of services and the assessment process continued within a context of political accusations. One of the key challenges they faced, was finding appropriate educational opportunities for young people because of the language differences between East and West Arrernte people. The Eastern Arrende families support the intergenerational learning of the Detour Program and more work needs to be done to support the West Arrernte families.

The issue of the cultural heterogeneity of Aboriginality is sometimes not taken into consideration when planning development initiatives.

Recreational activities favoured by young Indigenous people: 'Things which make us happy'.

Children aged 11–15 participated in this painting project entitled 'Things I can do to make me happy'. The idea was to produce a wall mural based on their brainstorming on all the things that made them happy. Participation in workshops on mental health was organised by Gap Youth workers. First the concepts of what made them happy were discussed after a workshop held by The Gap to address the depression which young people suffered after the suicide of a young 13-year-old girl. The things, which made them happy, were discussed and then drawn on paper. A researcher from Alcohol and Other Drugs discussed the process of drawing for wall murals with participants and youth workers and in a second session the concepts were finalised. A decision was taken along with the participants as to which drawings would be transcribed for the wall mural. Older boys and some youth workers applied some of the drawings to the wall. Participants identified the following recreational themes: music, dance, hunting kangaroos, using a bush vehicle for hunting, culture in terms of traditional food and in terms of world view, sport, particularly football and cricket. The older children mentioned being a success and this was seen in terms of educational success and earning money. The Indigenous youth worker carefully supervised and each participant was given plenty of room to paint their icon or drawing. Whilst the others waited their turn they played on the trampoline, squirted water at each other, or watched, making comments. The participants seemed more enthusiastic about applying the different coloured paints, once the outlines had been transcribed from their own drawings. Many of the participants seemed to have little confidence initially at drawing what made them happy, but as time passed more original contributions were made. One of the participants, however, felt very confident to paint. He said that his entire family was artistic. It was clear that squirting water and playing with water hoses is a favourite activity. None of the children at the Gap Youth Centre said that they used the public swimming pool very regularly. Energetic activities such as kicking footballs and jumping on the trampoline were other favourites. Participants were unanimous in their dislike of school. Most said that the only thing they liked was sport at school. Teachers were said to make them sit for long hours at school. Across the age groups from under 10 to 15 and above they complained of problems with family associated with violence and drinking. Also that family did not understand them. Parents don't listen to children and also abuse them. One young boy volunteered that his father gave him a black eye. This was found embarrassing by some of the other participants, as there was much forced loud laughter and the subject was quickly changed.

The needs of young Aboriginal people are far from homogenous. It is clear that the young people have widely different life chances and outlooks depending on their social circumstances.

### 6.1.2.4 Disabled Residents

'The world is fraught with access difficulties which have to be negotiated, with great difficulty by those representatives of the disabled who can articulate issues ... . For those who are less articulate and less empowered as a result of language barriers access is more difficult to achieve. Further, many feel powerless because they have had limited opportunities for employment and have little confidence.

'It is a citizen's right to go out, not to feel isolated and to use public transport, not just a special bus ... or special services'.

According to a spokesperson for mental health, people with mental disabilities also suffer particular problems: 'plucking up the courage to speak to someone at CentreLink or at the library can be a great challenge. Awareness of the interior world of people suffering from depression or anxiety is essential. A harsh word or "shaming" can make the next interaction even more threatening'.

*6.1.2.4.a Services.* Services for young people who are physically disabled 'are not well met and families tend to leave town'. 'Access problems for wheel chairs and walking frames'. 'Need more help for renal patients'. Many of the respondents talked of special needs and reflected a lag in understanding that services need to be generic.

### 6.1.2.5 The Elderly

The elderly feel threatened, as do younger people. Some elderly people fear young people and Aboriginal people and they tend to be isolated within their homes. Going out at night is considered risky because of the perceived danger to themselves and their property. 'It is a very dangerous, dirty and violent place ... I will not go out at night', said an irate elderly female resident. One of the reasons for this is that the areas for parking are considered to be ill lit.

'Reasonable, services, but hard if you have a disability'. More recreation 'bring back the waltz, the locals should not just provide entertainment for the white collar mob—an annual dance for Alice Springs is needed, why do older people always have to make their own fun?' Senior Citizens and Migrant Resource Centre serve different groups and not enough integration. 'Not much other than bingo and organized trips'. 'More home care is needed, but the Red Cross is good'. 'Old Timers services are excellent'. 'Fine if you are fit, everything is here'. Services to the aged who do not speak English well or at all need to be considered.

Overall the main concerns of the residents and service providers were for more home supports to enable people to remain at home for as long as possible and for accessible transport to enable them to participate in recreational activities and to access a range of services. Another major concern is the lack of specialised palliative care for people living at home. The concern of elderly Indigenous informants include access to clinical care within the town camp environment and more support for home nursing (e.g. washing blankets) and ensuring that the housing designs are suited to people with a physical disability. Other general concerns were for their personal safety and about the divisions within the community based on a range of factors: age, income and culture being predominant.

*6.1.2.5.a Recreation.* Recreation appears to be divided along the lines of those who are Australian born who make use of the Senior Citizens Centre and NESB people who use the Migrant Resource Centre. Some felt that the Senior Citizens Centre was 'only for locals or Australian born citizens'. Some elderly migrants said that it is 'a place for people who like drinking and bingo' or 'if you like doing pottery'. An elderly 93-year-old said that 'as a coloured person' he could not belong to the Senior Cit's Centre (as it is known) because 'they wouldn't want me there' and said besides he did not know where it was, despite living in Alice all his life. His main point of contact was Old Timers, but he said he was a member of the RSL.

According to the coordinator of the Senior Citizens Centre, it provides services to 130 members and on any one day they have about 30 people making use of the Centre and 50 to 70 people come to the monthly dinner. The activities are individually priced.[7] The coordinator said that the costs although low were 'possibly too much for some'.

A workshop at the Centre provides men with the opportunity to do handy work and repairs. In this way men can be persuaded to link

[7] Costs range from $3.50 for water aerobics, to about $2 for ceramics. The cost of the bus although free, requires a donation of about 50 cents. Lunch costs about $1.50 to $7 for the monthly formal meal. The annual fee is $7.50 per year.

socialising to doing 'work in the shed' which is normally a solitary pursuit of men at the bottom of their garden.

'Only English speakers use the club' the coordinator said, 'the Migrant Resource Centre provides for the migrants'. The Organisation is run by an Executive Committee that is elected every 3 years and an 'ordinary committee' that is elected every 12 months. 'No Indigenous people belong to the club'.

*6.1.2.5.b Care and Respite for Carers.* The Carers Respite Centre for the Central Australia and Barkly Region serves remote and urban communities and according to the brochure the functions of the organisation is to assist: carers taking a break and to establish links with other communities as well as counseling and advocacy. According to a trained nurse (who assesses community need) the scope of her work has narrowed slightly as a result of NPY Women's Council taking on some of the areas where she used to work, which helps. Their role 'is to network with other agencies and communities to address needs'. ... One of the missing links in the system is the lack of rehabilitation. A Rehab Unit is located at the hospital but does not appear to meet the needs of carers who come to town to be with patients. Carers and recipients sometimes need a break from each other, the role in remote communities is usually by aunts, grandmothers and certain daughters. Men particularly uncles address cultural concerns on their nephews behalf but not the physical care. Sometimes there is a reluctance to care for a frail young person, because if she dies then 'payback' may be an issue.

Home care services are provided by NGO organisations but they cannot meet all the existing needs of the frail and elderly. The Australian Red Cross provides home care services defined as follows in their documentation:

> People who due to various health related problems, are unable to cope in their own homes. The objective is to promote independent living. The fees range from $3.00 per hour for clients in receipt of a pension, then on a sliding scale of $12.50 per hour according to the combined family income.

The following services are provided: 'House keeping, clothes washing, shopping, gardening, home maintenance, personal care and respite care.' Also walking frames and wheel chairs are provided. With the referral of a doctor, meals on wheels are provided to people unable to prepare their own meals and at the cost of $2.50 for a 3-course meal

A spokesperson for the Red Cross, stressed that the key issues are isolation of elderly people who do not have access to the support networks of family members.

Many elderly feel unable to participate in social gatherings such as Anzac Day celebrations because they find organising taxis stressful and worry about how to return home. They also worry about using public toilets because of difficulties dressing and undressing quickly. So many small details become important in an elderly person's life. There were so many people this year who said they could not attend unless they were personally transported.

Social isolation of the elderly is partly due to physical problems and partly social. Many elderly people have stories to tell about the early Alice and these are being lost. According to a spokesperson for the Red Cross, oral history projects could involve elderly people and could help ensure that they are respected and valued as citizens. This informant said:

> 'I know an elderly woman and her husband who came to Alice in the 1920s on a camel and she was looked after by Aboriginal people and she lived in a humpy (sic, a colloquialism for a house, originally a derogative term). She has the most wonderful photos on her wall ... I am leaving town and I really worry that this is something that will be lost.'

*6.1.2.5.c Housing and the Elderly.* A non-government organisation provides transport and some day care for both Indigenous and non-Indigenous residents that could be used more according to a spokesperson, but many elderly feel that it is an admission of not being in control of their lives. Others say that they will not use services because people are cliquey.

According to the ABS 1996 data for Alice Springs there are 499 males and 508 females identified as being in non-private aged care, a total of 1,007 out of a total of 2,373 persons aged over 50, that is, 42.4%. Specialised care such as respite care and palliative care are areas in need of further development (through the Association of Pain Management and the Red Cross) and resourcing.

The director of nursing at a service for Indigenous elderly in need of respite care said:

> 'Without being idealistic', to use her words she 'tries very hard to provide an environment where the basic needs of people are met. Always having a blanket and having meals three times a day is some compensation for not having family around, although "the Oldies" tend to get disoriented when they come to town.' The residents spend their time sitting outdoors, facing the fence in comfortable chairs covered in gauze to protect from flies. They talk little, partly because of language differences and partly because of disorientation. Their view is the fence outside the hostel. According to the informant caring, attentive staff can minister to their body but not their spirit and their longing 'for country'. This is one of the reasons for delaying treatment despite the provision of a sensitive caring environment and material comfort. Elderly patients have a feeling that when they come to town for treatment they are not going

> to be able to return to country. There is a sense that before the physical body dies the spirit will die. Some of the elderly get distressed, some withdraw, and about 6 different language groups are represented so perhaps this makes conversation less easy. People go home to die, they go home to visit and the relatives are encouraged to give them all the emotional support they can, without judgement about the way they have cared for dressings, I can fix that ... but not their emotional needs. There are about 38 patients and 28 nurses and 10 other staff. This is necessary, because they require a great deal of care. Considerable networking occurs with other organisations in town, Royal Flying Doctor Service, Remote Carers Respite, St Mary's. The Hostel is about to move to be located near Old Timers. Outings include going for bus rides to the centre of town, sitting on lawns, going on picnics for lunch. The public sector could assist by providing a meeting place for the elderly. Many come in to town and they would benefit if they could all meet together outside in a central place. The venue for meeting others could be central to give elderly people a chance to come to town. Perhaps it could be as easy as providing a package lunch (donated) once a month at an arranged time.

The following key concerns were raised in an interview outlining the issues raised by elderly people who make use of the aged advocacy service:

> Infrastructure: The rural town is changing and the planning character is ugly. We need big humps to slow down the traffic and the traffic islands aren't enough, we need pedestrian crossings. Pavements are uneven and how about a few pedestrian crossings? Services: Home care nursing services aren't easy to accesses it depends ... People need more gardening assistance. Once you access the services you may not be able to keep them for long. Recreation: Clubs are full of cliques and strong class groupings. Also the problem is crime and a sense of fear to go out at night and a lack of resources for outings because the pension is limited. Accommodation: People who cannot afford to pay their rent in Housing Commission Houses. For instance: 'An old man went to hospital and he was behind with his rent. He lost everything. When he came out we had to find him accommodation and furniture ... some people stay on in hospital and frail care because they have nowhere to go. Where are they placed on the Housing Commission list when they are debtors? The options are relatives, a nursing home, going back to country or the river'. 'Where do the elderly tenants go when government housing is sold off?' Getting out: They cannot afford to eat out, just make ends on the pension and this is also the case for those on superannuation. The infrequency of buses at the weekend adds to this. Also their ability to get to the cemetery is an issue. Taxi subsidy schemes are not given to people in hostels and public transport is not accessible for people with disabilities. Alienation: Its no longer a small place where you know everyone.

According to a staff member one of the ways the advocacy service is used is to wage campaigns against neighbours and this can be as a result of noise or dogs, cats or simply mutual dislike. It can lead to involvement of the police and tax office as a result of passing on information.

### 6.1.2.6 Migrants

The challenge faced by migrants is to re-make their maps of the world by creating a new life in a new place. Access issues faced by elderly migrants at the Migrant Resource Centre were discussed by migrants present at morning tea and banner-making. They were from Germany, Sri Lanka, Italy, Malaysia and South America. The value of a small organisation in creating a sense of community was apparent. The sewing appeared to be an excuse for morning tea and conversation. Most of those present knew one another and said that the MRC was described as excellent because it is a small organisation and they have close contact with the staff. The English language skills of a Spanish-speaking woman from South America were minimal but she participated by nodding and gesticulating. Her son came to take her home after the sewing class. She numbered and named all her family for the benefit of all those gathered at the table. Most of her family was in Adelaide and in North America. She received tuition at home from one of the volunteer aged care workers who also shared the task of banner-making. One of the aspects of Alice Springs, which long standing residents complained about was that it had grown from a small, intimate place to a place where people did not know every one else as they had in the past. There was also discussion about the fact that some of the younger people they had met had moved away from Alice. D listed all the young people he had met and how they had moved interstate and overseas.

A sense of isolation was perhaps the only issue highlighted as a problem for people without families in Alice. D said depression was a problem for elderly with no interests. 'They (older people) can get into trouble like that'. The elderly migrant group included a male who had been a German prisoner in Alice Springs. After the war he stayed on in Alice for 50 years. His wife had died 2 years previously and despite a great interest in Alice Springs history, he felt lonely at home with his collection of books and documents and his cat for company, so he joined in the sewing morning just for company. Despite requiring a wheelchair, at the time of the research he was able to cope at home and had not required much more than Meals on Wheels and the occasional home help. He used to care for his wife at home who had suffered from muscular dystrophy. The other problem (associated with isolation) is the public transport system which does not cover all the routes they wished to use and which required long waits in town for return journeys.

Most said that they thought the Red Cross provided a good home service and that they thought it appropriate for this service to be delivered by them.

The popular perception is that the majority of people in Alice spent their time having barbecues and drinking and recovering from drinking!

There was a sense that Alice could 'do with societies for special interests' and that 'people came to Alice with interests, they started up good clubs and then left town'. They explained that there were too few people to set up many clubs for specific groups and that the Italian club did not survive, although there had been two attempts to keep it going. The clubs were perceived to be used by 'drinkers and bingo players and those who liked doing pottery'. Although it was acknowledged that the Araluen Art Centre was good, it was considered to be under-utilised.

A particular source of annoyance was the way in which the elderly were disrespected. For instance a friend of D's sat down at a shopping centre and was told to 'move along' by a security guard. She was feeling faint and was waiting to recover. She ordered a cool drink and only then she was allowed to stay. D said he had 'mentioned it to the Mayor' and that although he 'knew they were doing it for security reasons, the security guards could go too far'.[8]

Most of the participants seemed reasonably content about security in Alice. Drinkers were mentioned as a major problem but there was a sense that drinking is a general characteristic in Alice Springs and one associated with a number of problems.

Migrants talked about the quality and availability of food, for example an Italian woman stressed the poor quality of vegetables in supermarkets. A nutritionist and a diabetic educator gave a talk on the causes of diabetes and how to select healthy food in a supermarket. One of the women present had been diagnosed with diabetes and was interested, but the other participants (both men and women) seemed to feel that they knew how to plan a healthy diet and contributed knowledgeably to the conversation. The atmosphere was one of enjoyment and easy familiarity as the health session was planned around a meal served halfway through the proceedings. It was suggested by a nutritionist that perhaps tinned fish was not as bad as they initially thought, but the Japanese couple were pleased to hear about an arrangement for buying fresh fish through a network of Philippine women who had started the service. The knowledge of participants from Japan, Korea and from Eastern Europe on healthy eating was considerable, even if some of the technical information on diabetes

[8] The informant did not make the connection that shop owners in the mall wished their facilities to be used only by consumers, not by people who wished to escape the heat and enjoy the air-conditioning without actively consuming. People on low incomes (young, elderly and Indigenous people were discouraged from using public places).

was new, the basic principles of eating natural, unprocessed foods and avoiding sweets and processed foods was understood. Specific concerns were about fresh fish and fresh vegetables and the conflicting information they were receiving about the high cholesterol content of eggs. One participant said that she knew that eggs were fine provided they were eaten in a balanced diet.

## 6.2 *Drawing Together the Themes*

Social problems are as much a product of the way we think and conceptualise issues as they are the result of social structures and the way in which people participate in decision-making. Apparently discrete problem areas are in fact linked in complex ways, as conceptualised in Diagrams 6.1 and 6.2. If this is understood then coordinated development can occur that is based on partnerships with a real sense of the systemic links that can build or erode life chances and social capital (as it is perceived in terms of diverse sections of the maps on reality). The challenge is finding ways to co-create some shared visions and to also allow for diverse visions, in the interest of democracy and the need to sustain diversity as a jump lead of human creativity. A rich range of ideas is better than sameness, with the caveat that shared vision is vital on matters of human and environmental justice. This can only be achieved through ongoing dialogue.

The questions that people frequently ask are 'Why are things the way they are in Alice Springs?' and 'Are things really so different in comparison with other places?' The barriers to achieving service outcomes in the areas of health, education and employment for some people in Alice Springs are due to a number of reasons. The answers lie both in the present and in the past. The impact of colonisation on Indigenous people needs to be considered in all analyses of the current status of social health in Alice Springs (Menzies Annual Report 1999).

Indicators of well-being demonstrate that positive outcomes in some sectors of the population have yet to be achieved. The reasons span a number of interlinked areas: social–cultural, demographic, geographic, economic and political. It is no accident that geography has played a role in shaping social responses the world over. The evolution of Indigenous ecology began over 40,000 years ago and provided a sustainable way of life, if not economic progress in the ways measured by capitalist economics. The history of colonisation and social marginalisation has been documented. According to ABS census data, Alice Springs is characterised by high levels of unemployment amongst Aboriginal people (16.3% in 1996), high levels of social dis-ease that create an industry for many professionals. Health and community services (13.3% in 1996) and government services (8.3% in 1996) are one of the main economic areas. Retail follows at 14.9%. Despite the tourism potential of eco- and cultural tourism, culture and recreation makes up only 4.2%. This is an obvious area for development. The other area is arid zone technology, however this is outside the scope of this discussion. Working in harmony with the environment, however, rather than conquering the environment is a challenge well worth addressing. Historical, social and environmental factors are relevant to problem definitions and approaches to addressing problems.

The collaborative approach to governance is the focus of this discussion. It is relevant not only to thinking, but also to practice. A collaborative praxis to governance can only be established on the basis of systemic thinking that spans barriers. Governance approaches that assist participation are matrix networks that are based on shared foci of concern (McIntyre-Mills 2000).

Alcohol misuse is an effect of a legacy of colonisation and marginalisation and a cause of social ills. Higher mortality and morbidity rates in Alice Springs and the Northern Territory are outcomes associated with violence and road deaths as well as diseases directly and indirectly linked with alcohol and the associated poor nutrition (as a result of both spending money on alcohol as well as the unavailability and very high cost of food in remote communities). The ramifications of the abuse of alcohol and other substances need to be understood as being systemic in their causes and effects. Substance misuse is the immediate cause of many social ills. The causes and effects become a cycle of damage to individuals, families and communities resulting in ongoing and intergenerational poverty because of the modelling of behaviour to the younger generation and the sense of cultural loss, meaninglessness and dependency on welfare (Diagram 6.1). The cycle is one of marginalisation, alcohol misuse and further marginalisation. A broad-based systemic approach has been advocated by recent NT public health policies in line with the World Health Organisation's Approach (1995),[9] but the policy needs to be translated into practice.

To draw a link, for example, between the alcoholic despair of an Indigenous 14-year-old and globalisation is not so far-fetched when one considers the impact first of colonisation, then loss of land, family and culture of origin in the case of the so-called stolen generation and some

[9] In a paper entitled 'Alcohol policy and the public good', the WHO concluded that programs are only effective if they are broad-based and if they address the wider social context.

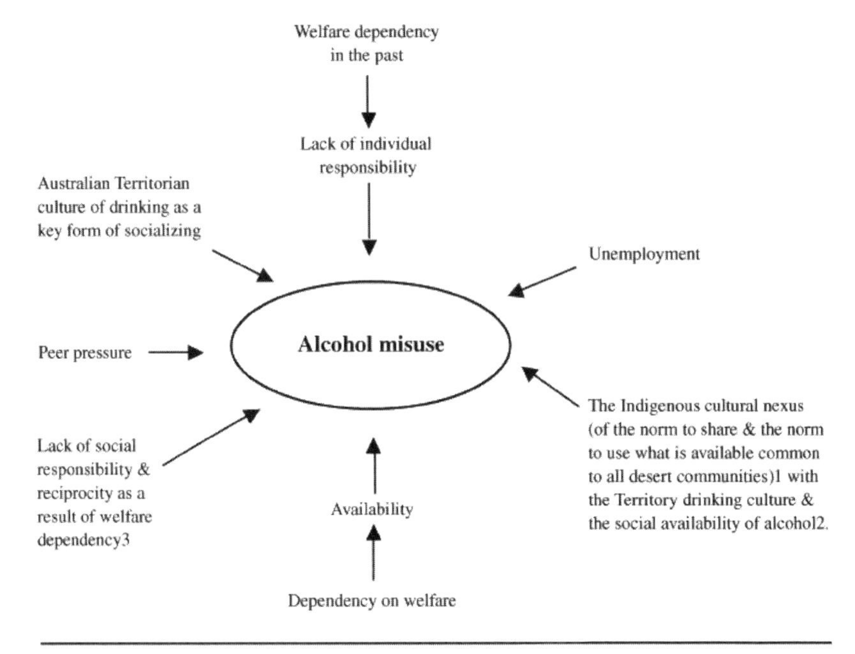


1 See CARPA Newsletter March 1998
2 See Rowse 1998
3 Pearson 1999

*DIAGRAM 6.1* The Context of Alcohol Use

form of welfare dependency in the case of the majority of Indigenous people. And later the changing nature of welfare from universal to residual rights to the social wage and the changing nature of the economy and work. The mismatch of cultural values and work skills causes anxiety and the lack of respect (resultant from the cash, work, status nexus, Pixley 1993) or a sense of 'being shamed' when people negotiate in non-Indigenous cultural contexts. Added to this, the culture of drinking in this remote area has been institutionalised in the dominant Australian culture (not only as a way to socialise at work, at public functions, but also to celebrate and mark social events in private life (see Lyons 1990) at each stage of the life cycle (see Van Gennep's 1960 notion of rites de passage). Further, it is also a way to cope with problems. Having a drink is seen by some as a way to treat or medicate most problems.

Linked with the so-called 'grog culture' is the commodification of culture by non-Indigenous players and paradoxically (as an effect of colonisation) by Indigenous people themselves to pay for grog (Diagram 6.2). The commodification of Indigenous culture (digiridoos that are not in fact

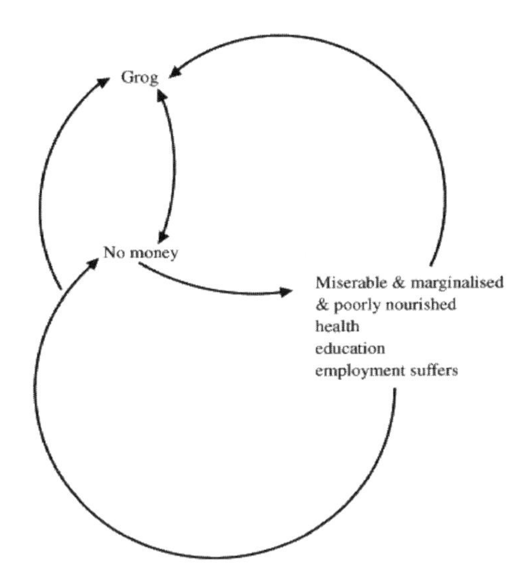


*DIAGRAM 6.2* Cycle of Alcohol Use and its Impact

Indigenous to this area) and commercialised dot paintings is an issue for those who feel excluded in their own place. Nevertheless, doing a painting for 'grog or grog money' at the time of the research was not uncommon. This chapter discusses the socio-political and economic legacy that is perceived as a sense of disease and sense of dis-location by those with limited life chances.

The activities of all organisations in Alice Springs take place within a wider framework of policies. Stakeholders at local, territory, commonwealth and international level attribute different layers of meanings and values to the same human and environmental issues and interventions. Further, planning for development requires a strategy that addresses three clusters of patterns along a demographic continuum with the life chances of a developed population, transitional population and a less developed population. The developed population pattern with low birth rates and low death rates can be defined overall as showing patterns of health associated with developed socio-economic societies. The transitional and less developed populations have higher mortality rates and shorter life expectancy showing the patterns of health associated with less developed socio-economic societies. Health patterns are linked with poor nutrition and risk-related behaviours associated with difficulties in negotiating competing cultural systems. There is also a sense of social, political and economic marginalisation amongst those with these demographic patterns. But nevertheless the pattern needs to be seen as a continuum of life chances linked with age, gender, race, culture, level of income and health status. The indicators of unemployment, suicide rates, incarceration and wide use of alcohol across all age groups and across both the Indigenous and non-Indigenous population are indicative of the need to build broad-based social prevention programs. A time frame of immediate survival, day-to-day, hand-to-mouth is usual amongst all people who do not have economic capital and have very limited access to social capital in the wider community. Thinking about the long-term implications of hazardous life styles is also not a priority. This is an area that has been well researched by Helman (1983). Sections of the highly mobile non-Indigenous population are also not coping (as indicated by the high levels of alcohol use and domestic violence across the population).

Nationally the Northern Territory has the highest youth suicide rate, the highest rates of public drinking, the worst health status in terms of injury rates, nutritional disorders such as diabetes, respiratory diseases and other health problems linked with the developing population profile. The NT and Alice Springs have been associated with mandatory sentencing and the highest imprisonment rates nationally. These factors have drawn national and international outcries by international human rights agencies and the United Nations.

The growing disparity in social health between Indigenous and non-Indigenous people has contributed to social disharmony and could continue to do so in the future, unless participatory designs focus on a wide range of social health and crime prevention measures that take into account multiple narratives and attempt to co-create shared areas for action. This impacts on 'the community' as a whole and makes it less likely that common life goals will be addressed. As Banathy (1996) has stressed co-creation can only occur when the conditions for co-creation are achieved. He calls this generative dialogue. The creation of a collective consciousness was the first challenge, of PAR and could only be achieved within pockets of the community, because of the low levels of shared social capital associated with a sense of powerlessness, marginalisation from grass roots decision-making, broken lines of communication and poverty. Poverty is linked with age, culture, language, gender, marital status, education, language, employment, level of ability and disability. Compound disadvantage occurs, for example through language difficulties being linked with poor levels of health and low levels of literacy (Diagram 6.3).

The class structure in Alice Springs can be measured by indicators of unemployment, incarceration, homelessness, employment indicators of occupation type, level of income, ownership of property and level of education. Unemployment is directly linked with crime rates in other parts of

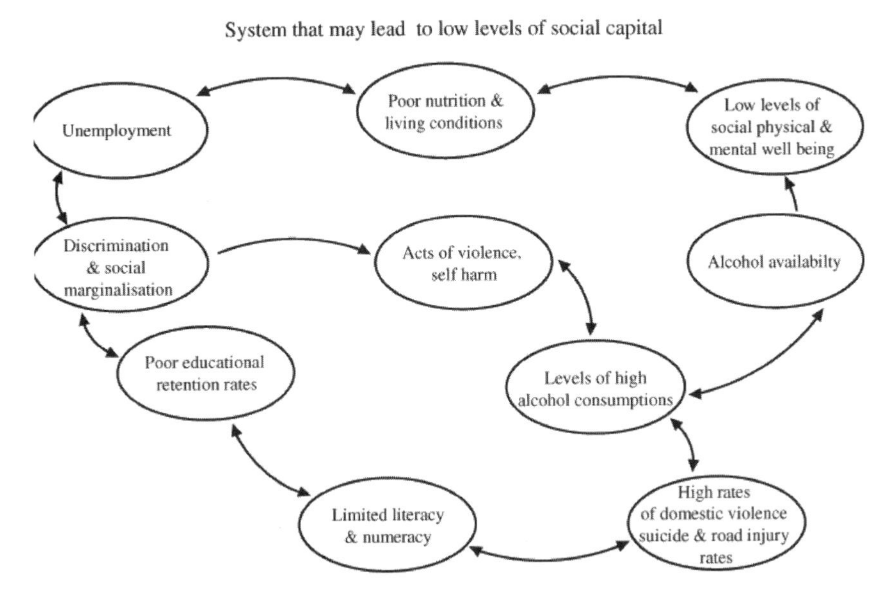


*DIAGRAM 6.3* The Impact of Alcohol on Social Capital

the world and there is no reason to consider that the NT is an exception.[10] According to Street Ryan and Associates (1998) low skilled jobs are disappearing which means that the unemployment rate amongst Aboriginal young people will only be addressed by better preparation for a skilled service industry and by creating specific opportunities. The government's introduction at the time of the research of mutual obligation policies for the unemployed need to be considered in the light of the many problems which people face and the fact that job placements favour those with a wide range of educational and life skills suited to a capitalist society. In addition, a culture of despair and cynicism exists as a result of the lived experiences of limited life chances. The high number of unemployed young Indigenous people who have dropped out of school poses a social and economic challenge and undermines the possibility of building both social and environmental capital. The pass rate for year 12 at the time of research was almost negligible in Alice Springs. Education and employment[11]

[10] Local statistics on crime by demographic characteristics were not available at the time of the research from Northern Territory Government.

[11] Human capital theory refers to investing in human capital and argues that by not investing the economy misses out on the skills of people and this is an opportunity cost.

considerations need to be based on an understanding of the different ways in which groups of people make sense of their world. The high crime rate attributed to young people has led to more social control, such as mandatory sentencing introduced in 1996 and revoked on 21st August 2001 as a result of sustained protests across the Northern Territory and criticism from leading Australian public figures in the field of law and a wide range of other human services.[12] This has led to some changes in the sentencing of minors and the funding of further diversionary programs.

Building social capital through a holistic web of measures to promote well-being of both young people and their families would be a positive and proactive form of social intervention, which will not necessarily be more costly to implement than the previous system of mandatory sentencing and Zero Tolerance policing[13] (given the cost of mandatory sentencing to the community in excess of $54,000 per prisoner annually).

One of the pro-active approaches could be the way we develop pathways in civil society to improve employment opportunities. The introduction of changes in the conditions of work (relations of production), types of commodities and changes to the very fabric of cultural identity as a result of social change have far ranging ramifications for those affected by the social and economic changes at a global level. The types of economic activity include: (i) knowledge production for sustainable living.

*Continued*

> 'One problem with applying "human capital theory" to indigenous education research and policy development is that as Schwab (1996) shows in his study of participation patterns in higher education, Indigenous people's behaviour does not "conform" to the theory's predictions. Rather than take courses, for example, which would "maximise their future returns," many Indigenous students choose to study in programs which prepare them to return to work in their own communities, "to work with their people," as they put it, rather than preparing them (for better-paid) employment which would isolate them, not only geographically but also culturally from their kin and homelands. The decisions pertaining to education by Indigenous peoples, Schwab concludes, may have much less to do with individual calculation of private rates of return than with individual calculations of cultural costs' (p 15).' (Boughton 1998: 10).

[12] Central Australian Youth Justice Coalition 2000.

[13] Stone, S (1998) Ministerial Statement Reclaiming the streets: Zero tolerance policing and the Northern Territory (August). The policy of Zero Tolerance policing introduced in August 1998, which can be summarised as being introduced to target specific 'problem areas'. It follows the 'broken window' philosophy; namely that if there is an opening people will take it. Once people realise that there are no openings the crime statistics will decrease.

(ii) localised, traditional approaches to traditional industries in primary, manufacturing and service work without information literacy and a local, rather than a global approach. (iii) Welfare dependency on the social wage and those struggling to find employment and to shrug off the legacy of what Rowse (1998) called the dependency created by rationing: 'white flour, white power' in the history of colonisation. (iv) Traditional Indigenous communal lifestyle. The 'cash work nexus' (Pixley 1993) is central to definitions of citizenship. The majority of Aboriginal people have difficulty in obtaining work opportunities. This was particularly so after obtaining citizenship and the notion of minimum wage associated with citizenship led to many pastoral workers becoming unemployed. Historically property ownership was a basis for the right to vote. Ironically dispossessed Indigenous people could not vote until 1967. In some sections of Aboriginal communal society, commodities are merely shared items, albeit reciprocity underpins sharing. Also the way in which land and culture is understood in capitalism and in Indigenous communal society differs. Land is not so much a commodity as a source of cultural well-being in communal Indigenous society. The context of land rights is the context in which the organisations within the public, private and volunteer sectors operate.

## 6.3 *A Historical Legacy of Colonisation and Marginalisation*

The life chances of Indigenous people can be mapped along a continuum from ownership of business and property interests, employment, underemployment and destitution. Alcohol and poverty play a role in property crimes in Alice Springs. Property laws were implemented in such a way that the judiciary was subject to the laws of the Territory. These limit the right of the judiciary to assess each case and to determine the appropriate punishment for the specific case.

Indigenous people have had the vote since 1967, but sections of traditional and transitional population perceive that they are still regarded as second-class citizens within their own land. Also successful Indigenous capitalists in Alice Springs are regarded with some ambivalence by both Indigenous and non-Indigenous informants and the need for this group to develop social responsibility towards those Indigenous citizens who are less successful. Negative attitudes to Indigenous people using public spaces and services by non-Indigenous people was stressed as one of the factors leading to the need for more recognition of Indigenous identity (language and land). The need for some separate and specific services to meet health, education and legal needs was stressed.

A section of the underdeveloped and transitional population, mostly Indigenous, feel that they are marginalised and they feel that they can make little impact on bringing about change. The causes of social ill health and social disadvantage are not unknown.

> '[T]he poor health of Aboriginal Australians is not due to any absolute lack of knowledge about the causes of their ill health, but to the fact that they have had limited access to health resources and knowledge because of poverty and educational disadvantage. In the past there has been limited understanding of Aboriginal health issues by funding agencies, compounded by inadequate knowledge and training of health advisers and providers. Unfortunately the poor state of Aboriginal health has also been perpetuated by disagreements about what should be done and how, who should do it, and who should pay for it. This lack of consensus amounting to a modern Babel ... is only now beginning to be resolved' (Menzies Annual Report 1999: 9–10).

Added to this analysis of health services per se is the problem of the lack of understanding that the responsibility for improving health outcomes is not solely the responsibility of the health sector, but of many sectors.

## *6.4 The Local, National and International Context of Policy Decisions*

The wider policy environment within which local players operate will be briefly outlined because the local case study needs to be considered within the wider context of the human rights, legal and policy environment. Local activities take place within a wider framework of policies at national and territory level, for instance: the role of legislation to control crime, police the rights of citizens and redefine Native Title and self-determination provides the contexts in which human service providers and citizens operate. At the time of the research, for example, the International Human Rights Committees associated with the UN considered that human rights abuses occurred in the Northern Territory. Mandatory sentencing at that time led to 'the Territory being placed into the national and world-wide spotlight of organisations including Amnesty International and the Human Rights Commission.'[14] International attention was focused on changes to the Native Title Act by the UN Committee on the Elimination of Racial discrimination and found that the Commonwealth Government had discriminated against Indigenous peoples.[15]

[14] Alice Springs News 1998 25th Nov vol. 5 No 43.
[15] Land Rights News March 1999 page 14.

Criticism was also directed at the non-support by Australian Government of the terms 'self-determination' which it had suggested should be removed from the Draft Declaration on the Rights of Indigenous peoples and replaced with the term 'self-empowerment or self-management'. The language of rights is one which is value laden and communication to co-create acceptable definitions is important for democracy.

### *6.4.1 Divergent Policy Environments at Different Levels of Governance*

Two examples of policy at the time of the research are analysed at the Territory and national level. The first is crime prevention and the second is public health. The two are systemically linked through the promotion of social well-being and social capital.

'The Pathways to Prevention' policy approach was introduced by the National Crime Prevention Strategy in 1999. It is argued that crime prevention is only possible through a broad integrated approach to addressing risk behaviour and building up 'protective factors'.

> At a broader level, protective factors can be enhanced by strengthening the capacity of a community to intervene positively in the lives of children, and by building facilities or social structures that support involvement and attachment, that help maintain a civil society rather than an oppositional culture. A general message is that we need to keep exploring what links previous conditions to later behaviours. There are many possibilities, such as the presence or absence of social strategies or social competence, the presence or the lack of a sense of shame, or the presence or lack of forward thinking, but it would be wise not to emphasise one at the expense of others. A further caution for program designers is that some of these factors may operate in different ways depending on gender and ethnic group ... (Pathways to Prevention 1999: 30).[16]

At the time of the research it was suggested that this approach to crime prevention at commonwealth level, for example, could be considered as an alternative to mandatory sentencing for the reasons detailed below, namely the social cost to individuals and to the community as a whole.

Public health policy at that time received much government support as its mission for 2000 and beyond. But the planned holistic service delivery faces the challenge to become integrated and holistic despite the competitive tendering contracts from government that works against integration because of competition across departments and organisations.

[16] Pathways to prevention: Developmental and early intervention approaches to crime in Australia 1999. National Crime Prevention publications. Attorney General's Department: Canberra.

The rhetoric of prevention and health promotion can only be implemented in an environment where organisations are prepared to work and communicate across sectors in an integrated manner: how do you move service delivery away from public delivery through bureaucracies and smaller organisations to competitive privatised delivery?

Subscription to the rules of global markets has led governments to balance their budgets at the cost of welfare spending. The implications of this global shift from universal to residual (limited) welfare will be discussed below. In this context of rationalisation,[17] the proposed Territory Health policy shift from specialised programs to generic programs in reality meant that specific services (in gender and cultural terms) would no longer necessarily receive as much funding as previously. The trend towards privatisation of services and outsourcing is likely to continue as government's face the pressure to compete in global markets and trade barriers are unlikely to be restored as markets move from a global to a regional focus. Within this broader context at the local level, small human service agencies respond to the centralising[18] and devolution policies of

[17] At a National Government level policy is shifting from universal to residual welfare. To quote and paraphrase: the example of the shift of funds from Medicare to the $300 rebate is a case in point.

> If more middle class people can move from Medicare to private health cover, then more will be available for those who need it. In reality that has not been the case elsewhere, if we consider the USA where the public health systems are far worse than in the UK who maintain a public health service … . 1.6 billion has been allocated to funding the rebate in the context of cost cutting and rationalising through privatisation, downsizing and outsourcing (Baum 1999).

[18] Devolution of responsibilities. The NT has the highest rate of population growth (2.9%) of any jurisdiction in the nation, according to Reed (Oct 1998). In line with the 'Planning for Growth' policy outlined in a speech by Treasurer Mike Reed (Oct 14th 1998) the centralisation of services will allow savings in the public sector that can be used for improvement of services. The review needs to be undertaken within the context of the Budget Speech on 28th of April 1998 that indicated the process of Planning for Growth: 'Major reviews are in progress in education, health and PAWA and Correctional Services all oriented towards a better service at a lower cost.' It is argued by NT government that savings will be achieved by economies of scale by bringing together like services and allowing organisations to concentrate on core business. The implications of privatising basic services such as Telstra and the PAWA could have challenging implications for the remote areas surrounding Alice Springs in terms of user cuts and access to vital services. ACOSS has cited (Nov 4th 1998) that all-Social Welfare services show an increase in the number of people using services and an inability to cope

Commonwealth and State government, by narrowly guarding their resources and trying to justify their separate existence that is understandable given the diverse needs of citizens.

The challenge of meeting social health needs is made more difficult as a result of the two opposite tendencies at local and regional level, namely: a push towards centralising services at Territory Government level and a pull by many agencies at the local level to 'protect their patch' from the rationalising process. This local approach to compete for turf was highlighted by service providers at the CARPA[19] Conference hosted by Aboriginal Congress and at a series of Territory Health Planning Conferences.

### *6.4.2 Addressing Values and Governance*

Organisations and individuals compete for scarce funding and to retain special services to meet specific needs. Further, different values underpin service delivery and leads to a wish to retain separate identities. For example: harm minimisation approaches for drugs and other alcohol versus control or services geared to primary and promotive health care versus curative care. Added to the withdrawal of funding from the welfare state and a style of operating which has for years favoured thinking within departmental or discipline-based compartments is the development approach based on the expertise rather than on the personal knowledge[20] of people with 'lived experience'.

Human service organisations the world over have been structured as bureaucracies and have communication styles that could benefit from more open, flexible and dynamic team approaches to addressing multifaceted social issues.[21] Complex development concerns could be better addressed by approaches that integrate sectors and disciplines when

*Continued*

with the needs of clients. The socio-economic context reflects the way in which Welfare States have moved from universal to residual welfare rights. This is reflected in the ministerial statement of the Treasurer The Hon Mike Reed MLA on 14th of October 1998. He discussed the review of Government that would involve 'planning for growth'. which involves maximising effectiveness and efficiency in services, outsourcing and trimming to core business. The centralisation of information services for NT government is one example of cost cutting.

[19] Central Australian Rural Practitioners (CARPA) Conference.

[20] Polanyi, M (1962) *Personal Knowledge*. London: Routledge and Kegan Paul.

[21] Flood, R. and Romm, N (1996) *Diversity Management: Triple Loop Learning*. Chichester: Wiley.

undertaking problem-solving. For example, unemployment, low retention rates in schools and social marginalisation could best be achieved through integrated development policy and decision-making approaches that work across organisations. In Alice Springs, besides the historical legacy, the global shaping of welfare delivery and the mobility of the population of service providers and users, there is another factor, namely the small town dynamics in a community with a diversity of political, social and cultural values. Non-cooperation across organisations can be the result of personal, family, organisational and political antipathies (often because of different values towards culture and development). Non-participation can also be a tool of resistance used by those who consider themselves to be less than powerful. The organisational context of human service program and infrastructural delivery is often not considered in terms of development philosophy and design. The philosophy of development is often written up as a mission statement and then left as the document that is considered as 'window dressing' without much relevance to day-to-day operations. If there were a change to increase the diversity of decision-makers this could be a step towards participatory management. Even if this did occur the management structures also need to change from segmented (according to function) to open networks and matrices (that reflect the holistic systemic nature of reality). Boxed categories are useless for problem-solving because problems are not discrete; they are situated in complex social, cultural, political, economic and environmental systems.

Participatory management involves all stakeholders because their contributions to problem-solving are valued as a resource. This is not merely motivating and good for production levels, but also good for management, because people's 'personal knowledge' (Polanyi 1962) helps enhance creative thinking and reduce risks in decision-making. Problem-solving in human service organisations using open systems and non-hierarchical approaches to collaborative problem-solving are the way forward.

This is vital to address the new conservatism associated with globalisation, managerialism, residualism in welfare and development and the control of social research. Scientific management approaches tend to regain popularity in times of economic recession and conservative government. Even if the shift away from participatory styles is not complete, hybrid forms exist with the emphasis shifting according to personalities and socio-economic factors. We need to avoid moving back to these forms as they undermine the notion of valuing the environment and people as ends in themselves and thus to ensure that conservative nationalist responses are resorted to by people who feel that they have limited access to resources. The valuing of people and the environment as resources need to be supported, because the opportunity costs of neglecting them are disastrous in social, political and economic terms.

Service providers frequently attempt to address the effects of social problems without a thorough understanding of the nature of the problems and the research process (that is part of a complex web of structures and processes). Often service providers try to address issues without understanding the systems that continue to produce the same effects. Policy research needs to address not merely the effects of social problems, but also needs to prevent social problems through participatory planning that addresses management and governance.[22]

Good management and good governance is about ensuring that the values of democracy imbued in the Australian constitution are implemented at the local level. Much could be learned about the ways of seeing and perceiving the world through engaging in a dialogue. Listening and learning from one another not only contribute to reconciliation, but also enriches transcultural understanding and enhances our understanding of cultural systems. Planning for development requires a process of including representatives of all interest groups. In this way an overview of perspectives and hopefully a sense of ownership can be achieved, rather than achieving an outcome which is top down and exclusive. The choice of the appropriate methodology and methods is essential. Working together in partnership across cultures, age, gender and differences in physical and mental health is the goal as is understanding the links within and across sectors and disciplines. An analysis is of value only if it is used for problem-solving to address identified problem areas.

Enhancing governance capacity is as much about achieving representation through appropriate voting mechanisms[23] within appropriate subsections, as it is about achieving participation through enhancing access through being information literate and theoretically/politically

[22] See Jamrozik and Nocella (1998).

[23] *Type of voting system*

This is integral to the discussion. The voting options are: (i) *Single vote first past the post* which is the 'simplest alternative but it denies the voter the opportunity to participate in the election of all vacancies in multi-member contests. It will also, in most situations but particularly where there are large numbers of vacancies, produce markedly undemocratic outcomes' (Heatley 1997: 9). (ii) *Exhaustive preferential voting (EPV)* which 'requires voters to indicate their preference for candidates contesting individual parliamentary seats (by writing a sequence of numbers on the ballot paper). When ballots are counted the returning officer totals the first preference votes cast for each candidate. Where none has an absolute majority and can be declared elected, ballots cast for the least successful candidates are recounted and distributed (upwards) among the remaining candidates according to the second preference marked upon them.

literate. This involves:

- Knowing about the legal system, the social, political and economic shapers locally, nationally and globally.
- Ability to make appropriate decisions based on systemic problem-solving skills.

*Continued*

If at the end of this procedure no one candidate has an absolute majority, it is repeated (including third and then subsequent preferences) until one candidate emerges as the victor with more than half of the vote. … .' (Stewart and Ward 1996: 233–234). (iii) *Proportional voting*: 'The variant most suitable in the municipal context (and in use in New South Wales and Tasmania) is the Single Transferable Vote, employing 'quotas' and transfer values … The basic advantages of proportional representation are claimed to be that it most accurately converts voter choice to membership and that it enhances the capacity for representation of significant minority interests. "If minority" candidates can secure a quota on the primary count, they are automatically elected but, unless they get an appropriate flow of preferences, even those who gain a relatively high share of the primary vote (but less than a quota) are not always guaranteed success under proportional voting. Considering the variation in the membership of the wards/undivided electorates … and in the size of the election fields, the impact of the general use of proportional representation in the NT is uncertain except that it is likely to encourage the formation of electoral groupings … and the more disciplined use of how-to-vote cards …' (Heatley 1997: 9–10).

Members of the public cannot rank all the people with equal knowledge or interest, this leads to 'donkey votes' these donkey votes are critical in marginal seats and this can lead in general elections to policy discussions to advise voters where to place their preferences. In local government elections it just means that those who lose out have their preferences redistributed. This is unfair as far as three members of the public are concerned (albeit for different reasons). It was argued that 'people get in on the coat tails of other people that is on preferences rather than on primary votes. This means that very popular people can miss out simply because they are not placed high on preferences by the majority. Minority groups such as: rural people, Aboriginals (*sic*), Migrants, Young people. There was a sense that the status quo would remain, because previous criticisms in previous reviews had been ignored.'

As Dean Jaensh, a political scientist and commentator has pointed out, this result could mean nothing other than maintaining the status quo. Kramer a participant in a discussion on this issue introduced the following example of the way 3000 electors voted for one vacancy:

| A | B | C |
|---|---|---|
| 1001 | 1000 | 999 |

C can get in on preferences.

- Ability to create pathways for people to access resources and services.
- Ability to create appropriate organisational and communication structures to ensure that people can participate in designing their own futures.

This is only possible however if people have the confidence and the will to participate. This is an area that needs particular work as a result of the history of marginalisation and intergenerational welfare dependency. It is also the biggest challenge to understanding rights and responsibilities. It may be useful at one level to use self-determination along separate lines as an argument. But non-engagement can allow others to determine one's life chances and one's future. The crude saying 'you must be in it to win it' applies as much to the lottery as it does to the fact that people need to be actively part of the decision-making process if they do not want to have future designs excluding them, or including them in the same dependent roles as before. Participation based on co-creating a future is

*Continued*

The present voting system justifies the status quo and returns to power those who reflect the interests of the status quo. A more diverse group of people is needed to represent other people as well as those who are currently represented. According to a government official, the electoral office can count any variation of voting system and can also have split voting systems half-proportional and half-preferential. Different systems could be associated with different wards, half of Alice could be ward-based another not. The options are open to creating a unique system for Alice. All the members of the public present said that they had problems with Exhaustive Preferential Voting and would like to see a change which preferably did not involve the introduction of wards but which made it possible to allow minority groups to be represented. It appeared that there is a drive to move away, from preferential exhaustive systems to proportional voting system. If this does not occur then a ward system would be preferable. In 1995 McDonald Brain and Associates conducted an Electoral Review comprising an expo, a survey, a public meeting, telephone survey and 16 submissions from the public. The results of this review led to the following conclusions: a system of wards would represent geographical areas, but in Alice Springs minority interests do not overlap with geographical areas. Minority interests could be served by maintaining the current system of no wards, but introducing proportional voting. Support for this approach is evident from the results of the 1995 review. 'It is clear from the consultative process that there is no significant support for changes to the municipal boundaries, for changes to the number of Elected members or for the introduction of a Ward System. Alistair Heatley conducted a review in July 1997 commissioned by LGANT which comprised a review of the literature, an examination of the 1996 results, written submissions from the public and municipal councils, a questionnaire administered to all candidates who had stood for election in 1996–1997 and consultations with representatives of LGANT.

not however about winning or losing a debate, or a pool of funds. It is about achieving social and environmental harmony and justice. In terms of governance at the local level a proportional system of voting was deemed appropriate in an electoral review[24] of the system and that a ward-based system should be avoided to prevent further divisions in the community.

The values of the players are essential determinants of the way things are and the way in which things could be done. As community development leaders, managers and change agents we need to know the values and assumptions of where our organisation is located, where we are located, where the stakeholders are located and where we want to go and why, based on co-created designs. Ethical responsibility is required (McIntyre 1996) based on eco-humanistic thinking tools (McIntyre-Mills 2001).

The problem however is that quality decisions are linked with the input values of human service staff and where there are competing values (in terms of understanding the nature of development) and the process of development it becomes particularly difficult to bring about change because it requires not merely training, but a shift in values. The philosophy of planning by, with and for people involves valuing performance in terms of effectiveness (processes and qualitative outcomes) as much as efficiency (outputs counted in terms of numerical items). This requires an enormous shift in organisational thinking. It involves taking into account narratives and their relevance to co-created problem-solving.

[24] An electoral review in July 1997 was commissioned by LGANT which comprised a review of the literature, an examination of the 1996 results, written submissions from the public and municipal councils, a questionnaire administered to all candidates who had stood for election in 1996–1997 and consultations with representatives of LGANT. According to Heatley (1997): 'Local government elections are conducted largely under the provisions of the Local Government Act, the Northern Territory Electoral Act and the Local Government (Electoral Regulations)'. 'In Alice Springs, where the issue of wards has been most salient, opinion was more or less evenly split ... .' (see Heatley 1997: 14). According to Heatley (1997) Exhaustive Preferential Voting (EPV) allows candidates who ranked highly on the primary counts to be eliminated and encourages donkey votes or informal votes as a result of very long lists of candidates. Heatley (1997) concluded that EPV is not suited to diverse communities. A study of the previous reviews on behalf of Council indicates that the EPV system has been debated since at least 1996. The Minister for Aboriginal Development, Local Government and Housing and LGANT concluded as a result of research undertaken by Heatley (1997) that EPV needs to be reconsidered as this 'could well have the effect of broadening community representation ...'.

The challenge is that whilst the changes are occurring there is a sense of having comfort zones in thinking and practice removed. New faces appear in organisations with new ideas, old familiar faces go and the lines of management change. Just when issues are resolved and progress seems to be imminent, another direction is suggested and the morale of staff (who may or may not see the so-called big picture and if they do, it may be quite different from the picture of the managers, plummets). This is where the soft systems mapping (Checkland and Scholes 1990) of ideas, perceptions and values can be of assistance in understanding the different ways in which reality is perceived.

Another challenge is that it is popular amongst both the politically conservative and progressive to blame the lack of unity in Indigenous leadership but with little understanding of why it has occurred. Some Indigenous leaders have taken on board capitalist culture and others have not and face different goals in the area of health service delivery. Further, cooperative efforts have as yet not been established through current efforts of government-led reconciliation, because of a sense amongst some interest groups that a lack of rights and sense of self-esteem (that could be obtained from political self-determination) remain barriers. Managers of some Indigenous organisations have a clear sense that social health can only occur when Indigenous people are active citizens with a sense of rights based on a sense of nationhood and a sense of responsibility based on the feeling that they have a stake in society. Owners of Indigenous business have succeeded within a capitalist society that remains alien and threatening to many people with thinking patterns characteristic of traditional, egalitarian societies. Their perceptions of commodities and the nature of work are different as discussed below.

To achieve civic pride through generative communication requires a quantum leap, based on trust and real participation across a range of sectors and members of the community in participating in 'mainstream' as well as Indigenous community organisations. Developing a separate Indigenous power base has been necessary to move rights to their current level. There is less enthusiasm for generic services when there are great differences in culture and life chances of citizens as far as service providers and users are concerned.

Nevertheless the servicing organisations in Alice Springs have not achieved socio-economic development outcomes for many Indigenous people as evidenced by the statistics. Overservicing leads to dependency. Involvement in decision-making is a more positive goal for development geared at enhancing the life chances of citizens.

### 6.4.3 *Resistance to Changes and Fear of Mergers and Cuts in Services*

The world over people have a vested interest in the status quo. The community services and health industries support many professionals and there have been and continue to be cuts to the size of the industry as post welfarist (Jamrozik 2001) cuts continue. Some question the extent to which there is a real willingness to bring about social change in a community where poverty is an industry. Resistance to change is due as much to fear of job loss as it is due to cynicism about the motives for change. Although the rhetoric of the Ottawa Health Charter of 1986 has been taken on by most of the public sector, the organisations, the management style and the thinking remain discipline- and sector-oriented. There is considerable tension in asking managers to make a shift in their thinking when funding cuts make them wish to justify the existence of their own organisation (and portfolio). It is perceived that working 'systemically' (across sectors) and 'generically' (as opposed to specialising) in terms of strategic plans is all very well as far as managers and staff are concerned if it does not lead to identifying individuals and specific organisations to be 'surplus to requirements' because services overlap!

Citizens need to have a sense that they have a long-term stake in a socially just and sustainable society. Trust and hope lead to active participation in society. Without trust and with a high degree of cynicism felt by middle management, staff and the users of the services there is little enthusiasm for top-down political decisions about development, no matter the language or the idiom in which it is expressed. Services need to be designed to meet the specific needs of people with different life chances in terms of age, language, gender, culture, education or specific health needs. But even so there is potential for cooperation to ensure that duplication and wastage is minimised and finding ways to enhance the multiplier effects of working systemically.

### 6.4.4 *Volunteering and Post Welfarism*

Post welfarism makes the reality of asking people to do more in the guise of civic responsibility problematic. People are suspicious of the motives behind what is seen to be the rhetoric of cooperation and community participation. Unless people organise at the local level, change cannot occur.

But if community involvement is successfully implemented and volunteering rather than paid work is prioritised, rationalisation of paid staff and funding cuts could eventuate. The perceived potential result could be the further erosion of community services. Local volunteering needs to be balanced by ensuring that volunteering leads to the creation of employment pathways rather than undermining workers rights and awards. Thus volunteering can either build or undermine social capital depending on the way in which it is implemented.

## 6.5 Social Indicators of Well-Being and Life Chances

The challenges are a growing population of young Indigenous people that have life chances that are limited by a host of socio-economic factors. Active steps need to be taken to redress the educational trend for low literacy and numeracy outcomes.

### 6.5.1 Socio-demographic Factors[25] Associated with Identity and Meaning

Alice Springs has three continuous demographic patterns (developed, transitional and less developed). Planning for development requires a strategy to address a diverse and transient population that addresses clusters of demographic characteristics along the continuum of life chances of a developed, transitional and less developed population. The continuum of life chances is linked with age, gender, race, culture, level of income, level of physical and mental health.

*The less developed population pattern* has high birth rates, high fertility and high death rates with health status not dissimilar to less developed socio-economic societies. A transitional pattern has health associated with poor nutrition and risk related behaviours associated with difficulties in negotiating two cultural systems and a sense of social, political and economic marginalisation. Sections of the population are not coping with change and feel out of control.

The high suicide rate in both adult males and young people is indicative of mental health problems. There is a need for both individual intervention (anger management and means of expressing grief) which are not stigmatised and accessible without being labelled as counselling

[25] The data which follow are sliced in many ways so that the Indigenous and so-called ethnic population are addressed separately but also integrated into broader generic demographic categories such as youth, families or the elderly. In other words data can be analysed in many ways depending on the issues on which we wish to focus planning and policy.

per se which is often regarded as an acknowledgement of weakness or failure and hence 'not cool' in terms of youth culture and 'very unmasculine' in terms of male traditional culture as well as social intervention (enhancing participation across the sectors). The high rates of injury and poverty associated with substance abuse are problematic and it is likely that losses from gambling could also be an issue. Alice Springs has very low unemployment rates overall, but within the youth sector it is much higher and within the aboriginal and aboriginal youth sector it is even higher. A time frame of immediate survival, day-to-day, hand-to-mouth is usual amongst all people who do not have economic capital and have very limited access to social capital in the wider community. The suicide rate amongst young people is also indicative of a sense of not having a stake in society and a very strong sense that their life chances are unfair. Services and activities need to be presented in positive ways for instance skills training and recreational learning that is fun, accessible and not stigmatised.

Thinking about the long-term implications of hazardous life styles is also not a priority. This is an area that has been well researched by Helman (1983). Some Indigenous people identify mainly with extended communal households and in terms of the extended family, some identify themselves mainly with the nuclear family whilst some consider that their identity as a nation within a nation is a first step in self-determination, so that they can shrug off notions of assimilation (because they feel that they have little stake in the way in which the nation is currently defined). Some felt they have global or wider national and international links and use the Internet, social networking, travel and social movement approaches to address their need for self-determination and social justice. Some have a sense of place that is local, and a sense that they have very little voice in it. These Indigenous people are marginalised from the Internet and the global environment. The challenge is to consider rights and responsibilities in this fast changing world and to not forget the diversity of needs in planning for the future. Nationally the Northern Territory has the highest youth suicide rate, the highest rates of public drinking, the worst health status in terms of injury rates, nutritional disorders such as diabetes, respiratory diseases and other health problems commonly linked with a developing population profile (in terms of social health and demographic characteristics), rather than a developed population. The opportunity to address the needs of young people needs to be seen in the light of the social, political and economic costs of allowing a young group of people to grow up marginalised from mainstream society. In the context of high unemployment there was a diminished number of jobs for low-skilled workers at the time of the research and a sense that young Aboriginal people have to come to grips not only with adolescence, which is a usual time for rebellion against one's elders, but a sense of rebelling against both traditional and Western, capitalist culture, a double challenge for already marginalised young people. These are some of the linked factors that shape social health and well-being. At the time of the research high incarceration rates for young people in the Northern Territory and Alice Springs was also a threat. More supported accommodation for young people and a sobering up shelter for those aged under 18 years of age who binge drink at the weekends was identified as a particular need by human service providers.

The high rates of injury are associated with alcohol abuse and the use of inhalants: road accidents and homicides, domestic violence and incarceration rates are associated with a sense of social marginalisation. High-risk hazardous behaviour associated with alcohol is indicated across the population and the use of intravenous drugs is increasing. Future areas of concern are sexually transmitted diseases such as HIV.

*A developed population pattern* is characterised by low birth rates and low death rates overall. Working couples lack the support of family and long-term friends to support child rearing in particular. Families who have moved to Alice Springs for employment reasons are isolated from their extended family and from friendship networks. In these circumstances parenting is particularly difficult because limited opportunities exist for childcare and emotional support from family. Extensive childcare during the day and night is necessary to enable parents to meet their employment and recreational activities. Further, the highly mobile population means that friendship networks are also short-term for many.

The high cost of living needs to be considered as a possible threat given the higher costs of food, childcare, housing and rent. Parents make informal arrangements to look after one another's children, for both day and night care. Some make more formal arrangements than others do, for instance establishing cooperatives where services are traded as opposed to paying for services. Some rotate childcare and/or pay a childcare worker to assist. Some informants stressed the difficulties they faced as carers of young children without a support network. They also highlighted the lack of changing spaces for babies around town, as the facilities in the public library were not well known.

Home support services to the elderly and more child care services for parents of young children and after school care could be a growth area for the private, public and voluntary sector.[26] Young people complained of a lack of variety in recreation, lack of access to facilities at night because

[26] Overall those who are over 50 have increased in the last decade from 10.5% to 16.6%, but only 3% of NT residents are over 65 years of age, compared to 12%

of the lack of a safe, affordable bus service and lack of access to shady, cool public space where they could meet friends and talk without being regarded as disruptive. Another concern expressed by young people is that they wish to have a safe place to do their roller blading and skating.

Funding the social wage and other services could become a challenge in the future as the baby boomers start to retire and as the dependency ratio rises, as a result of lowered fertility rates and a skewing of the population. As the population ages there will be greater need to address issues such as home care support, frail care and palliative care. Professionals tend to depart after retirement age. In terms of planning, attracting professional service workers needs to be balanced by encouraging them to stay on until after retirement age. The reality is that mobility is a defining characteristic and even with incentives it is likely that Alice Springs will continue to be a service centre for the region and a place where many spend short periods of time, namely 2, 5 years, that is, the length of a contract. The lack of continuity in professional knowledge is a particular problem and it partly explains some of the repetitiveness in relearning the local issues (not merely statistical knowledge but the detailed qualitative understanding of a place that is built up over the years) and the fragmentation in service delivery. The population is still relatively youthful in Alice Springs with a high percentage of children, young people and young parents and (although below the national average) a growing number of people over 50 (Table 6.1). The life chances of the population vary widely and the challenge is to meet the diversity of needs.[27] Overall

*Continued*

of the rest of the Australian population. 49.9% of those who responded on Census night are male and 50.05% are female (1996 Census ABS, quoted by the General Practitioners Survey 1998).

[27] The NT overall has 44% of its population under 25 years of age. In the Aboriginal population 61% are less than 25 years of age. According to the ABS, the Indigenous population comprises 27% of the NT population, two thirds of whom live in remote communities, outstations and cattle stations. The remainder live in urban areas. In the remote parts of the Central Region the percentage of young people less than 15 years is slightly higher than in the municipal area of Alice Springs. An analysis of age demographics shows that the over 50s age group in Alice Springs has increased, but only 10.5% of the population are over 60. Only 3.2% of the Indigenous population are aged 60 or more, compared with 8.2% overall in the NT and 15.9% Australia wide. The median age of Indigenous males and females is 22, whilst the median age overall is 30. There is slightly more variance in family type in Alice Springs than in the Central Region balance and a lower percentage of people speak English in the central region balance than they do in Alice. The most common household type in Alice across the collector districts excluding the Town Camps is the single

*TABLE 6.1.*
*A Profile*

Regional characteristics for a 500 km radius including Tennant Creek

| Selected characteristics | Male | Female | Persons |
|---|---|---|---|
| Total persons | 24,323 | 23,995 | *N* = 48,318 |
| Aged 15 and over | 18216 | 17933 | 74.81% |
| Aboriginal | 7666 | 8148 | 32.72% |
| Torres Strait Islanders | 22 | 46 | 0.14% |
| Australian Born | 19040 | 18931 | 78.58% |
| Born Overseas (Canada, Ireland, South Africa, UK and USA) | 1769 | 1692 | 7.16% |
| Other country | 955 | 882 | 3.8% |
| Speaks English only and aged 5 years and over | 13895 | 13387 | 56.46% |
| Speaks language other than English and aged 5 years and over | 5284 | 5531 | 22.38% |
| Australian citizens | 20679 | 20361 | 84.9% |
| Unemployed | 750 | 569 | 2.72% |
| Employed 10308 | 8531 | 38.98% | |
| In the labour force | 11058 | 9100 | 41.71% |
| Not in the labour force | 4489 | 6304 | 22.33% |
| Unemployed rate | 6.8% | 6.3% | 6.5% |
| Enumerated in private dwelling | 21100 | 21085 | 87.30% |
| Enumerated in non private dwelling | 3222 | 2915 | 12.7% |
| Persons enumerated at the same address 5 years ago | 9344 | 9312 | 38.6% |
| Persons enumerated at different address 5 years ago | 9585 | 9409 | 39.30% |
| Overseas visitors | 1,240 | 1344 | 5.34% |

*Source*: ABS 1996 Regional Statistics Report 1362.7 and ABS profile for 500 km radius of Alice Springs based on 1996 data.

*Continued*

family. Parenting is difficult for those families in Alice who do not have extended family on whom to rely for childcare whilst they are working. Many stay only whilst children are at primary school level and leave when children require secondary schooling. Educational status is particularly problematic, because educational outcomes for sections of the community and retention rates are not competitive with standards at a national level. Appropriate models for learning are however being developed locally to address these concerns. Couples tend to strive for dual incomes. Parents, particularly women have the responsibility of dual roles, mother and worker. In the Town Camp areas the challenge faced by Indigenous extended families is the difficulty of accommodating and providing care and safety in an environment affected by the large number of visitors (both family and strangers) who make use of the Town Camps for a host of reasons, detailed below. In terms of health status indicators the population growth has slowed which has implications for decision-making for the future[28] and changes have occurred in the demographics of Alice Springs,[29] despite the continued growth of the NT population overall. The total number of persons in Alice Springs in 1991

*Continued*

the residents in the NT and Alice are below the national average and in many respects are on a par with health standards in developing countries. Indicators such as life expectancy, infant mortality and morbidity rates are associated with poor nutrition and poor coping behaviour (which in turn is indicated by high levels of hazardous drinking, domestic violence, injury and suicide rates). Adult individual incomes in Alice are higher than for the Central Region balance. A considerably lower percentage of people are within the labour force in the Central Region balance than they are in Alice Springs. Forty one percent of households in Alice Springs earn more than $52,000 per year compared with 27% throughout Australia (Street Ryan 1999). This section of the population buys homes, whist those earning below $35,000 tend to rent. The median housing loan repayment in 1996 was $867 per month. The median weekly individual income is over $460 and the median weekly household income is $896. The cost of living is higher than in some other less remote areas, therefore those on lower incomes have limited amounts of disposable income after paying for essentials. The median income of young people aged 15–25 years was $192 in 1996 that is above the national median of $181. But the higher living costs associated with living in remote areas needs to be factored in to assessments.

The unemployment rate in Alice as of June 1999 was lower than for the national average (4.5% compared with 7.6% throughout Australia). Alice is a destination for professionals and business people seeking opportunities that are not available elsewhere. Unemployment is clustered amongst young people and in particular Indigenous young people, but for those with contacts and skills unemployment is less of an issue. The workforce participation rate is 73% for Alice that is higher than the national average of 60% throughout Australia. This is largely because of community development employment programs (CDEP). In reality the workforce participation rate within the Indigenous population is about 445 whilst for the non-Indigenous population it is over 80%. CDEP programs in Alice however ensure that the employment rates for Indigenous people are higher here than in the more remote parts of the NT. This is an area that needs more attention given that the rate of youth unemployment is likely to grow as a result of higher Indigenous birth rates and as yet disappointing education outcomes and a lack of available jobs to match existing skills.

[28] According to informants this is possibly because of out-migration by families seeking wider education opportunities once young people reach the age requiring secondary and tertiary education and at retirement age.

[29] There is debate about the actual growth rate because the ABS figures for the census and the estimated figures between the census dates included visitors in the earlier figures and more recent figures make a distinction between residents

was 25,585[30] whilst in 1986, the usual population numbered 20,683.[31] Eighty four percent of the 27,092 persons who responded on census night in Alice Springs were Australian citizens. The tourist average per night was 1,658.[32] If one also considers the number of people from the surrounding population accessing GP services, specifically, a further 2,083 people can be added which makes a total of 30,833 people using services in Alice Springs.[33]

O'Kane (1999)[34] quotes Wakerman's (1999) estimate of a total regional population of service users of 45,672 in an area of 827,470 square kilometres. This is also supported by the Territory Health Services Annual Report (1997/8: 96) that estimates 45,000 people live in the Central Australian catchment area, as well as numerous interstate and overseas tourists who make use of the Alice Springs Health services. Others include so-called 'transients', local and interstate visitors (3,746 comprising 13.8%) and overseas tourists (1,140 comprising 4.2%) as well as contract workers associated with the Australian and American Military base.[35]

Some service providers believe that in 1999 Aboriginal people made up 20% of the population in Alice Springs. The percentage of Aboriginal people for the Northern Territory as a whole is 24.7% and for Australia as a whole is 2%. The very definition of 'resident' and 'visitor' needs

*Continued*

and visitors. For instance: the average annual growth rate for the decade from 1987–1997 is estimated at 1.7%. Between 1991–1994 the population is estimated to have increased by 2.4%, from June 1993–June 1994 the percentage changes was 0.7%, from 1994–1997, the increase is estimated at 3.7%. The overall growth rate of the population is estimated at 2.7% from 1996 to 1997 when the population increased by 673 persons out of a total of 25,713 residents, according to ABS estimates. The latter figure has been disputed by some locals.

[30] See table 1 of 1991 Census (Census Counts for small areas), page 20.

[31] Census Data for 1986 Profile of Legal Local Government Areas Usual residents counts for NT Catalogue No 2476.0.

[32] General Practitioner's Community Profile (1998).

[33] General Practitioner's Community Profile (1998).

[34] O'Kane, A. (1999) Service required by residents with a mental illness in Central Australia. *Paper presented to the 31st annual Public Health Association Conference: Our place, our health: Local values and global directions*. Darwin.

[35] At the time of the research 820 personnel were based at Pine Gap with an estimated increase to 930 in the future. If one adds spouses it makes about 1800 which is 6.6% of the population (Public Relations Information Officers at the Defence Administration, 1998, personal communication). The 1996 census lists 1,002 persons employed in Government Administration and Defence (7.8%) of the population.

consideration. If one's dreaming site is located in town the tag 'visitor' could make little sense to an Indigenous citizen. The definition of transient is also problematic given the greater regional stability of Indigenous people.

> The end of the training allowance scheme, the advent of award wages, the decline in the pastoral industry, access to welfare entitlements and improved transport services all added up to an escalation in the movement of Aboriginal groups to Alice Springs in the mid to late 1970s. This in turn put pressure on Aboriginal housing services and camping areas (Drakakis-Smith 1980a,b; Collman 1988 in Coulehan 1997) and highlighted the 'needs basis' for Aboriginal special purpose leases and urban excisions for town camps in Alice Springs.[36]

Migrants of NESB comprise 7.2% of the population. Cultural diversity and youthfulness are the defining characteristics of residents. The median age is 30. The range of languages spoken (see ABS Census for 1996: 8) reflects this diversity: 20,145 speak English only and the rest of the residents and overseas visitors speak a host of other languages. Migrants and refugees—this diverse section of the community comprises at least 57 nationalities according to the Migrant Resource Centre and makes up 7.2% of the community (1,800) of whom 3,923 are NESB residents. Of these 10.8% speak English very well and 2.9% speak it poorly or not at all. At least 10,180 (37%) of the population was under 25 years of age according to the 1996 census data. From 50 onwards the numbers of residents decrease as a result of out-migration. If we compare this to 1986 we can see the following pattern: in the last decade the over 50s age group have increased from 10.5% to 16.7% (Total N = 27,100) and the percentage of young children, has changed very little (27.5%–25.8%). The percentage of adolescents and teenagers has dropped slightly. Young adults and the middle-aged have maintained more or less the same demographic profile if we compare 1986 (42.5%) and 1996 (41.4%).[37]

[36] Coulehan (1997) Paper prepared by Dr Kerin Coulehan c/o School of South East Asian and Australian Studies Northern Territory University Darwin NT.

[37] Service provision needs to compensate for unemployment, lack of extended family support faced by mobile workers and for the special needs of Indigenous parents and young people. Current service needs are for the young prior to and during years of primary, secondary and tertiary education and for a range of relatively young households and families. The challenge is to decide whether more resources should be channelled to those who are tending to leave as retirement approaches.

According to the 1996 census, unemployment rates were at 19% for 15–19 year olds and 13% for 20–25 year olds. The number of Aboriginal young people is increasing. 24.9% of the population was aged less that 15, compared with 21.6% of the population for the rest of Australia. Of the 1,535 18–24 year olds living in a household, 47.5% are either with partners or with parents. 4.4% are dependent students and 27.7 % are non-dependants. Of the 514 living in non-family households, 5.8% are living alone, 13.5% are living in group households. A total of 84 young people claimed to have no usual address on census night out of a total of 5,879 persons.

#### 6.5.1.1 Religion

Anglicans, Catholics and Uniting Church are the three most common Christian denominations. Buddhism, Islam and Judaism are the most frequently mentioned other denominations and make up 4.8% of the responses.

#### 6.5.1.2 Family Structure and Household Groups

The census data for 1996 indicate that as far as marital status is concerned, 48.9% (10,068) of the total population are married. 11% of the population experience relationship problems, for instance 3.9% (819) are separated and 7.1% (1,461) are divorced. 3.6% (757) are widowed and the majority of these are female. 36% (7,471) never married, but the data includes 15–24 year olds and later marriages are usual. One family households make up 83.4% of all households in 1996 compared with 79.1% in 1986, but the nuclear family is now made up of de facto parents and often comprises blended families as a result of divorce and re-marriage. Single parent households have increased from 6.2% in 1986 to 8.5% in 1996. This is significant. Couple households with dependent children are still the most frequent family types. In total there are 2,611 of which 16% are de facto rather than in a so-called registered marriage. The majority of nuclear families have children below 14 years of age. Parents with young children and without extended families are a significant group in Alice Springs as a result of the changing structure of families and an increasingly mobile workforce. The number of sole parent families is increasing and the number of families with younger children under 15 has not changed significantly since the last census. The average size is 2.8 persons per family. This is above the national average of 2.7% A range of new family types plus households of non-families need to be considered. 1,643 lone person households[38] and 449 group households. In total there are 802 one-parent families[39] and 878 lone parents.

*6.5.1.2.a Indigenous Population.* The Indigenous population is diverse (as indicated by the range of language groups discussed below) and has some diversity in life chances, but the life expectancy is overall much lower than for the rest of the population (Table 6.2). According to the 1996 census:

> Indigenous households are more likely to be multi-famiy, increasing in size progressively from capital cities to urban, rural and remote areas in terms of the average number of families, persons, adults and dependants per household. Overall the adults are younger, have lower levels of education and are less likely to be in employment than non-indigenous Australians. They are twice as likely to contain sole parent families and less likely to contain couple families … .

### *6.5.2 Socio-Economic Indicators*

An overview of the life chances of people who live in Alice Springs and who use local services would be incomplete without an analysis of socio-economic factors. The Alice Springs geographical community can be understood in terms of: (i) the relations of production and (ii) perceptions of commodities, culture[40] and consumption. The introduction of changes in relations of production, types of commodities and changes to the very fabric of cultural identity has far ranging ramifications for the residents of Alice Springs (who are affected by the social and economic changes) at a global level. Relations of production can be understood by means of a typology of the relations of production in Alice Springs including: knowledge production for sustainable living, localised, traditional approaches to traditional industries, welfare dependency and traditional Indigenous communal lifestyle.

[38] ABS Community Profile, Page 26.

[39] According to ABS household type by family type by number of people usually resident, page 26 of Community Profile, but according to the table showing relationship in household by age, page 22 of ABS Community Profile.

[40] Cvetkovich, A and Kellner, D (1997). 'Both the classical liberalism of Smith and classical Marxism see capitalism as a global economic system characterised by a world market and the imposition of similar relations of production, commodities, and culture on areas throughout the world, creating a new modern world system as the capitalist market penetrates the four corners of the earth …'.

**TABLE 6.2**

*Selected Person Characteristics by Indigenous Origin by Sex for Alice Springs*

| | Indigenous | | | Non-Indigenous | | | Not Stated | | | Total | | |
|---|---|---|---|---|---|---|---|---|---|---|---|---|
| | Male | Female | Person | Male | Female | Person | Male | Female | Person | Male | Female | Person |
| Total | 1.867 | 2044 | 3911 | 10472 | 10352 | 20823 | 631 | 586 | 1217 | 12.970 | 12.982 | 25952 |
| 0–14 | 656 | 701 | 1357 | 2467 | 2388 | 4855 | 116 | 136 | 252 | 3239 | 3225 | 6464 |
| 15–44 | 959 | 1024 | 1983 | 5181 | 5343 | 10524 | 351 | 302 | 653 | 6491 | 6669 | 13160 |
| 45–64 | 199 | 244 | 443 | 2234 | 1974 | 4208 | 131 | 100 | 231 | 2564 | 2318 | 4882 |
| 65+ | 53 | 75 | 128 | 590 | 647 | 1237 | 33 | 48 | 81 | 676 | 770 | 1446 |
| Median Age | 22 | 22 | 22 | 32 | 31 | 32 | 32 | 30 | 31 | 31 | 30 | 30 |
| Dependency ratio % | 61.2 | 61.2 | 61.2 | 41.2 | 41.5 | 41.4 | 30.9 | 45.8 | 37.7 | 43.2 | 44.5 | 43.8 |
| Attending educational institution | 537 | 651 | 1188 | 2496 | 2730 | 5226 | 19 | 30 | 49 | 3052 | 3411 | 6463 |
| Never attended school | 63 | 72 | 135 | 20 | 14 | 34 | 0 | 0 | 0 | 83 | 86 | 169 |
| Australian Aboriginal Traditional Religion | 11 | 7 | 18 | 0 | 0 | 0 | 0 | 0 | 0 | 11 | 7 | 18 |
| Speaks Australian Aboriginal or Torres Strait Islander Language aged 5 years & over | 656 | | 708 | 1364 | 22 | 11 | | 33 | 3 | 0 | 3 681 | 719 |
| 1400 | | | | | | | | | | | | |
| Speaks English only 5 years and over | 903 | 998 | 1901 | 9029 | 18089 | 81 | 74 | 155 | 10044 | 10101 | 20145 | |
| Employed CDEP | 50 | 38 | 88 | 0 | 0 | 0 | 0 | 0 | 0 | 50 | 38 | 88 |
| Employed other | 386 | 411 | 797 | 6333 | 5458 | 11791 | 59 | 44 | 103 | 6778 | 5913 | 12691 |
| Unemployed | 96 | 76 | 172 | 268 | 216 | 484 | 3 | 0 | 3 | 367 | 292 | 659 |
| Total | 532 | 525 | 1057 | 6601 | 5674 | 12275 | 62 | 44 | 106 | 7195 | 6243 | 13438 |

*TABLE 6.2 Continued*

| | Indigenous | | | Non-Indigenous | | | Not Stated | | | Total | | |
|---|---|---|---|---|---|---|---|---|---|---|---|---|
| | Male | Female | Person | Male | Female | Person | Male | Female | Person | Male | Female | Person |
| Not in the labour force | 615 | 716 | 1331 | 1318 | 2213 | 3531 | 48 | 72 | 120 | 1981 | 3001 | 4982 |
| Unemployment rate % | 18.0 | 14.5 | 16.3 | 4.1 | 3.8 | 3.9 | 4.8 | 0.0 | 2.8 | 5.1 | 4.7 | 4.9 |
| Participation rate % | 46.4 | 42.3 | 44.3 | 83.4 | 71.9 | 77.7 | 56.4 | 37.9 | 46.9 | 78.4 | 67.5 | 73.0 |
| Median individual income | 173 | 192 | 184 | 548 | 370 | 457 | 519 | 264 | 403 | 508 | 343 | 423 |
| Separate house | 1145 | 1326 | 2471 | 6915 | 7034 | 13949 | 263 | 263 | 526 | 8323 | 8623 | 16946 |
| Improvised home | 35 | 38 | 73 | 167 | 124 | 291 | 6 | 5 | 11 | 208 | 167 | 375 |
| Other private dwelling | 300 | 338 | 638 | 2531 | 2414 | 4945 | 253 | 239 | 492 | 3084 | 2991 | 6075 |
| Total | 1480 | 1702 | 3182 | 9613 | 9572 | 19185 | 522 | 507 | 1029 | 11615 | 11781 | 23396 |
| Non-private dwellings | 388 | 343 | 731 | 857 | 779 | 1636 | 110 | 78 | 188 | 1355 | 1200 | 2555 |

*Source*: ABS Census Data 1996.

Knowledge workers may contribute to primary, manufacturing and service work however they also understand Internet technology and are able to access 'global culture'. They are salaried professionals but the nature of contract work has changed their bargaining position. Mobility is used as a means to advance their career, but has also become a necessity for many. This section includes the business and management class, which has kept up with changing technology. Highly mobile knowledge workers are globally focused:

> The 'cosmopolitans', professionally oriented people, and the 10 per cent of the population with a university degree. These are up to the minute men and women, with international experience and interests, working in the knowledge professions, in the realm of culture, arts, finance and high technology; they make up the new class of so-called symbolic analysts.[41]

The challenge is to involve those on short-term contracts to contribute to the social fabric of the town and to be seen not merely as 'takers', but as 'givers'. People are perceived to come to Alice Springs to get away from city lifestyles, to pursue a career or to make money. When they are ready they leave and focus on the next place and the next work contract. In this way the long-term residents tend to lose out because the community has a high turnover and discontinuity in a wide range of relationships: professional, business and friendships.

Those who are local in focus (rather than global in outlook) and not information literate have a sense of comfort in the taken for granted and familiar. Locals value the character of home and 'rely on its borders to protect them from external turbulence and distress ... they represent the antithesis of the internationalist ideal ... They are enthusiastic nationalists ... they are the losers in globalisation'.[42]

Those with a local, inward focus include people with limited levels of education. Alice Springs attracts people who wish to remove themselves from challenges and to withdraw. Tourists, short-term 'blow ins' and new comers who wish to introduce changes to the social fabric of the town are regarded with suspicion.

'Traditionally ... Australia—the Australia of suburbs and small towns—has been rich in the intangible qualities of collective life: collaborative action, trust, mutual responsibility, and all those things that jargon

[41] The New Nation Nicholas Rithwell on 'A national divided: Peoplehood resides not in the big picture but in the details of community life'. Weekend Australian 27th–38th March 1999.

[42] The New Nation Nicholas Rithwell on 'A national divided: Peoplehood resides not in the big picture but in the details of community life'. Weekend Australian 27th–28th March 1999.

labels with the tag of social capital. ... Where social capital is high, people care for a local park; where it is low, they dump their household rubbish in it ... Social capital is the sense of community belonging made practical' ...[43]

In this analysis Aboriginal and non-Aboriginal cultural categories are not the only consideration, instead Aboriginality can also be a proxy for relations of production and class. Both Aboriginal and non-Aboriginal people fit into the categories of knowledge production for sustainable living, localised traditional approaches to traditional industries and welfare dependency and articulate with the categories according to the way in which production, commodities and cultural opportunities are experienced.

The changes in markets have had an impact on local commodity prices that have fluctuated dramatically. The primary industries such as the pastoral industries (beef, lamb) and the mining industry (gold) were the least profitable they have been in 20 years at the time of the research. Farms and mines were at their most vulnerable. The locals who make a living from these areas were being encouraged to make changes dictated by global markets. For some this was particularly difficult. Later the market boomed, so that diversification seemed less necessary until the next dramatic market move.

Manufacturing was an area of the local economy that was identified by informants as a weakness and that the provision of cheaper local goods for the local market was worth considering. For example making use of recycled materials from the Town Dump was identified by informants as resources as that could be used more sustainably at the local level.

The high turn over of mobile service workers in the sectors of health and education in particular who leave at the end of their government service contracts (between 2–5 years) when they have a greater insight into the local community is a weakness. The lack of representation by Indigenous people in human service provision to provide continuity is in part due to lack of appropriate training and in part due to the lack of appropriate employment pathways. This could be addressed by providing case management approach to ensuring that people get the right cluster of skills, work place training and life skills. Some of those who are locally based, long-term residents feel out of touch with the wider world.

#### 6.5.2.1 Development, Choices and the Social Wage

The difficulty of making ends meet is faced by (i) Migrants and refugees have expressed particular concerns about the cost of fresh food (Migrant

[43] The New Nation Nicholas Rithwell on 'A national divided: Peoplehood resides not in the big picture but in the details of community life'. Weekend Australian 27th–38th March 1999.

Resource Centre, Healthy Living Seminar for Refugees, October 1998) and Service Providers (such as the Women's Information Services, NT Health Services). (ii) Single parents (usually women) with dependants are statistically amongst the lowest income group and are likely to be affected most adversely by any rise in the cost of living. (iii) The needs of the single elderly, despite being a very small percentage of the population need consideration in order to prevent out-migration. Living costs were rising as a result of the introduction of the GST. This could become a particular challenge in the NT where living costs were already higher. The most vulnerable demographic groups (namely those with less developed and transitional characteristics) need to be given particular attention.

Indigenous people on Town Camps and in the urban areas need to be seen as a diverse group, not as a completely homogenous group. Although many share similar life chances as a result of the shared environment, they have different levels of education and have incorporated the aspirations of mainstream Australian society to varying degrees, namely: education (albeit differently defined in some instances) for their children, a secure job, material comfort and respect within the community.

Although traditional Indigenous culture strives to share all commodities, many see that their self-determination could involve: 'getting off welfare'. The role of dependency on the social wage defines the current situation of many Aboriginal people simply because they are unable to get out of the cycle of poverty. For some this goal is desirable, but for others to understand why it is necessary or desirable requires a certain frame of mind and a certain capitalistic understanding of economics and finances. Also self-determination from the point of view of many residents is the right to declare a locality a dry zone and to concentrate on creating a safe space for family whilst discouraging some family members who do not share their vision. For many the issue and the vision is broader and hinges on obtaining land rights as a means for achieving socio-cultural and political progress.

Rowse (1998: 40)[44] looks to the past to explain the way in which the delivery of rations to Indigenous people has led to dependency. He argues that they were at best a means to 'civilise', at worst 'a means to control and manage Aboriginal people'.[45] Either way they were a top down

[44] Rowse, T. *White Flour: White Power: From Rations to Citizenship*. Cambridge University Press.

[45] 'The movement of people to sources of food is traditional and the movement to missions and pastoral stations in pursuit of food made sense, albeit the acceptance of rations was prompted by ambivalence: Desire for commodities and a fear of the new forms of power which they confronted which undermined relations between kin and lead to a destruction of a way of life'.

approach to development and that 'the agenda was certainly not of their own writing'. If one sees 'the gift'[46] as a means to obtain reciprocal gains, 'namely compliance or access to land, then it implies that the power relationship was equal which of course it was not'. Rowse emphasises that 'effort and reward' are however well understood in traditional societies and this was effectively undermined by the system of rations without work. Dependency was therefore created by taking away the right to self-respect by taking away their responsibility to address their needs and not to be dependent on others. But moving along to today, the relevance of the exchange remains, the suggestion made by those who were invited to contribute to the NT Face of Alice in Ten Years Project (2000) was that Aboriginal people should collect their social wage away from the town centre in their remote communities, once again echoes the notion that the rights as citizens are not so great for those who do not earn a cash wage as they are for those who do. The dependency relationship on commodities is a central aspect of the issue.

The goal for some Aboriginal residents to achieve economic independence is currently compromised by the fact that spending on alcohol and the effects of alcohol-related problems, erodes household income. Domestic violence and the dangerous, noisy environment that is not conducive to sleep or study currently affects some residents. The lack of employment opportunities makes this goal unattainable for many.

The user pay philosophy could be balanced by an understanding that the cost of setting up a cycle of debt through automatic deductions is also problematic as it causes a sense of hopelessness and despair as the cycle of debt increases (personal communication 1999, Alice Springs).

[46] Marcel Mauss (1990) elaborated a theory of reciprocity in '*The gift*'. Rowse emphasises that in a communal society the notion of gift is quite different. Possessions are owned communally and it is a right and a responsibility to receive and share commodities. The notion of gratitude is not the same. The notion of receiving money as a social wage or the dole is perhaps seen as simply part of the natural scheme of things. The concept of property is also quite different in traditional communities. Rowse applies this argument to the way in which unequal exchanges of commodities can lead to changes unexpected by those who do not understand the dynamics of power. White flour, beads, sugar (and alcohol) were 'traded' which rendered people vulnerable to white power as a result of dependency. This is the exchange that underpinned much of the 'trade' by colonists.

### 6.5.2.2 A Close up on Employment, Unemployment and Poverty

*6.5.2.2.a Employment.* Over the past decade the workforce participation rate in Alice Springs has been consistently higher than that nationally with a consistently higher percentage of full time positions (Street Ryan 1998).

*6.5.2.2.b Household Income.*[47] The 1996 Census demographics quoted by Street Ryan (1998)[48] highlight the fact that 59.5% of the population earn above 36,000 per household and the median weekly income per person is 463 dollars. According to this report the average size of the family is 2.8 (versus 2.7 nationally). The number of professionals and paraprofessionals is 41%. This means that in salary terms more than half of the population is better off than other Australians are. However the isolation from extended family means that limited opportunities exist for childcare and emotional support from family.

The other important point they make is that home ownership is not high, despite the high incomes. This may be because people tend to rent because they do not intend remaining in Alice Springs long-term. The street Ryan Report details that overall the median weekly *individual* income is over $460, but the median weekly individual income for Indigenous residents is about $200. The median weekly *household* income is $896. The cost of living is high; therefore those on lower incomes or who have high mortgage payments have limited amounts of disposable income. The median housing loan repayment in 1996 was $867 per month. The average weekly earnings in the NT for all employees was about $576.40. The average for males was $676 compared to $469.40 for females, due to more women working part time.

Consultation with service providers[49] and an analysis of survey data, indicated that some couples with dependent children (despite in many cases having dual incomes) do not have large amounts of disposable incomes after paying for mortgages or rental, high food bills and power bills. This could become more of a concern when the GST is implemented.

[47] ABS census data for 1996.

[48] Street Rayan and Associates: Alice Springs Economic Profile and indicators October 1998.

[49] Kim Ruthann, Psychologist, Women's Information Centre and spokes persons from non government organisations.

*6.5.2.2.c Volunteering.* Nobel (1991)[50] conducted a survey in South Australia, which indicated that much volunteering is associated with fund raising and sporting activities, rather than activities immediately linked with the provision of welfare services. This appears to be the case in Alice Springs where there is a mobile workforce. Professionals work on contract for a few years and concentrate most of their time in paid employment and recreational activities. Volunteering in welfare areas did not feature very highly in Alice. The participation rate in the NT as a whole is high, however, according to Harrison (1996: 71). During the 12 months (prior to June 1995) 23.4% of the population aged 15 years and over provided some form of voluntary work. This amounted to 5.2 million hours. This is second only to the ACT.

*6.5.2.2.d Employment through Community Development Employment Programs (CDEP).*[51] This program developed by Nugget Coomes is a policy that has supporters and detractors, because it can be seen as having positive and negative aspects (Rowse 1998: 205–222). Positive aspects are that the CDEP gives opportunities to town-based Aboriginal people with limited employment opportunities. According to the Tangentyere Council Annual Report 1993–1994 'A survey undertaken before CDEP commenced found 85% of men under twenty-five had no job and received no Social Security money. CDEP has created jobs … [t]his has a positive effect on town camps'. The CDEP program remains

[50] Nobel, J. (1991) *Volunteering; a current perspective.* Adelaide. Volunteer Centre of South Australia.

> 'If work for the dole participants were included, our jobless figure would shoot up to 14.3%, by far the nation's highest, and more than six points above the Australian average of 8.2% … In Central Australia the picture would be even worse—close to 15% would be officially unemployed. More than 8% of the Territory's "civilian" workforce of 93,100 are on the ATSIC funded CDEP scheme, a total of 7,524 people, including 1,616 in the Alice Springs region. The CDEP participation rate in the Territory is 26 times greater when compared to the nation's 0.31%. CDEP participants, almost all Aboriginals, generally get the equivalent of the dole—an average of $183.24 a week and are required to work 15 hours per week. The allocations are made by the ATSIC regional councils, in response to applications, direct to the communities, which are responsible for the implementation of their schemes … Some organisations—notably Arrernte Council … are progressively incorporating CDEP into business ventures …' (Alice Springs News 1998 September 16th) This is an initiative which could act as a model for further development.

[51] This section relies on and draws from Rowse 1998: 205–222. The perceptions quoted are based on experimental learning.

the major means to obtain employment. It allows for self-determination of communities who can develop opportunities for all within a particular geographical community. The wages are not however competitive with mainstream society.

CDEP is perceived to be problematic and has been described as menial work by informants. Whether a task is perceived as: *'Worthwhile environmental health work' or 'dirty work'* depends on the point of view of the informant. As a participant in the CDEP women's program said:

> 'I do dirty work, things like cleaning bathrooms and toilets, to ensure hygiene, also drains. I qualified in a program for health workers at Batchelor (a college).' Picking up papers is done by CDEP, leaving papers is seen as creating work for people.

*6.5.2.2.e Unemployment.* The unemployment rate in June 1999[52] is lower than the national average (4.5% compared with 7.6% throughout Australia). Unemployment is clustered amongst young people and in particular Indigenous youth. Low skilled jobs are disappearing which means that the unemployment rate amongst Aboriginal young people will only be addressed by either better preparation for a skilled service industry or by creating specific opportunities associated, for example with environmental protection and tourism. The introduction of strict mutual obligation programs for the unemployed, whilst they may appear sensible for many people, need to be considered in the light of the many problems which people face and the fact that job placements favour those with a wide range of educational and life skills suited to a capitalist society.

According to the 1996 census, unemployment rates were at 19% for 15–19 year olds and 13% for 20–25 year olds. 16.3% of the Indigenous population are unemployed, but if one considers that the dependency ratios are higher, that is, by 20%, that is more people in households are dependent on the breadwinner, then this is more significant. Also the number would be higher if the government run training programs were excluded from the calculation. The reality is that the only form of employment is the CDEP program and training programs that currently do not lead to permanent employment of either a full- or part-time nature. The existing job networks in Alice Springs that are run as businesses are linked with the public sector. The problem is that only those who are seen as highly employable are likely to be easily placed, because economic and efficiency concerns are paramount. Unless there is an ongoing allowance to ensure that mentoring and training can occur within the workplace only 'the cream' will be given attention. A manager of an employment agency said

[52] Northern Territory Labour Market Report June 1999.

that the reality is that many of the young people need *work readiness* training and cadetships or mentoring programs to ensure that work habits and routines can be learned and re-inforced.

At the time of the research the data indicated that Indigenous people in Alice Springs score higher in terms of social dependency and lower in terms of social well-being indicators. According to Whyte[53] the majority of those who are long-term unemployed in Alice Springs are Indigenous people. According to estimates by King (of the Canberra Centre for Social and Economic Modelling 1999):[54]

> Somewhere between 1.3 million and 2 million individuals in Australia are below the Henderson line. The Henderson inquiry based its poverty line for a family with two children and an employed father on the basic wage, which at the time was derived from an analysis of household budgets, plus child endowment for two children. Its report described it as 'so austere as to make it unchallengeable. No one can seriously argue that those we define as poor are not so'. Today the line for such a family is set at 464 a week and for a single person it is 247 … . People on social security have incomes close to the poverty line so that with any change in circumstances they can move below it. (king 1999).

The current divisions based upon difference of race, class and culture, (which are largely proxies for one another in the Alice Springs context) need to be addressed through a range of social measures to enhance employment opportunities and thus to address the divisions between those who are employed and those who are uneducated and unemployed, according to a manager at Centrelink.

*6.5.2.2.f Key Issues.* Extreme depression is associated with a sense of not coping with the demands of paying for services (power and water, the telephone, taxis, food). Depression and a sense of 'learned helplessness' (Renwick 1999, personal communication) based on the psychological assessment of users of public sector services). This concept refers to a sense that people have very little control over their lives. The intergenerational impact of welfare culture has been identified as another key factor by Rowse (1999)[55]. Hopelessness and associated alcoholism, violence and suicide were raised as being a matter of urgency. The sense of hopelessness is associated with trying to break out of the spiral of debt and welfare dependency. This was summed up by some of the younger Town Camp

[53] 1999, personal communication, Manager Centrelink, Alice Springs.
[54] Poor Fella our Countrymen: The new forgotten people. Weekend Australian Feb 27–28 pages 22–23.
[55] 'White flower, white power'.

residents that paying for services such as rent is not the same as paying off a mortgage and becoming independent. Basic needs such as safe housing and a healthy diet also remain unaddressed. Food security is a concept introduced in The Health and Welfare of Australia's Aboriginal and Torres Strait Islander People's Report[56] that highlights the fact that this basic need is a source of concern for people in remote areas where food prices are high, their incomes low and their ability to manage finances limited by low literacy and numeracy and by the misuse of alcohol. Transience and homelessness also add to vulnerability. Transience in this instance refers to the movement of Indigenous people across the land. Some of the people who are homeless in Alice Springs have homes in remote areas and for a range of reasons choose to be in Alice Springs. Others are stranded because they cannot afford the cost of transport home. Resources are less accessible to those without a permanent address. With the introduction of 'strong rent' policy for visitors, the options are limited. The provision of a 'return to country' bus service alleviates some of the problems.
The following impact on unemployment:

(i) Historical factors: 'The repeal of the Wards Employment Ordinance in 1971 ... saw the introduction of award wages for Aborigines, increased wage costs for Aboriginal labour and a commensurate drop in the number of Aborigines in paid employment (Brendt and Brendt 1987). Missions and pastoral properties were no longer able to afford the number of Aboriginal workers on award wages and welfare entitlements became the means of support for most Aboriginal people. In Central Australia, the introduction of award wages coincide with a decline in the pastoral industry, which in turn saw many Aboriginal stockmen and their families voluntarily move off or be evicted from pastoral land, move to Town Camps in and around Alice Springs and have to rely on welfare entitlements' (Coulehan 1997: 7).

(ii) Cultural considerations contribute to unemployment. Indigenous people do not share the Western welfare notion of the so-called 'deserving poor' who cannot work and 'the undeserving poor' who ought to work. Such distinctions make little sense in a communal society where values can be quite different because if material items are available in traditional society they ought to be shared and thus there is no notion of shame associated with consuming items that are given. Receiving material items holds no shame in traditional society because these items are not seen as commodities for exchange or profit. Children in remote

[56] The Health and Welfare of Australia's Aboriginal and Torres Strait Islander People's Report 4704.0.

Aboriginal communities around Alice Springs and within Town Camps grow up where it is the norm for many people not to have a job. Generations of unemployment in Aboriginal communities remain a challenge. Low incomes derived from social security and CDEP programs are no insurance against poverty even if strict budgeting is adhered to. Lack of numeracy, literacy and high cost of foods and higher living costs in general in the NT along with higher levels of drug usage (particularly alcohol) pose a particular threat to health and living standards.

(iii) Practical issues associated with the lack of transport (private and public) contribute to unemployment. For instance as an informant who lives in Inarlenge Town Camp pointed out: 'I got up early and washed and waited for them to pick me up, they (the welfare service bus) have not come to collect me …'

#### 6.5.2.3 Perceptions of Commodities, Culture and Consumption

People with different life chances perceive commodities very differently. Access to commodities differs dramatically according to level of income and type of occupation. The latter confer status differences that belie the notion that Australia is a classless society. Alice Springs is a community characterised in terms of different levels of access to employment and commodities.

Access to commodities is one of the key determinants of life chances. Class divisions are built upon access to economic, social and cultural capital (as per Bourdieu 1992) by virtue of historical social position (linked with class, race and gender) and/or level of education. A range of class positions exists along the spectrum of development within both the Indigenous and non-Indigenous population. The residents of Town Camps happen to be Aboriginal, but they share a position of being unemployed or CDEP workers. Aboriginality from the point of view of this group is a proxy for a disadvantaged class and citizenship position.[57] Others in Alice Springs are marginalised by virtue of a range of criteria, for instance: age, education, culture, level of ability, gender. To be an Indigenous person and resident of a Town Camp, is to experience yet another level of disadvantage. Indigenous residents who live within the community rather than in Town Camps have the advantage of being able to control their lives to a greater extent, even if this is merely being able to limit some of the worst aspects of violence associated with strangers and visitors using Town

[57] They have been described as 'citizens without rights' (Chesterman, J. and Galligan, B. 1997).

Camps as places to drink. One of the strategies for achieving greater control on camps is simply to move away. This is the goal of many young people. Commodities and culture can be perceived or treated in terms of their dollar value (commodified) by people in the business sector. Indigenous people who adhere to a traditional lifestyle see commodities very differently. Commodities are not seen to signify a monetary value. Food is shared and used immediately in desert communities. As stressed in a CARPA[58] newsletter this different notion of commodities needs to be borne in mind when considering alcohol usage by people who are confronted with items in abundance. If people have experienced scarcity and it is customary to share whatever they have in communal society, then immediate consumption of items in abundance can be understood. Alcohol is an effect and a cause of social ills. Alcohol is an effect of social marginalisation, low self-esteem and low self-respect. Alcohol is in turn associated with and directly related to further environmental health problems, such as the modelling of violent behaviour to young children and the break down of the fabric of social life.

The diverse nature of Aboriginality has been widely discussed by both Aboriginal and non-Aboriginal scholars and the debate, developed for instance in Hollinsworth and the reply to Holinsworth[59] is as follows: culturally, Aboriginality is lived and expressed *contextually*, because the sense of reality and the experiences of the urban, the rural, the tertiary educated, the minimally educated, men, women, the elderly and the young can be very different.[60] These tensions in the definition of culture

[58] CARPA Newsletter 1998.

[59] Hollinsworth, D. Discourses on Aboriginality and the politics of identity in urban Australia *Oceania* 63 1992 pages 137–154. Replies to Hollinsworth by Beckett, J., Atwood, B and Mudrooroo Nyoongah Wiping the blood off Aboriginality: The politics of Aboriginal Embodiment in Contemporary Intellectual debate. Lattas, A. Australia *Oceania* 63 1992 pages 155–167. Coagulating categories: A reply to responses by Hollinsworth 168–171.

[60] For some Aboriginality is perceived to be impermeable and unchangeable. It persists against all odds. For others it is about resistance. It is used as a political vehicle for lobbying for the rights of an essentially marginalised group of people who share in common their sense of a dispossession of land and family ties as a result of government policies of the day. For some, who have experienced intergenerational unemployment, poor health, high levels of domestic violence, poor levels of education and high levels of incarceration and the loss of a sense of control over their lives, a culture of despair exists. Re-focusing on the positive aspects of tradition can be a means of moving beyond a sense of dispossession and despair.

are evident locally and are reflected in Annual Reports. For instance in the Central Land Council Report (1998: 59) in reference to the Aboriginal and Torres Strait Islander Heritage Protection Act of 1984 it states that 'the scope of the Act applies to any area or object of significance to Aboriginal people, whether they live a traditional lifestyle or not'. The report implicitly recognises that culture is evolutionary and contemporary.

On the other hand, according to Nellie Camfoo in the Aboriginal Areas Protection Authority Annual Report (1998):

> Some people don't fully understand this Law that we Aboriginal people still follow. They only think of Australian law, made in Parliament. You can vote in elections to change the Parliament, the politicians and laws, but Aboriginal law can never change ... and we can't vote to change ... the elders must make sure the right decisions are made about sacred sites ...

The link between Aboriginal Law and Culture is explicitly stated.

Successful communication to enhance reconciliation requires an understanding that culture is defined differently by different stakeholders. This has implications for policy and practice. Culture can and is used as a resource for survival and self-determination and needs to be considered in terms of specific social, political and economic contexts. Human beings use cultural maps that explain their own sense of the world in terms of personal identity, destiny and place. When these are shaken by social changes, the maps no longer explain reality. A sense of inner or interpersonal conflict occurs in these contexts (see Giddens 1995: 193–194). In some instances when working with traditional rural Indigenous people the basic building blocks of reality do differ dramatically. Their view of the nature of the world and their perception of being custodians of nature and their perception of so-called 'success' in a social and economic sense differ dramatically from those who have experienced life in urban contexts. Thus culture is a vehicle for expression and can have many faces. The fact that culture can be defined differently does not lessen the social importance of the varied definitions. The cultural differences amongst Aboriginal Australians and between Aboriginal and other Australians cannot be dismissed as being unimportant, nor can the issue of cultural definition be submerged within the concept of multiculturalism. Aboriginality is about self-determination and for this reason there is a strong need for difference between the concepts and the distinction is of relevance to planning and development. The links between current social problems and the history of colonisation cannot be forgotten. The results are spelt out in alcoholism, cultural despair and a sense of real powerlessness as demonstrated by the social health indicators. The Central Land Council strives for Land Rights as a way to obtain citizenship rights that they consider are not realised adequately for Aboriginal citizens.

Culture can also be reduced to the lowest common denominator of mass culture which is the way in which culture can be used by some sections of the public and tourist industry: plastic boomerangs and mass produced digeridoos. Thus if culture is seen only as a resource by some citizens and not as a form of identity it can be reduced to cultural artefacts and rituals in shops and restaurants. Ecotourism, which is respectful of people and their environment, could be a more viable approach. Culture can be used as a resource for survival and self-determination and always needs to be considered in terms of specific social, political and economic contexts. Human beings have and use cultural maps that explain their own sense of the world and their sense of personal identity and place. The causes of social ill health are a loss of land that has class, and status implications and loss of identity associated with being required to negotiate multiple cultural frameworks requiring competitive, individualistic behaviour and egalitarian group behaviour. It is however impossible to type-caste or stereotype Aboriginal people or their culture.[61] People need to be understood in terms of their life chances (associated with gender, level of education, occupation and urban or rural life experience and class position). Nevertheless this opinion is not necessarily shared by traditional Aboriginal people who define culture as an absolute associated with kinship, land and law. As insiders they do not see that culture is learned and that it can and does change for many people. For traditional people using culture for survival in a world perceived as hostile, culture is immutable, not a set of guidelines or a resource for living.

Bowden (1994)[62] emphasises the way culture is used by young people for resistance both against their parents and the wider society. This is a source of great difficulty for young people. Injury rates, incarceration rates, low literacy and numeracy rates indicate social outcomes.

[61] Hollinsworth, D. Discourses on Aboriginality and the politics of identity in urban Australia *Oceania* 63 1992 pages 137–154. Replies to Hollinsworth by Beckett, J., Atwood, B. and Mudrooroo Nyoongah and Wiping the blood off Aboriginality: the politics of Aboriginal Embodiment in Contemporary Intellectual debate Lattas, A. Australia *Oceania* 63 1992 pages 155–167. Coagulating categories: A reply to responses by Hollinsworth pages 168–171.
Keene, I. (1988) Being Black: Aboriginal cultures in 'settled' Australia. Canberra, Aboriginal Press.
O'Keeffe, K. (1988) Aboriginality: Resistance and Persistence. *Australian Aboriginal Studies* (1): 67–81.

[62] Bowden, M. (1994) Will they be Aboriginals as we are today? A case study of the education of Arrernte youth in Alice Springs. *Masters thesis* submitted in partial fulfilment of the requirements of the Degree of Master of Education at the Northern Territory University.

To sum up the socio-economic challenges are to address the global market proactively by researching the impacts of increased privatisation and ensuring that the negative impacts on welfare are addressed by keeping in place and developing the supports required by people who are currently marginalised. Staff who have a sense of loyalty and continuity to an organisation and the people best serve the delivery of human service. This applies as much as it does to schoolteachers as it does to social workers and community nurses. Encouraging innovative and creative thinking through open and flexible organisations that can produce value added goods, rather than only commodity-driven goods. Acknowledging that the market sectors are now linked and that thinking systemically is rewarded. Divisions and categories are not useful for problem-solving in global economies. Flexible, responsive thinking is needed for organisations to survive in a fast and changing environment.

The cost of basic services such as power and water could be limitations. The challenge is to address the differences in life chances in terms of:

- The economic aspects: the relations of production under capitalism and under communal society.
- The socio-cultural aspects: the notion of commodities being private as opposed to being communal. Commodities are understood as private items of accumulation, status or profit in terms of capitalism. But in some sections of Aboriginal communal society, commodities are merely shared items. The way in which culture and land is understood in capitalism and in Indigenous Communal society differs.
- Service delivery that reflects the values of the user group needs to be based on broad-based participation in the planning, implementation and monitoring processes. Service users need to be addressed not only as consumers, but also as participants in a development process.
- Maximising access to opportunities for economic and social development could be achieved through teaching computer skills and exposing people to the wider world through the Internet and media. Basic literacy and numeracy are essential for job readiness. As are providing portals to the local and wider arena through sport, arts and cultural events that builds self-esteem and recognition.
- Much could be gained in terms of reconciliation by engaging in a dialogue that celebrates the diverse culture of Alice Springs. It is this diversity and the unique beauty of the landscape that draws tourists and travellers to Alice Springs. Listening and learning from one another is the first step in reconciliation. Culture and landscape are also two major economic resources for Alice Springs. The potential for Indigenous people to celebrate their culture in a way that they determine, and to share their stories is as yet undeveloped.

- A culture of despair and cynicism about work opportunities exists as a result of lived experiences of limited life chances and survival in difficult environments. According to Shepard (1998) economic development needs to address the challenge of youth unemployment in Alice Springs. Their research data indicated jobs growth in the professional sector, in financial services and skilled clerical jobs, but jobs for labourers and unskilled clerks were disappearing. The young population of future school leavers need to be prepared to compete in a changing job market. 'School retention rates are low and there is a marked disparity between the retention rates for Indigenous and non-Indigenous students'. The retention rate for Aboriginal students' in the NT in 1997 was 9.2% compared with 54.5% for non-Indigenous students ... . (Alice Springs Profile 1999: Retention rates in particular are influenced by a host of factors, one of which is the extent of support young people receive from their home environments.) The high number of unemployed young Indigenous people who have dropped out of school poses a threat to their life chances.
- Education and employment[63] considerations need to be based on an understanding of the different ways in which groups of people make sense of their world. An opportunity exists to harness the goodwill in the business community, expressed by the Chamber of Commerce to create employment pathways for young people that are employable because they are 'job ready' (NTCOSS forum, Alice Springs 1999). The challenge is to ensure that the goodwill and the initiatives are coordinated. Young people who are not job ready need to be case managed to enable them to obtain the necessary skills and training from basic numeracy, literacy, people skills and work culture, time management and self presentation. The skilled service sector is an area of growth, but the education sector does not retain young people for long enough nor do the standards of education across all the population groups prepare the young unemployed for the kind of employment needed for the future growth of the economy.
- The arid zone economy could be a central focus of sustainable development: living in harmony with existing resources and creating business, educational and economic opportunities locally, nationally and globally. We need to think of the way in which arid zone technology can use the environment sustainability by using recycled grey water and planting as many Indigenous plants as possible to maximise shade and reduce heat of tarmac and cement. Barrett Drive is known as 'broken promise drive' because the tail of the caterpillar-dreaming site was cut off during the development.[64]

[63] Kieran Finnane Alice Springs News vol. 16 No 16 May 19 1999: 3.

[64] The extent to which house and rental prices will be sustained given the drop in the number of residents is debatable.

The challenge is to help people in Alice Springs to address the differences in terms of economic aspects: the relations of production under capitalism and under communal society. The notion of resources being private as opposed to being communal. The way in which commodities are understood under capitalism and in traditional communal society is not the same. But it is important to understand that Aboriginality spans the full class spectrum and cannot be stereotyped. It is however true to say that life chances of the majority of Indigenous people are still clustered at the most disadvantaged end of that spectrum.

- To date the major growth industries have been linked with population growth. Street Ryan (1998) have identified that the major growth industries in Alice Springs are service-oriented and driven by population growth, for instance property,[65] business and human service industries.[66] The development of both service and information industries is appropriate because for too long Australia emphasised primary industries and exported raw materials to which others added value. Vulnerability to the rise and fall in commodity prices has compounded this approach to development. A study in 1985[67] flagged tourism as the 'driving force for economic development. A recession in the region or in the global economy could impact on this. In the 1999 study conducted by Street Ryan tourism is also stressed to be a major industry that appears to be 'showing worrying trends'. The NT Tourism Commission, for example, estimates that in 1997 there were 227,000 visitors to Alice Springs, a decline of 17% on the previous year and well below the peak of the early 1980s of 360,000. Against this the number of international visitors has increased by 15% to now represent around 41% of all visitors, and both the average length of stay and average expenditure per visit have increased ... . Further, the Tourism Forecasting Council (TFC) predicts much stronger growth in international visitors than with domestic tourism ...' (1999: 42).

## 6.6 *Physical and Mental Health Status Indicators*

> Costs to Health Services are rising faster than any other Government service. In this situation, there is a very real risk that the health budget will consume

[65] Street Ryan and Associates (1998), Economic Profile.

[66] Shames Harley (1985), Base align 82 & 83 Community Needs Study.

[67] Human Rights and Equal Opportunity Commission (1997) Bringing them home: A guide to the findings and recommendations of the national Inquiry into the separation of Aboriginal and Torres Strait Islander children from their families. Sydney: Human Rights and Equal Opportunity Commission.

> all the discretionary resources available to us. This is due to a combination of population growth, increasing standards as technology progresses and, sadly, the inescapable fact overlaying all of this, 25% of the population are Aboriginal people with such poor health status that they currently account for over 50% of health outlays (Reed 1998: 13).

Indicators such as low life expectancy amongst the Indigenous population, higher infant mortality and morbidity rates are associated with poor nutrition and poor coping behaviour. The latter is indicated by high levels of hazardous drinking, domestic violence, injury and suicide rates (Table 6.3). The results are spelt out in alcoholism, cultural despair, a sense of powerlessness linked with mental ill health, a sense of 'loss of country' and 'a loss of culture' linked with low self-esteem. Living within two different cultural frameworks means learning to survive within both competitive, individualistic frameworks and within egalitarian, group frameworks which poses particular stresses. In order to address mental ill health the different way in which it is defined by Indigenous people needs to be understood and a holistic, whole of community approach is required to prevent illness by working across sectors and disciplines.[68]

> The state of Aboriginal health is the factor which is having the greatest impact on demands in the health system. Aboriginal people in remote town and town camp communities carry the burden of the worst health status in the NT and the rest of Australia. 'By Australian standards as a whole and by the standards operating amongst the indigenous peoples of other western countries, the rate at which Aboriginal health is improving in the NT is not good. Over the last decade the gap between Aboriginal and non-Aboriginal health has shown little sign of narrowing ...'[69]

The NT has the highest mortality rate across all the states for the period between 1992–1996.[70] Lung cancer and stroke are causes of death, which are increasing in NT whereas in other parts of Australia the opposite trend is occurring.

> 'Australian mortality rates overall have declined steadily over the past 15 years. Non-Aboriginal mortality in the NT is about the same as the rest of Australia, but for one of the areas where there has been mortality due to suicides and mental health trauma associated with suicides. The mortality rate for Australia is 6.7: 1,000 on average, but it is 10.8: 1,000 for the NT.'[71] Aboriginal mortality

[68] The NT Government Health Policy, NT website on public health.
[69] Source: Australian Institute of Health and Welfare 1998: 10.
[70] General Practitioner's Survey 1998.
[71] Mackerras, D. (1998) Evaluation of the Strong Women, Strong Babies Strong culture Program Menzies Occasional Papers Issue No 2/98. Menzies School of Health Northern Territory Australia.

*TABLE 6.3*
*Years of Potential Life Lost before 65 for Aboriginal Adults in NT*

| | Men | Women |
|---|---|---|
| Motor vehicle accidents | 17 yrs | 7 yrs |
| Homicide | 8 yrs | 11 yrs |

*Source*: Central Australian Division of General Practice (1998).

rates are about three to four times higher than non-Aboriginal rates. Aboriginal male mortality decreased from 1979 to 1991, though at a slower rate than for Australian males generally, so Aboriginal men have become relatively worse off than before. Aboriginal female mortality has risen 20% in that time (*op.cit.*).

The main causes of excess deaths in men were heart and circulatory disease, followed by respiratory disease and injuries (motor vehicle, homicides and accidental falls) for women, firstly heart and circulatory, then respiratory, then cancers, diabetes and other endocrine disease renal failure and other genital and urinary tract disease. Cervical cancers are more common amongst Aboriginal women. Statistics show that Central Australian Aboriginal people are admitted to hospital much more frequently than other people and overall have higher mortality, and lower birthweight babies than people in other regions of the NT.[72]

[72] A meeting was held in November to respond to the suicide by a young 13-year-old. The discussion was about ways for young people to cope with sadness. It was suggested that they make use of counselling services available to them. As a participant I suggested that we explore 'ways of coping positively when we are sad'. Often risk-related behaviour is an outcome of sadness and feeling unable to cope. The group dynamics did not work well because there were too many people including adults who were strangers. The adults included professional youth workers, counsellors and a psychiatrist. It was clear that young people did not want to speak much in public about matters that are hard to speak about at the best of times. It was also clear that they wanted to engage in fun activities and that whilst doing something else they would talk. The focus needs to be on something other than themselves as individuals but instead to talk whilst being engaged in some other activity. At the meeting an Aboriginal elder spoke at length about how he personally coped with much sadness in his life by looking at nature and talking with his family and friends. He said that they must look forward to good things in their lives, like visiting the sea or going on trips. He emphasised that even overseas holidays were not impossible, that he could make things happen. The faces of the young people became really animated at this point. It underlined that what young people are looking for is

*6.6.1. Mental Health*

The full extent of mental illness and preventative, accessible services within the community have been assessed in terms of the National Mental Health Strategy (Tsey and O'Kane 1999) of the Menzies School of Health. The majority of residents of Alice Springs with a mental illness and using mental health programs are part of a stable, Indigenous population (40% had been living here for 20 years or longer). In Alice Springs the non-Indigenous population is more transient and tends to move if they become ill, according to this report. The needs of Indigenous urban residents and those from remote communities need to be considered. (1999: 123). In order to assess the extent of mental illness a range of indicators need to be considered; the high rates of alcohol consumption, the level of domestic violence, the level of suicides, the extent of unemployment amongst Indigenous young people in particular and the number of people who are homeless. Quality of life and life chances contributes to the use of drugs and alcohol, anxiety and depression, for example. Three suicides occurred on one day in Alice Springs (Centralian Advocate 24/06/98) including a 26 and 47-year-old male, and a 13-year-old girl. Addressing mental health issues using a community psychology approach, rather than only one-to-one counselling is a priority.[73] This needs to be explored as mental well-being has a major impact on the lifestyle of the whole of Alice Springs. According to the Centralian Advocate[74](24/03/98):

> Since June 1997, 10 people have taken their lives in Central Australia including 5 under 20—in Alice Springs ... Territory Health Services child and

*Continued*

an opportunity to be happy, to explore and to fulfil themselves. They need hope. They also made it clear that their parents are often unable to help them. Often parents are part of the problem in dysfunctional families. Wright (1999, personal communication) has reinforced this point.

[73] Centralian Advocate 24/03/98.

[74] The Youth at Risk Committee was set up in response to suicides earlier in 1998. According to their spokesperson: 'hundreds of young people are at risk ... either from their own damaging behaviour, including petrol sniffing and alcohol abuse, from violence directed at them by others, or from living in an environment where violence is an everyday fact of life.' Together with staff from Alcohol and Other Drugs and the Gap staff, it was decided that painting could be used as a vehicle for young people's voices to be heard. They could conceptualise ways of coping with sadness and share their positive stories with other young people via wall murals.

> adolescent psychologist Kim Rathman said feelings of loss of control and having 'no way out' were present in both adult and youth suicides. She had been in Alice Springs for 14 years and suicide had never been as big a problem as it was now, Ms Rathman said ... Teenagers often gave out 'little hints' something is wrong ... Signs a teenager may need to talk to someone could include apathetic behaviour ... risk taking like drinking and driving unsafely and talking about suicide ... . Ms Rathman said people were often reluctant to seek professional help, because it seemed like admitting failure ... .[75]

Current statistics are inadequate and attempts are being made to develop procedures to identify mortality and deaths associated with suicide more accurately:

> Data on deaths from self-inflicted injury have limitation, and there is little information on the link between mental disorder and suicide. Mortality data for suicide are obtained from State and Territory Registrars of Birth, Deaths and Marriages and are usually based on records provided by the Coroner's Court. Identification of suicide cases by these methods may be undercounted, as some cases cannot be clearly identified through coronial procedures ... Efforts are being made to link mental health case registers with death registers.[76]

Death rates from suicide are also relevant to mental illness ... The suicide rate for the 0–14 year age group has ranged between 0.8 and 0.4 per 100,000 during the period 1979 to 1996 (AIHW Mortality Database). The highest rates occurred in the mid to late 1980s ... ., and again the year with the latest available data 1996 (and 1997) ... .[77]

> According to ABS (1997) the high number of female suicides in the NT needs to be noted as unusual in comparison to other indigenous and non-indigenous suicide trends interstate.
>
> 'The number of suicides in 1997 was the highest ever recorded (2,723). This represented a 14% increase on the 2,393 suicide deaths registered in 1996 and a 24% increase over the 1988 figure of 2,197. The 15–24 year age group accounted for 19% of total suicides in 1997, with a corresponding age-specific death rate of 19.1. However, the age-specific death rate for suicide was highest in the 25–44 year age group ... Male suicides still outnumber female suicides by nearly four to one. In both the 15–24 and 25–44 year age groups. Between 1988 and 1997, the standardised death rate for both male and female suicides increased by 9%.' [78]

[75] Australian Institute of Health and Welfare 1998: 107.

[76] Australia's Institute of Health and welfare (1998) Australia's children: Their health and wellbeing 1998: 86.

[77] ABS Causes of Death 3303.0 1997.

[78] Cited in *Understanding mental health in Central Australia* (part 1 and 2) CARPA Newsletter vol. 28: 26–32. According to O'Kane and Tsey (1998: 26) the definition of mental illness is the starting point for all research to assess the extent of the mortality and morbidity associated with mental health problems. There are many disagreements about definitions. Wider definitions are essential for preventing social causes of mental health.

An ABS[79] survey cites that at a national level 1 in 5 Australians (18%) had a mental disorder at some time during the 12 months prior to the survey (ABS 1997: 5). According to ABS[80] as people age the prevalence of mental illness decreases. In Alice Springs the population is demographically young, hence there is no reason to expect that the statistics for specifically Alice Springs will be lower (particularly in the 18–24 year old category) than elsewhere. It is very likely that mental disability in Alice Springs is higher than the national average given the following factors:

> the comorbidity (more than one illness) associated with heavy alcohol intake, the impact of low school retention rates, high youth incarceration rates, high levels of unemployment for Indigenous youth, high levels of domestic violence, the transience of the non-Aboriginal population, the lack of support for young Aboriginal people particularly from remote communities and the lack of extended family supports for mobile non-Aboriginal workers.[81]

According to the High Support Needs Project (1996)[82] which was commissioned by NT Health to assess 'needs and evaluate options for service delivery for disability and mental health clients with high support needs and challenging behaviours', there are about 220 people in the NT whose needs are unmet by current services because of their challenging behaviour. They include:

> Adolescents and young adults with intellectual disability and challenging behaviour, children with severe and multiple disabilities, adolescents and young adults with petrol sniffing or substance abuse-related brain damage, people with disability and high support needs living in or originating from remote Aboriginal communities.

The group is not large in size when compared with the overall population of people with disability but they consume a large amount of the resources (1996: 1). It is anticipated from preliminary research conducted by O'Kane (1999) of the Menzies School of Health that the incidence could be higher across both the Indigenous and non-Indigenous population in Alice Springs.

[79] Cited in the *Mental Health and Wellbeing Profile of Adults Australia* (1997) Report 4326.0.

[80] Tsey, K., Scrimgeor, D. and McNought, C. (1998) An Evaluability Assessment of Central Australian Mental Health Services, Menzies School of Health Research, Alice Springs.

[81] High Support Needs Draft Final Report Territory Health Services (1996) compiled by the Ernst and Young Consulting Team.

[82] Australian Bureau of Statistics (1997) The Health and Welfare of Australia's Aboriginal and Torres Strait Islander Peoples Report No 4704.0.

The National Aboriginal Health Survey (NATSIS)[83] stresses that the conceptualisation of issues can be very different. Health status is not merely about physical health but about a sense of 'strength and worthiness'. To an Aboriginal person 'strength and worthiness are drawn from being free in one's own country and what is health beyond strength and worthiness?' (Smith and Smith 1995: 28 cited in ABS report 1997: 41). Nationally only about 2% of Aboriginal informants described their health status as poor, because they associated health with what happens in a clinic or in a hospital, not with how they felt or their wellness. In the NATSIS report people were asked if they were happy with health services (ABS 1997: 43) and only 10% said that they were not happy (ABS 1996 cited in ABS 1997: 43). The ABS report warns that this result should be considered with caution.

In an interview with a staff member at a mental health service it was stressed that the definition of mental illness is too narrow and not sufficiently responsive to cross and bi-cultural definitions. Language and concepts can be a barrier to mutual understanding. Aboriginal informants define illness and health in holistic terms and cannot see the point of narrow definitions. They see history and social circumstances as underpinning their life-chances in terms of physical and mental well-being. For example: Arrernte Council for instance stressed that they provide services to many who can be classified as 'part of the stolen generation' mental health issues associated with family breakdown need to be recognised.

It was stressed that wellness and ongoing support services 'need to be the base of a pyramid of services moving to pre-crisis support, crisis response, resolution and referral, community based crisis stabilisation and hospital acute care'. The wide support base is at the bottom of the pyramid and the narrowest at the top. In other words more preventative and health promotion work needs to be done. Building a wider base of support in partnership is possible once there is a sense that citizenship rights extend to promoting well-being in the community rather than dealing with the effects of problems which have escalated as a result of unemployment, isolation and domestic violence.

Alice Springs is a place where the turnover of residents is very high and thus the transitory nature of life in the community is a contributory factor to mental health. Transience adds to the usual stresses of living in any community. But in Alice Springs the stresses faced by the Indigenous population add particular problems. The availability of mental health services in Alice Springs is limited. According to Tsey, Scrimgeour and

[83] GP survey 1996.

McNaught (1998: 4): At the time of their research it (Mental health services, a branch of Territory Health Services) comprised:

> an 11 member community mental health team servicing the Alice Springs community; a staff of 18 providing residential and out patient care at Alice Springs Hospital; 2 forensic workers. A 3-member bush mobile team servicing all the remote regions South of Barkly; a child and adolescent worker; a team of 2 based in Tennant Creek which services the whole of the Barclay region; 3 community-based Aboriginal mental health workers; 1 full-time manager and an administrative support person ... the total population is about 50,000 people, a third of which is Aboriginal living predominantly in remote communities. In urban areas the population is 80% non-Aboriginal, but in remote areas 70% of the people are Aboriginal. ...

### *6.6.2 Adult Health*

'Non-Aboriginal adults experience the same good health as young and middle-aged Australians generally, with the exceptions of higher injury rates in young adults, especially males, and higher alcohol and tobacco consumption rates. Sun-related cancers are also significant ... . By contrast Aboriginal adult health is bad and actually deteriorating ... especially amongst women ... due to lung cancers, ischaemic heart disease, smoking-related respiratory disease, diabetes and renal disease. Currently over 100 people are on renal dialysis in both Darwin and Alice Springs. This could rise to 350 in 10 years time', according to the Menzies School of Health.[84]

Overall Aboriginal people score higher in terms of social dependency and lower in terms of social health indicators.[85] Longer life expectancy amongst some sectors of the population will result in more caring required in the home context and a shift away from nursing homes to care within the community.[86]

### *6.6.3 Child and Maternal Health*

According to the NT Women's Health Policy: 'the dual roles of many women results in the need for more child care options, which are currently being reduced. They are too tired to do much to look after

[84] Indigenous social-health and well-being indicators from the National Aboriginal and Torres Strait Islander Survey, Social Atlas 1997.

[85] NT Women's Health Policy: Preliminary focus groups.

[86] Used by many living on Town Camps. It induces a state of intoxication. If combined with alcohol it appears to enhance the effect of drinking and vulnerability to violence.

themselves.' The well women programs need to be adapted for urban women because they have been largely for remote women (*op cit.*). Cancer prevention through breast screening is a high priority area and there is a need to promote cervical screening. The remote well women's screening program promotes a holistic approach to women's services THS Annual report 1997/8: 65.According to the NT Women's Health Policy:

> sexual health is not as yet a priority although some communities have 20–30% STD rates which results in: infertility, complicated pregnancies, pelvic inflammatory diseases, paybacks which result in domestic violence ...

At the time of the research the Aids Council had only recently begun to serve women's user groups and an Aboriginal Women's Health Unit called Alukra is also available to women and is funded by the Central Aboriginal Congress. They supply specialised services to women, such as dialysis. Tangentyere Council also has a Women's Committee that addresses social health issues. The risk of alcohol use and other drugs used by women (antidepressants, cigarettes, pitjarie[87] and alcohol) also need to be considered. The extent to which the socially disadvantaged are more prone to alcohol abuse than the rest of the population need to be considered. Both D'Abbs (1998) and Gray (1998)[88] argue that the costs of alcohol are generated and shared by the whole community. The use of alcohol and other drugs is an area that continues to need attention.

Malnutrition and infectious diseases remain the major causes of chronic illness and death in Aboriginal infants according to the NT Health Services. Some programmes are effective in preventative and promotive programmes to improve the health of pregnant women and thus the birthweight of infants.[89]

[87] Dennis Gray (1998) Paper presented at Alcohol Availability Workshop on 2nd Nov.

[88] Mackerras, D. (1998) Evaluation of the Strong Women, Strong Babies Strong culture Program Menzies Occasional Papers Issue No 2/98. Menzies School of Health Northern Territory Australia. According to Mackerras (1998) 'the maternal health of non-Aboriginal mothers is good and the infant mortality is 5 per 1,000 live births. Maternal deaths for Aboriginal mothers occur at a rate of one every two years or 30 times that of Australian mothers overall and still births have declined from 34 per 1,0000 in 1976 to 20 in 1993, but the rate for the rest of the population is 6.5 per 1,000 births in 1993. Infant mortality has fallen from 120 per 1,000 live births in 1966 to 28 in 1993. This is 4 times higher than the Australian infant mortality rate. 12% of Aboriginal infants were of low birthweight in 1993 that is twice that of non-Aboriginal babies.'

[89] Gleeson, B. (1999) Beyond Goodwill: The materialist view of disability. *Social Alternatives* Vol. 18 no 1: 11–17.

### *6.6.4 Morbidity, Disability and Access to Services*

Disability precludes participation in society and this has been addressed through legislation by lobbyists who set aside the notion of good will and talk in terms of rights.[90]For many spending a few days each week on a dialysis machine is a way of life that has dictated their living in town and not with their families in a remote area. It is also a way of life for many on Town Camps. Indigenous informants on Town Camps recited the days and times they went by bus for dialysis treatment.

#### 6.6.4.1 The Disablement Process

Limited epidemiological data on disability exists, hence the decision to include a specific question on disability in the census in 2001.[91] According to the publication 'The Health of Australians' (1998): the three dimensions of the *disablement process* are 'impairment, disability and handicap'[92] which refer to a level of ability and quality of life.'

Just over 4% of Australians are classified as profoundly handicapped, requiring assistance for self care, movement and communication. Mental ill health or head injury are the most likely causes. The total prevalence of handicap is 14.2%. Those with profound handicaps have multiple impairments.[93] In 1993, 18% of the population of Australia was defined as having a disability (note that by 1998 it was 19.4%). According to this study,[94] "people aged 65 years and over account for only 36% of Australians with a disability. In excess of 70% of people with a disability reported that it was primarily due to a physical condition, 18% were affected by a sensory condition and 12% have mental, psychiatric or intellectual problems. These data counted each person only once, not the prevalence of conditions. The national statistics on mobility handicaps are a baseline for estimating the local needs in the absence of complete data sets." In Alice Springs there is no reason to think that the number of disabled is lower than 18% as suggested by the Australian overall figure. The standard of health in the Indigenous population suggests that

[90] According to the Australian Institute of Health and Welfare Report (1997) The Disability, Ageing and Carers Survey (1993) had adequate detail on disability but an inadequate sample of Aboriginal people.

[91] According to the World Health Organisation's Classification (WHO 1980).

[92] Australian Institute of Health and Welfare Report (1997).

[93] Australian Institute of Health and Welfare Report (1997).

[94] Spokesperson for Aboriginal hostels who manages the unit at Renner Street, Eastside (Personal Communication 1999).

diabetes-related, respiratory-and-heart related conditions, plus higher than usual injury rates could lead to a figure close to if not higher than the suggested national percentage (Australian Institute of Health and Welfare Report 1997). The age at which Indigenous people need renal care for kidney failure is getting younger.[95] It was suggested that this was because of the age at which young people start drinking and combining drugs with very poor diets. Some as young as 15 years have to begin dialysis. Children start binge drinking at the age of 11–12. The informant stressed that the cost of food in remote areas: $7 for a tin of corn beef and $12 for a chicken, make living difficult and lead to 'booking down' that is writing up debts in the local shops. The cost of food in remote areas leads to poor nutrition, which in combination with excessive amounts of alcohol leads to diabetes and renal malfunction. This contributes to poor health and the need to access health services in Alice Springs (Table 6.4). Disability and access are particularly relevant for Indigenous groups who have much higher rates of diabetes and renal failure and a higher number of surgical procedures that affect mobility.[96] 'Aboriginal and Torres Strait Islander people in the NT are twice as likely to be users of disability services (Black and Eckerman 1997) and make greater use of Home and Community Care (HACC) services at younger ages.' (Jenkins 1995 in Australian Institute of Health of Welfare Report 1997) (*op cit*.).

By realising that citizens and visitors with a disability and special access needs, include the elderly, the frail, parents and young children in prams as well as wheel chair users, the size of the population of users could be in excess of 20% according to Rob Nankerville of The Disability Advocacy Service (personality. communication. 1998).

#### 6.6.4.2 Physical Barriers to Mobility

A citizen's group to me the need for a generic accessible infrastructure and public transport system for all Australian citizens because:

- "The high level of diseases such as diabetes, cardiovascular diseases and injuries in the NT is well known.
- The high percentage of parents using prams requires a generic service, which is accessible.

[95] Australian Institute of Health and Welfare Report (1997).

[96] The life chances of citizens in Alice Springs vary greatly. The changing, mobile population of Alice Springs includes tourists, regionally stable but locally mobile Indigenous citizens and local citizens. The life chances range from professional two income couples to the unemployed living in hostels, supported accommodation and those living in town camps and those without a roof over their heads.

*TABLE 6.4*
*Indications of Need in Alice Springs*

| | | |
|---|---|---|
| Disability Support Pension | In town | 1459 |
| | Remote | 766 |
| Mobility Allowance | In town | 38 |
| Carer Pension | In town | 41 |
| | Remote | 18 |

*Source*: Centrelink 1998.

- People who are frail, elderly or on frames and crutches or users of wheelchairs also need accessible infrastructure.
- Members of the Indigenous community suffer illnesses associated with ageing at an earlier age.
- Local and international visitors expand the number of people who need to access services safely."

## *6.7 Health Services*

The main challenges for all service providers in Alice Springs is to address the needs of: (i) a diverse population of Indigenous people (both transient and urban based) who have overall health and development indicators which differ in terms of life chances from the rest of the population. (ii) a section of the residential population who are well educated and have a high standard of living and a section who face a range of socio-economic and health challenges, (iii) the fluctuating needs of the tourist population.[97]

Clinical care was considered to be a strength. The weaknesses included: (i) lack of access to health promotion services and community support to create a safe community. (ii) Lack of accommodation for out of town service users and transport to meet the needs of patients from remote areas, assisting patients to manage when they have returned home, (iii) lack of continuity in case work. Opportunities identified included: (i) appointing more case work coordinators to manage pathways and ensure that appropriate treatment is found and followed through. Managing treatment is not easy for those from remote communities and those with limited understanding of bureaucratic systems. (ii) Developing services to promote well-being and prevent domestic

[97] Health Worker at General Practitioners Association, personal communication 1999.

violence, frail elderly abuse and youth suicide. (iii) Promoting access to education and pathways to employment. A congregation spot or waiting space is needed for people who wait throughout the day to access services or to visit. This could be a point for community education and casework coordination. Ongoing dialogue, not once-off consultation is needed to build mutual understanding. Particular emphasis needs to be placed on health promotion and preventative care (particularly healthy nutrition, numeracy, literacy and budgeting). Also there is a need for coordinated care and case management (according to Deborah Fry)[98] to avoid losing people in a mass of referrals within a foreign cultural environment. The point is relevant for all Australians who have limited ability to access information and limited sense of personal autonomy (or a sense that it is possible to control their own lives). A sense of being able to control information and manage appointment times, finding transport and organising personal life to accommodate a time framework requires a number of skills which many people find a struggle if they have limited literacy and numeracy.

The threats identified included the lack of effective integration of services to address systemic health problems of alcohol and other drugs, poor mental health, high suicides and high injury rates. The lack of palliative care services,[99] management of alcohol and other drugs prevention of sexually transmitted diseases.

[98] At the time of the research palliative care and pain management services included the service of the Red Cross that had a minimal 3-hr position for palliative care per week and needed further support. The Alice Springs Hospital had a 4-bed suite for palliative care with full-length windows, also space for relatives to stop over. Small touches such as soft furnishings and tea and coffee making facilities make it possible for some sense of home comforts.

[99] Thirty-eight of the total of 243 clients are aged less than 25 years (15.6%). 143 are aged between 26–49 years and make up 58.8% of the users. Only 7 clients were aged above 50 years. 178 clients were male and 68 were female.

| Needle exchange Statistics | Needles distributed |
|---|---|
| 89/90 | 410 |
| 90/91 | 728 |
| 91/92 | 4394 |
| 92/93 | 13484 |
| 93/94 | 15289 |
| 94/95 | 16117 |
| 95/96 | 18829 |
| 96/97 | 18501 |

*Source*: Aids Council's Annual Report (1998).

### 6.7.1 *Timing and Choice*

The way in which health services are used by consumers is different from the way they are intended in terms of a Western biological development model. It is clear from a review of health services that Aboriginal people make use of hospitals as a last resort and by the time they arrive at the Alice Springs Hospital their condition is often in an advanced state. Alice Springs is the service area for the remote communities supported by the Flying Doctor Service. The challenge is to develop more preventative services in remote areas.

The delays are probably as a result of distances that need to be travelled to services, but also because of a desire to use culturally familiar options first. According to Bowden (1999), administrator for Alice Springs Hospital, the strategic direction for the Territory Health Services Corporate Plan 1996/1999 is to use more preventative and health promotive measures to reduce the number of people resorting to clinical care. The majority of admissions to the Alice Springs Hospital are by Aboriginal people for the treatment of advanced diseases. The stays are therefore longer and more intensive. This pattern of overuse of curative services is repeated in most parts of the world where either there is a lack of promotional and primary health care and an emphasis on curative service delivery or there is a lack of close community involvement in the planning and implementation of primary health care community-based development as a result of having different frames of meaning and without shared understanding of common goals.

## 6.8 *Social Health Indicators of Poor Coping Behaviour, Alcohol and other Drugs*

Drug usage follows the national trend and includes the use of more than one drug simultaneously. The Indigenous population tend to combine alcohol and cigarette use. Women also use Pitjarie. Young people tend to use alcohol, petrol and other inhalants, but the likelihood of using a wider range of drugs needs to be considered. The need for a needle exchange service in Alice Springs has remained constant, because as the statistics demonstrate the usage of the exchange and the numbers of needles

[100] Other drugs excluding alcohol

| Heroin | Pills | Speed | Coke | Roids | Other |
|---|---|---|---|---|---|
| 33 | 26 | 75 | | 13 | 13 |

*Source*: Aids Council's Annual Report (1998).

distributed has continued to grow.[100] According to the Aids Council's Annual Report the use of heroin made up a high proportion of drug use at the time of the research.[101]

The most commonly used and the most readily available drugs are cigarettes and alcohol. The morbidity associated with smoking has risen in the NT. According to Gray, (March 1999, Public Forum Alice Springs) the cost of alcohol related misuse in Australia as a whole is in the region of $4.7 billion if one includes costs associated with drinking and its control. The costs incurred range from the associated issues of absenteeism, to injuries, domestic violence and other crimes associated with intoxicated behaviour. The use of other drugs (such as cigarettes and pitjarie) along with alcohol could compound the impact on overall health. If general statistics on injury related deaths are regarded as an indicator of alcohol misuse, it is clear that injury rates in the NT are higher across all residents living in the Territory. But Gray (1998) also spelt out the support for across the board alcohol restriction.[102] The systemic causes and effects of alcohol abuse need to be recognised as having biological and psychological ramifications for current and future generations. For instance:

> 'Studies overseas show babies of alcoholic mothers face alarmingly high odds of maladjustment, violence and jail when they grow up … . Australian authorities hardly know about it'. According to this study foetal alcohol syndrome is 'caused by alcohol abuse during pregnancy. There is no conclusive study on what amount of liquor constitutes an intake damaging enough to cause neurological damage …'.[103]

The impact of Foetal Alcohol Syndrome (FAS) is unknown in Central Australia according to a CARPA News Letter (No 27 March 1998). Prior to birth, drinking impacts on the health of the foetus. The effects are both physical and intellectual. According to this News letter, there is no data on FAS, but that if data from Western Canada and South Africa on Indigenous children provide any indication, for instance 46/1000 and 189/1000[104] respectively are affected. According to another source the Western Cape (South Africa) as many as 1 in 50 children are affected[105] on the Cape Flats

[101] Dennis Gray (1998) Paper presented at Alcohol Availability Workshop on 2nd Nov.

[102] The Weekend Australian March 11–12, 2000.

[103] May, P.A. Foetal Alcohol Effects among North American Indians. Alcohol health and research World Vol. 15 No 3 1991: 239–248 in CARPA News letter No 27 March 1998.

[104] Martin, G. McIntyre, J. Parry, C., Stewart, R. and Tolman, S. and Yach, D. (1987) Alcohol and Alcoholism. Supplement 1: 125–132.

[105] A demographic profile of victims and abusers According to the Northern Territory Domestic Violence Data Collection Project (1996 July–1997 July) 'In total

where alcohol misuse is prevalent. The Cape Flats has one of the highest crime rates and highest rates of unemployment in the world.

Indications of stress and poor coping behaviour are evident if one considers that the Northern Territory and Alice Springs in particular, has one of the highest rates of domestic violence[106] in Australia. In addition, the rate of alcohol abuse is also one of the highest in Australia across the

*Continued*

1,285 incidents of domestic violence were reported to 29 government and non-government agencies collecting data throughout the NT in the reporting period ... Of these 322, or 25% of reported incidents occurred in Darwin. The next highest rate was recorded in Katherine with 215, or 17% of incidents and 181, or 14% of incidents, were from the Alice Springs Region. ... In these incidents:
'98% of victims were female.
96% of offenders were male.
76% of victims experienced violence by their partners or ex partners.
71% involved victims aged between 20 and 39 years and 51% involved offenders aged 20–39 years.
70% involved emotional/psychological abuse, 67% involved physical violence and sexual abuse was reported in 10% of incidents.
76% of victims experienced physical violence when offenders were affected by alcohol or drugs compared to 67% in the total sample. Of all offenders 63% were affected by alcohol or drugs at the time of the incident.
55% of victims and 49% of offenders were Indigenous Territorians.
47% involved children being exposed to violence.
45% of victims had previously sought help for a domestic violence incident, with 32% of victims first seeking help from the NT Police.
4% or 45 victims reported having a disability. These victims experienced more violence than the overall victim sample.
3%, or 33 victims were pregnant at the time of the domestic violence incident.
55% of victims, out of 1,046 incidents, were reported as having lived in the Territory for more than 10 years, 16% for more than 5 years and 13% had lived in the NT for less than 1 year.
69% of offenders, out of 743 incidents, had lived in the Territory for more than 10 years, 11% for more than 5 years and 8% had lived in the Territory for less than 1 year.'

[106] Pring, A. (1990) Women at the Centre, Pascoe Publishing. "Evidences the violence in traditional society and the violence to children who were removed. In male dominated white culture women were considered 'lubras for assisting men to open up the Territory: Domestic servants by day and sex objects by night." The testimonies of many women about their traditional lives in the 1920s detail stories of injury and death. Rape and beatings were part of the rubric of life for women when they became part of the stolen generation (Sam, M et al 1991).

whole population. The rate of serious violent crime in the NT is five times the national average (Crime Victims Advisory Committee, 1995: 7). Domestic violence (DV) is an issue in the NT for both Aboriginal and non-Aboriginal women. The injury rates are (includes accidents, suicide and murder) 25% higher for non-Aboriginal women than Australian rates and even higher for Aboriginal women, according to the NT Public Health Mission and Strategy Document (1998). According to the NT Women's Health policy Preliminary focus groups, the number of NESB women and women with disabilities who are subject to DV is increasing. A 1994 report is quoted in the Central Division of General Practice, Strategic Planning Background Papers Feb 1998:

> the indications are that NT has the worst domestic violence problem in Australia and that the rate of domestic violence suffered by Aboriginal women in the NT are in the highest estimated range of Australian and foreign studies.

According to Wright (1998 who cites O' Kane 1994) the full extent of the incidence of domestic violence is unknown:

> The reliability of information regarding the incidence of sexual abuse is questionable due to a victim's lack of access to those who wish to hear (O'Kane 1994), not being believed (Central Land Council Report to the Senate Enquiry 1997), fear of reprisals from the perpetrator's family (O'Kane 1994) and confidentiality issues (Brady 1992). … . In addition, sexually abused women's ability to parent effectively is often minimised, being unable to place boundaries and recognise unacceptable behaviours (O'Kane 1994). Role modelling of ineffective parenting could be continued through generations (ASYASS staff minutes 1997, and case notes 1997/1998).

Aboriginal women are dealing with sexual assault by 'turning to alcohol and drugs to stop the pain' (O'Kane 1994) It could be possible that those children who are abused model the behaviour of adults (Engle 1996) in dealing with their pain and even those who are not abused model the drinking and drug taking behaviour of parents and peers. At a meeting of the Central Australian Inhalant Abuse Network (CAISAN) meeting in 1997 sexual assault and violence were acknowledged as possible causal factors of petrol sniffing. Role modelling behaviour could also suggest reasons why representatives at the Violence against Women Seminar (1993) noted that perpetrators of sexual violence in their communities were becoming younger.

Violence for dealing with disputes and disobedience to cultural law[107] and "grief was commonplace within traditional practices, though it

[107] Wright, K. (1998) Aboriginal Youth Violence a way of Life published in part in CARPA Newsletter Vol. 28 Nov. Cites the above references Langton and Sam et al 1991.

is recognised that grief and despair 'produces a never ending state of anxiety, more injury and more death' (Langton 1991: 378). "Many people put it down to either alcohol or our low socio-economic status today and the pressures that go with that. Others will say it has to do with the way this society moulds us and expects us to behave. Some will tell you that it is the result of oppression our people have experienced this past 200 years. All the above are quite valid factors that contribute to family violence—although they are not excuses (through Sam, M Black Eyes 1991: 9)."[108] Women identified the following as causes of domestic violence: 'alcohol, women's lack of self-esteem, not knowing that violence is "not how it has to be", lack of anger management programs for men in Alice, STD are a cause of DV and paybacks.'[109] Office of Women's Policy (1996: 11)[110] shows that 'the majority of domestic violence costs are

[108] O'Kane Anne (1994) Sexual Assault Services in the Alice Springs Region. A Report for the Alice Springs Sexual Violence Action Group.

[109] Office of Women's policy Northern Women's policy Domestic Violence Strategy: The financial and economic costs of Domestic Violence.

[110]

| Description of cases based on domestic violence legal services records 1/7/96–30/6/97 | |
|---|---|
| Clients | 168 |
| Aboriginal | 63 |
| Non-Aboriginal | 9 |
| Unknown | 13 |
| Total | 253 |
| Types of Abuse | |
| Emotional | 177 |
| Verbal | 198 |
| Physical | 212 |
| Sexual | 30 |
| Total | 617 |
| Restraining orders | |
| Withdrawn | 26 |
| Variations | 8 |
| Revocations | 1 |
| Total | 35 |
| Police | |
| Police involvement | 175 |
| No Police involvement | 43 |
| Unknown | 29 |
| Total | 253 |
| Police Assistance | |
| Satisfactory | 86 |
| Unsatisfactory | 21 |
| Total | 107 |

associated with the provision of social security (housing and financial) and other forms of financial support/subsidies for those having suffered after they escape from the violent relationship. This constituted some 48.2% of the total costs of domestic violence. This study estimates that the number of women affected per year in NT is 750 and the cost per case $11,812 thousand per year. (The projection is based on modelling and extrapolating from the biographical data of 32 participants.) According to the Domestic Violence

*Continued*

| | |
|---|---|
| Remote communities | 40 |
| Housing type | |
| Lands and housing | 52 |
| Private rent | 30 |
| Remote community | 76 |
| Employer provided | 0 |
| Boarder/lodger | 12 |
| Family | 13 |
| Owner/buyer | 16 |
| Caravan park | 3 |
| Transient/squatter | 1 |
| Town Camps | 56 |
| Other | 24 |
| Unknown | 26 |
| Locality of clients | |
| Alice Springs | 128 |
| Tennant Creek | 22 |
| Other Areas | 4 |
| Total | 154 |
| Referrals from | |
| Magistrates Court | 19 |
| Police | 55 |
| Refuge | 16 |
| Been before | 36 |
| Family court | 2 |
| Legal Aid Commission | 3 |
| CAALAS | 6 |
| Crisis Line | 0 |
| DVS counsellor | 0 |
| Newspaper | 0 |
| TV | 0 |
| Gov. Dept | 3 |
| Community org | 7 |
| Aboriginal org | 15 |
| Word of mouth | 21 |
| Alice Springs Hospital | 1 |
| Solicitor | 2 |
| Total | 190 |

Legal Services, alcohol is related in 188 cases and other drugs in 41 cases.[111] The number of Indigenous and non-Indigenous women and their children making use of the Women's Shelter is indicative of family dysfunction in Alice Springs. Access to emergency accommodation for women and their children are inadequate according to service providers in Alice Springs. The demand outstrips the number of spaces available at the Women Shelter. Service providers at the Salvation Army and St Vincent de Paul reiterated this. Another gap in service delivery is the lack of safe accommodation available for young people below the age of 15 years who are intoxicated. One of the main reasons cited for making use of the shelter is the desire 'to have time out from partners'. Alcohol was cited as one of the key factors for family dysfunction. Other indicators of dysfunction detailed are the numbers of cases of domestic violence reported to legal services and to the police. The Northern Territory has one of the highest rates of domestic violence nationally. The rate of serious violent crime is five times the national rate. Alice Springs also has one of the highest rates of alcohol consumption. Polydrug use is likely to become more of a problem if the current trend continues.

## *6.9 A Proposed Systemic Approach to Address the Causes and Effects of Alcohol and other Drugs*

Following Stafford Beer's (1974) work 'Designing freedom' the links in social issues need to be understood. It is argued that alcohol needs to be understood as an effect and a cause of social ills. The issue of alcohol abuse in a report entitled: 'What everyone knows about Alice' (Lyons 1990) stresses that the problems of alcohol abuse in Alice Springs are both a regional and a local concern. Residents from some remote dry communities come to Alice Springs to buy alcohol. The role of alcohol in Alice Springs and the region need to be understood systemically. Alcohol is an effect of social marginalisation, low self-esteem and low self-respect. Alcohol is in turn associated with and directly related to further environmental health problems, such as the modelling of violent behaviour to young children and the breakdown of the fabric of social life. Some of the other main issues are the lack of basic, well-maintained housing and the lack of safety in Town Camps.

Alcohol misuse is the immediate cause of many social ills, this is indisputable. But what is less well understood is that alcohol misuse is also an effect of social marginalisation. A broad-based systemic approach is

[111] As far as the World Health Organisation (1995) is concerned in a paper entitled 'Alcohol policy and the public good', it is concluded that programs are only effective if they are broad based and unless they address the wider social context.

advocated by the WHO's public health policies.[112] The causes and effects become a cycle of damage to individuals, families and communities resulting in ongoing and intergenerational poverty. The health of women and children is affected whether or not they consume alcohol directly. It needs to be stressed that a high percentage of Aboriginal women abstain from the use of alcohol. This is not the case for Aboriginal men. The lack of a sense of power to control their chances in life is part of the problem. More Aboriginal men than women are unemployed. Poor nutrition is associated with the misuse of alcohol and this exacerbates health problems. The principles for successful programs are ownership and control over the decision-making and 'research should be directed towards providing Aboriginal people with information that empowers them in their quest for self-determination and the provision of appropriate services'.[113]

Scrimgeour (1997) stresses that self-determination is not merely control of service delivery, but control in decision-making. The problem with the latter is that not everyone has the resources to make informed decisions and those who do participate do not necessarily represent all the stakeholders. Thorough processes ensuring broad consultation and participation are essential.

> Aboriginal Congress believes that continuing to place a strong emphasis on these types of strategies is wasting precious resources and failing to address the problem. ... Substance misuse and the social disruption it causes will undoubtedly continue unless our efforts to take responsibility for the problem are supported. (CARPA newsletter No 27 March 1998).

According to D'Abbs (1998):[114]

> It is important at an early stage to define exactly what is meant by alcohol problem in any community, because it means different things to different people. Possible concerns include public drunkenness, domestic violence, self harm (health, career and lifestyle), cost to the public (police, hospitals), crime, road deaths, etc'. 'Research shows that restricting the availability of alcohol reduces the harm that arises from alcohol use. ...

Gray (1998), a senior public health consultant specialising in alcohol related issues, spelt out the support for across the board alcohol restriction in Alice Springs at an Alcohol Availability Workshop.[115] 'The dollar

[112] The National Centre for Research into the Prevention of Drug Abuse, Curtin University.

[113] Peter D'Abbs (1998) Paper presented at Alcohol Availability Workshop on 2nd Nov.

[114] Unpublished notes of meeting.

[115] Dennis Gray (1998) Paper presented at Alcohol Availability Workshop on 2nd Nov. Waltja Tjutangku Palyapayi Doing good work for families (30/3/99).

cost of alcohol harm in Australia is estimated to exceed 4 billion dollars per annum'. Preventing alcohol misuse is about addressing the social context. According to Bernie Kilgariff: 'If social conditions could be improved, maybe we would see less public drunkenness'. The debate between detoxification and harm minimisation versus re-criminalisation via the 2-kilometre law is not well understood, according to a DASA spokesperson quoted by Alice Springs News (August 5th 1998).

Everyone has an opinion on alcohol, because alcohol affects the whole community. But it is seen as a problem of the other, not oneself. Indigenous drinking is public and non-indigenous drinking tends to be less visible in private homes, restaurants and clubs. Public drinking and private drinking are two aspects of the same problem. Alcohol was cited in focus groups with young people as a way in which young people have fun, but also the way in which they address feelings of depression and a way of coping with grief and a sense of powerlessness in life. There was a strong awareness that there were other activities which could make them happy as detailed in the wall mural painted by those who participated in a project entitled: 'Things which make us happy'. This was conducted in response to a workshop held at the Gap Youth Centre at which a wide range of health professionals attended after a suicide of a young 14-year-old girl. At the workshop it was stressed that alcohol was a problem for young people, because when they felt in need, their parents were often not there for them. Alcohol use is modelled by the older generation and used by the younger generation to address a sense of despair. The emphasis in the research by O'Reilly and Townsend (1999) was on entertainment, rather than on coping with problems. Alcohol is also used to 'medicate' feelings of sadness and depression, as far as young people are concerned aged 16–24.

Aboriginal Congress believes that 'substance misuse and the social disruption it causes will undoubtedly continue unless Indigenous efforts to take responsibility for the problem are supported'. (CARPA Newsletter No 27 March 1998). Suggestions made by informants ranged from prohibition to strict control to the support of more responsible public and private drinking by Indigenous people through education, provision of serviced clubs at which entertainment and food is provided and safe transport home. To address public drinking a multifaceted approach is advocated by The World Health Organisation and researchers at Menzies School of Health, Curtin University and human service providers in Alice Springs echo this. Merely raising the cost of cask wine to the level of other alcohol costs will not necessarily act as a deterrent to all serious drinkers, who will simply spend a greater proportion of their income on alcohol or shift to drinking other spirits. According to Central Aboriginal Congress

[117] CARPA News Letter No 27 March 1998.

in a policy paper entitled 'Substance Misuse in Central Australia' the misuse of all drugs can only be addressed by avoiding top-down approaches. Communities have to take responsibility together with governments. For this reason the idea to control the movement of Indigenous people would be anathema in terms of citizenship rights, but also inappropriate because it is not based on generative dialogue.

The most controversial of all the suggestions was the control of Indigenous movement. The plan proposed the return to previous measures to control the movement of Indigenous people.

> The proposed pilot program outlined in the Face of Alice in Ten Years Project Public Discussion Paper (1999) is 'to manage social security payments to improve social development'. 'This project is intended to pilot a modified management of social security payments to residents of bush communities. The purpose is twofold: to minimise some of the consequences on social issues arising from the drift to Alice Springs, and to maximise the potential for social development in bush communities.' This was described regressive in that it moved away from rights of all citizens to free movement and to operate within a cash economy. This proposal was workshopped by Waltja with people from remote communities.[116] 'Part of the 10 year plan is about anti social behaviour in small groups. Part of the proposal is to get communities to agree that cheques were to be collected out on communities only and not in town any more. They wanted some communities to agree to try this before making it a law. Indigenous Informants said the following: Good aspects 'living together, all family strong, no more drinking everywhere in town, money stays in the community, more jobs, houses get bigger, mothers and fathers cant take off with money ... . More young people will stay on their community ... good if there is a bank on the community, ... less domestic violence in town ... . Bad aspects'. If someone is sick and need to see Doctor in town where they gonna get money from? You can get lift, but what if they leave you behind in town, you're stuck, Booking up will be a problem, never any money to spend, only money to keep paying back. There are no banks or post offices on some communities, Do not send money in on the plane, food expensive on the community, it can't be just a white town, -anti social behaviour is not all caused by drunks, Poor solution to social problems—should be looking at the cause not just removing the problem from eyesight. More travelling with grog on the roads—more accidents. Further pressure on families in town to support individuals who won't have access to their cheques. It's a breach of human rights. There needs to an increase in Emergency relief funds to help families get back home and when they are stuck in town. Controlling movement ... Difficult to move to look for work ... taking away a right ....

Controlling access to social security payments, because one is Aboriginal and from a remote community infringes civil rights. The

[116] 'The face of Alice Springs in Ten years Project Public Discussion Paper' Northern Territory Government.

colonial and missionary approach has been revisited in this policy and resonates at best with a paternalistic past. Urbanisation is inevitable and the right to choose where one lives should be an inalienable aspect of citizenship. It is more than 'paternalistic' and it is worse than 'an infringement on an individual'.[117]

For some the debate of controlled drinking versus abstinence is a false dichotomy. For middle-class Aboriginal people policy prohibition does not make sense. But for others, who have experienced the violence and reality of living with alcohol, prohibition does make sense and needs to be a community decision. Gary Stoll, the ex-Superintendent of Hermannsburg Mission summarised his experiences (1998: 98)[118] of controlled drinking which did not work because the concept was inappropriate for the following reasons. 'First the notion of a two-beer limit at a wet canteen was regarded as a two-beer quota, not an option. People felt that drinking a ration was appropriate, rather than seeing it as a limit'. He explained that:

> Council laws were seen to be less important than loyalty to family and the law that family comes first. People see themselves as family groups belonging to particular tracts of land with their own leaders and ceremonies. They only have responsibility for what happens to their family and on their land. They have no right to interfere in the matters of other families from other lands … The rules of leadership and the ceremonial laws are all believed to have been given by the various creative beings at the beginning of time. Thus the authority of the law is vested in the past people, even legitimate leaders of clans, do not have any authority in themselves. Their value to society is in them (*sic*) 'knowing the law' and this knowing enables them to guide their particular clan to enjoy a good and peaceful life as was intended by the creative beings in question. Anything outside this system is difficult to cope with. Rules or laws such as those made by Council to cope with new situations can be ignored … . The authority of the law is vested in the past.

He concludes that: 'Given the high degree of association with traditional clans and traditional law in Central Australia, ways of addressing the issue in terms that are understood is essential …'

This was confirmed by the focus group discussion with women on town camps, where the use of bush camps have been cited as being helpful and are greatly enjoyed as a means to persuade people of the value of traditional foods and recreation, rather than alcohol as a means of enjoyment. 'Berries are better than lollies said an informant. I paint this food and tell the young ones … .' Older women expressed a sadness that

[118] The Jay Creek Reserve was originally set up as a place for the protection of families who were at risk because their male relatives were considered violent by pastoralists. Rowse in Strehlow Occasional Papers 1999: 65.

the old way of women meeting together did not occur more frequently 'a good meeting is something to remember ... to sleep outside, to find bush tucker ... even here (in the camp) is good it makes women strong ... it could help us ... .'

The challenge is to enable citizens to take responsibility for alcohol through building capacity to govern their own communities. There needs to be a shift away from blaming and a shift towards owning the problem. At the time of the research the emphasis was on fines and community policing and not enough emphasis on community and individual responsibility. Admittedly it is particularly difficult to exert individual choice in an Indigenous community environment. In this context community responses continue to make more sense, for instance:

> 'Control grog through fining the suppliers'. 'Prevent and control through community policing and ensuring that people and animals do not use the roads at the peril of themselves and motorists'. 'Limit the location of outlets to places away from schools and residential areas, which should be grog free areas'. 'Declaration of Community Council areas as dry areas if this is the wish of the members of specific communities'. 'Limit take away licenses. The result is drunkenness and violent behavior (assault and rape)'. 'Provide licenses for more formal public drinking places for Aboriginal people. These could operate as Serviced Clubs with transport home.

Citizenship means civil rights and individual freedoms. By building citizenship rights and responsibility and achieving outcomes in social, political and economic terms social capital can grow. A basis for building trust is communication based on attitudes of mutual respect, right to information and freedom of expression. Where people can test the validity of others ideas through debate, development opportunities widen. All citizens have constitutional rights to the provision of services to meet their needs in such a way that they can retain their human dignity and that a healthy environment is sustained. According to an informant who had been part of the social movement on alcohol issues for over a decade:

> For many people alcohol misuse requires a whole community response. Some argue that town camps should be declared dry. Others argue that the combination of safe alcohol free areas for rearing family and safe places for responsible public drinking (with transport home if sober or transport to sobering up shelters) could redress some of the existing alcohol problems.

Safe places for public drinking for all members of the public both Indigenous and non-Indigenous are required in the long-term. Drinking in the streets and other public places is a result of not having safe places to drink. If there were more social clubs, with a family orientation, where food was served and where activities other than drinking were organised, safer drinking practices could occur. Drinking areas out of town such as Jay Creek[119] at an abandoned mission outside town are used to ensure non-prosecution for drinking related offences. People from dry communities drink the grog they have brought from Alice Springs, before returning home. The debris of drinking gives an indication of the chaos associated with heavy drinking: 'A babies bootie, discarded broken bottle, twisted beer cans and a knife. Graffiti walls of an abandoned mission vent an angry, broken spirit'.

The hours of trading for liquor outlets with take away licenses should be limited; sales should be strictly restricted to those who are sober and of the legal age. Residents of Town Camps should have more support in keeping camps dry if that is what they wish. Some Town Camps may wish to be dry and others not. It is a fact that communal living in Indigenous communities requires that communal, rather than individual choices need to be made. It is impossible for people to study or to sleep when alcohol related noise and violence occurs. The way forward is through building opportunities through education and employment. This cannot occur when alcohol provides a stumbling block.

> It is rarely recorded that Aboriginal people have been the leaders in the debate for twenty years, there are many Aboriginal people who live very functional lives (and are thus invisible) and who bear the projected shadow of the entire community. This leads me to the last point. While the Indigenous issues are so well highlighted and documented, it takes the focus away from the entire community issues and allows the majority to stay in denial. This is such an important part of any reform. What I do not have are the figures that show what a responsible level of takeaway outlets would be for a town this size. If the takeaways were limited and there were more eateries with limited licenses, then this could be a way of having culture friendly environments, without the problems that go with clubs. None of the above can happen until this town has a vision of what it wants—one that includes a commitment to our most valuable resource, young people and the very vibrant Indigenous culture that keeps providing leadership in this place.

## 6.10 *Environmental Health, Access to Services and Quality of Life*

Improving the health of the residents of Alice Springs is not merely about providing curative services but improving the quality of the environment.

[119] That applies to public drinking within 2 kilometres of the town centre.

Promoting health is as much about housing, water, sanitation and access to employment and access to public facilities and a sense of belonging as it is about access to specific medical services. The NT government policy (1999) to achieve better health recognises that improvement in health can only occur 'through intersectoral collaboration and acknowledgement of the environmental not merely individual factors, which determine choices, maximising dollars spent and applying lessons which work'. Strategically the policy stresses 'prevention and health promotion through improving water; disposal of wastes, shelter and nutrition; supporting families and fostering cross cultural awareness'.

Safe housing includes the need for safe water, drainage, sanitation, rubbish removal and power supply in town camps. The lack of these basic building blocks of health promotion has a devastating effect. It has been well-documented that water-and-sanitation-related diseases are best treated through preventative measures alongside education in hygiene. Improving and maintaining these basic aspects are an essential starting point for promoting health and preventing disease.

The environments in which people live are very diverse in quality. The major environmental hazard is the lack of safety as a result of extensive violence associated with alcohol abuse. The damage to housing stock and the overall quality of an environment is directly linked with alcohol abuse. The issue of housing and hygiene is fraught. On average 6–10 people at least occupy each immediate family unit and so crowding becomes a problem. Added to this is the need to clean regularly a small crowded house with indoor cooking and washing facilities. If it does not occur because it is difficult to mobilise people off the floor in the morning or because cleaning materials (detergents and tools) are not at hand—because they have been borrowed or stolen or because a large quarterly power bill has to be paid or some other contingency has occurred which simply could not be budgeted for—then the house becomes very unhealthy. Open designs with separate areas for different functions could alleviate the problem.

Residents of town camps are being initiated into the capitalist system through user pays programs, but the problem is that a host of barriers need to be addressed along the way. The challenge for some is to create private, safe spaces on public town camps. The younger generation appeared to want more individual control over personal space. This is very difficult to achieve on a town camp, where many things are regarded as common property. Also the older generation find the move towards social interaction in nuclear type families challenging. They prefer communal interaction.

Physical, mental and social well-being is another challenge. Informants exhibited limited self-esteem and a sense of helplessness. A key point raised was that it was difficult to show young people the right ways when adults model drinking behaviour. One of the most difficult problems faced by residents of town camps is the perception that visitors 'trash the camps' and that uninvited people 'squat' on the perimeters of some camps. The notion of who has a right to be in camps is fraught because drinking and violence is considered by many in camps to be largely caused by visitors or outsiders. Control of outsiders is perceived as vital to promote functional camps. A camp can be 'rendered dysfunctional in a weekend as a result of outsiders visiting', according to the coordinator of the Housing Association.

The expression of independence by many of the marginalised men is largely limited to drinking, football and violence. Football allows for a sense of catharsis, solidarity and a vicarious sense of power and social agency, while the violence associated with drinking enables men to maintain control over a camp and to maintain their dominance. Positive initiatives include courses run by Tangentyere Council and Congress on life skills. Another is the Territory Tidy Town competition that awards the camp, yards and homes of residents with large cash prizes. The role of cleaning and tidying public space is very hard when residents have to deal with the mess and damage caused by strangers (who come into camps to escape the 2-kilometer drinking law[120] and the problems created by unwanted visitors). Cash prizes are an incentive to maintain camps and not to trash them. Most of the winners mentioned in Ayeye[121] are women or women and their husbands.

Once a household has achieved a functional level, with houses well-maintained and utilities in working order and possibly a stock of domestic equipment and some basic furniture, they can lose these commodities during a weekend of drinking and violence caused often by strangers or the extended family (one' own or related to someone else on the town camp). Another challenge is the depletion of social wage or CDEP wages through alcohol use. Community, rather than merely individual responsibility makes more sense in traditional culture and for the older residents, but self-respect for body and spirit and hope for the future are equally important and stressed by alcohol management

[120] Volume 5 Issue 1 April 1999.

[121] See Deborah Durnam (1997) Federation of Independent Education Providers Study.

programs. Breaking the cycle of violence and alcohol needs to be multi-faceted: from community counselling, to control of levels of consumption and alcohol availability to creating employment opportunities and hope for the future. These are effective ways for people to break out of the destructive cycle. The control of alcohol availability cannot be the major thrust without the sense that individual responsibility opportunities and life-chances are being addressed as well.

At that time control of drinking and of outsiders was considered vital to the promotion of functional camps. The blame is usually attributed to outsiders, rather than those who live in the camps. The local residents are daily confronted by the difficulties of dealing with the accessibility of alcohol. The challenge is to face individual responsibility and community responsibility and not place all the blame on the social availability of alcohol, albeit the strategic placement of alcohol outlets can only be seen as commercially driven, without consideration of the social impacts. The notion of the extended family is differently interpreted across the residents of town camps and for some (although not all) the sense of responsibility to extended family appears to be changing. At one end of the continuum there is a clear wish that limits be placed on what families are expected to do for other extended family members. At the polar opposite end of the continuum there is a sense that responsibility to family overrides all other responsibilities. Some young mothers emphasised that they wished to ensure that their children received the best chance of an education. They saw this as a priority above all other family responsibilities.

The exasperation expressed by the women at one town camp is indicative of the way in which the residents of other camps tried to control the consumption of alcohol illustrates the range of opinions on alcohol but also about the range of opinions on the responsibility to family. Family on some camps are acknowledged and welcomed even if 'they trash the place and cost them money'. The notion of rent being paid by visitors could be a dramatic turning point in the notion of family obligation to allow people to 'stop with you, simply because they are family'. It was clear that on this camp women of different age groups had different priorities some were lonely because of the change in the way in which the camp was organised in terms of nuclear family houses. An informant expressed it thus: 'The women stop with their husbands, not with the other women ... ' Another female, elderly informant stressed that she wanted more communal meeting around tables outside the houses on camps. She wanted clean river sand to be placed in the communal spots for seating and the Community Development Employment Program

(CDEP) to be used to get the men to make tables and chairs. The CDEP program is perceived as a way not merely to keep people employed and active, but also as a way of ensuring that tasks are performed around the camp by those employed as CDEP workers. Most informants said that they understood that the wages for CDEP and the social wage were the same and that there was little incentive to undertake CDEP work, because it involved doing other people's 'dirty work' around the camp.

Low levels of literacy and numeracy create barriers to accessing information and services and limit job opportunities.[122] As does the lack of telephones and computers also limit life chances, because access to education, recreation, employment and health services is curtailed. Different attitudes and values make communication difficult. The informants felt that they were always struggling to communicate with people who did not really understand their story and this had an impact on their struggle for self-determination, whether it is interpreted as: (i) The Indigenous culture of sharing and a culture of community, family rights and family responsibility, (ii) Return to land and country, (iii) Declaring a dry area, (iv) Attempts to control visitors and strangers or (v) Getting a 'real job'. (vi) Attempting to redefine the extent of family responsibility. Responsibilities are difficult to accept, not merely because of the frequently quoted welfare mentality or because of the high cost of living (both of which are aspects of the situation), but because the notion of rights and responsibility in terms of Western notions of citizenship, do not necessarily have the same meaning for all citizens. Self-determination is the goal but it is defined and conceptualised very differently by different age groups, people with different levels of education and depending on gender. Multiple definitions associated with cultural values and maps have implications for planning and practice. Participation in discussions that lead to co-created decisions is vital if meaningful development is to occur.

A perceived lack of trust and limited life chances adds to difficulties. Everyday is a struggle for survival. The challenge is to create space for people to negotiate their own sense of citizenship not merely in individual (age, gender) terms but in communal terms. For some Indigenous people living in a communal household, sharing and following communal values is central to their life. For others moving away from communal to an individual lifestyle is preferable.

[122] Labour Party Survey 1999 on food and the impact of the GST in the NT and in remote communities.

Higher mortality and morbidity rates in Alice Springs and Northern Territory as discussed above are associated with contextual issues. Violence and road deaths as well as diseases linked with alcohol and associated poor nutrition (associated with spending patterns, literacy, numeracy, the cost of food and services in remote communities as well as alcohol-related problems) are some of the major challenges that need to be addressed systemically. The approach to governance impacts on the quality of life of citizens.

Housing design needs to accommodate extended family and fluctuating numbers. Access to safe power and water supply and drainage is also an issue. The standard of sanitation varies across the camps from pit latrines to flush latrines, which are in varying stages of working order and need regular maintenance. Drainage of water, which has flooded through blocked drains, is ongoing and contributes to environmental health problems. The issue of hygiene is fraught. On the one hand, the building of separate houses is not merely symbolic of self-determination but is a basic indicator of development. On the other hand, the construction of houses are designed to be cleaned and lit by electric light and filled with appliances and furniture that are expensive to maintain and maintenance through domestic housework and technical repairs are skills which are not yet universal. Another central issue is the cost to maintain and repair houses. The lack of disposable income as a result of user pays policies and house maintenance policy (as a result of funding cuts to an Indigenous housing association's operational budget) are also of central relevance. The designs for housing are well known to either enhance or limit environmental health. Housing which meets the needs of communal living is essential. The informal and structurally inadequate housing at one town camp comprising tin sheds and small and large thatched structures provide separation of functions: sleeping, cooking and communal meals. Formal houses need to be designed to suit the needs of an extended family, not a nuclear family. Design committees need to consider the separate functions that the house needs to fulfil for a fluctuating number of people. Separate house designs may meet the needs of younger people more than older people who wish to maintain more communal living.

Limited collection of garbage leads to the increase in the rat population. The use of rat bait leads to children being poisoned. The rats eat the electrical power cables, which means that cooking and heating does not occur in some homes. The live wires hanging down are a health hazard. The rubbish is a health hazard. The cost of addressing the effects of the non-collection of rubbish would be lower than simply collecting the rubbish. For example on an informal housing settlement the problem of the lack of rubbish removal can be seen in two different ways: this problem is systemic. It is one that leads to rat overpopulation. They eat electrical wiring and this leads to the increased vulnerability of children in particular and inability to use electrical appliances. As a result there is no hot water for washing, food preparation, recreation or study. In this context family dynamics deteriorate and so do life chances pertaining to health, education and employment. The ability to pay for services is limited by challenges at a personal, interpersonal and social level. The web of disadvantage has many strands and solutions need to be systemic, across sectors and disciplines (as per the World Health Organisation's Ottawa Health Charter of 1986 and the United Nation's Agenda 21). This is a problem that relates solely to lack of responsibility. The consequences follow from direct causes. Payment leads to service. Non-payment to lack of service or reliance on welfare service and volunteers. The two different maps of reality are first, on multiple loop and second on single loop perceptions of cause and effect. The non-systemic thinking can lead to higher costs in the long-term.

In all the camps running water is available although the number of taps per household is low in some instances as a result of visiting family. Hot water is not available in all households because electrical power supply is erratic due to rodents eating through the electric wiring. The housing committee membership is used as a vehicle for creating a sense of self-direction and control. The notion of women and citizenship rights and responsibilities is particularly difficult because women bear the responsibility in many instances of being house boss, but having little ability to address the never ending problem of uninvited visitors who are said to 'squat' on the perimeters of camps. The move towards requiring residents to pay for rent and to pay for damages means that deductions are made automatically from social wage payments. This leads to residents having to make use of food parcels and vouchers from St Vincent de Paul. Centrelink expressed concern about the impact of debt in coping strategies of families and on levels of depression.

The high cost of food in remote communities is well known[123] and the cost of food is 'booked down'. Family members who come in to Alice Springs to visit and access services will be required to pay rent after two weeks. The move by the NT government and ATSIC to make town camp residents more self sufficient has been implemented via incentives such as

[123] Street Ryan 1998: 25.

IHANT (Indigenous Housing Authority of Northern Territory) funding of $1,700 per house if residents pay a minimum of $30 and all visitors pay a portion of their salary through Centrelink deductions. The reduction in funding a few years ago has not made an impact on camp residents until recently, because Tangentyere has been footing the bills for maintenance and repairs.

Improving hygiene at home can only occur through education along with a sense of being in control of a number of areas of life. If alcohol abuse occurs hygiene becomes a low priority. The need to have appropriate cleaning materials readily to hand is the other priority. For respondents living across the seven residential areas (excluding town camps), food hygiene in food outlets was a concern.

The number of 'camp dogs' has been reduced according to a spokesperson for local government. As far as control is concerned, there have been only three Aboriginal dog attacks since 1979. The only attack by an Aboriginal dog was on another dog, outside a camp area. The community perceptions expressed in the Rates Survey (1998) of ratepayers was that the control of dogs was mainly an Indigenous issue, for instance: 'Camp-', 'River', 'Creek', 'Aboriginal' and 'Indigenous'-dogs 'cause problems'.

Indigenous and non-Indigenous residents stressed problems with the control of rodents. Open drains were mentioned as a source of concern as a breeding ground for mosquitoes. Non-Indigenous residents expressed concern about the types of chemicals used to control pests. Water to Alice Springs is supplied from Roe Creek bore field located 14 km south west of the town. This resource is expected to be depleted over the next 15 years or so, although its life is being extended through supplementation from the town basin.[124] Non-Indigenous residents concerned about recycling set up a lobby group. The cost of recycling was investigated by the public sector and at the time of the research did not have a cost-effective way of managing household recycling through split bins (waste in one side and recyclable materials on the other). Issues such as landscaping, landfill, access to the dump were stressed and informants suggested the following recycling alternatives:

> Town dump should be open to scavengers in a section of the dump during particular hours and under management to ensure safety. Materials could be sorted to facilitate building using recycled materials. A market needs to be set up at the dump to sell items or give them away. This could also be run as a business with assistance from volunteers. Encourage local arts and crafts groups to use wire (to make decorative Central Australian artefacts, toys, sculpture), rubber tyres (to make sandals, to resurface roads, to decorate gardens, to make play areas [swings adventure grounds], sculpture, dye/paint

> the tyres to make decorative surfaces for traffic islands once shredded), using dot painting motifs and under the auspices of an Aboriginal Organisation Arts/Cultural. Encourage building contractors to charge less for work if they use recycled material and if they can have the material which they are removing from a building (wood, window frames whiteware, tiles etc.). Wood from the dump could be sold as firewood. Deposit on all bottles, tins etc. could lead to less littering. A lack of services beyond the Gap means people need to take their own litter to the dump and this could be 'a health hazard'. Set up collection points for members of the public to deliver their own sorted waste and for Contractors to move the waste from collection points in each of the 7 geographical areas to the dump. Ascertain if a user pays service could be set up to complement the collection from central points. Create job opportunities using the material at the dump for welding or carpentry training, for example. Use materials for building affordable or free housing. The project could be used as a vehicle for job creation and training. Find out the feasibility of using empty trucks to take recycled building material with them, it would be viable even if the cost of the material offset the cost of transport because it would impact on reducing landfill and making the manufacture/extraction of more materials unnecessary.

Participants (excluding the town camps) were on the whole satisfied with the quality of litter control. They consider that the use of low security prisoners is a good idea and that there should be more inducements not to litter though fines and through ensuring that people who litter are required to clean up. Litter and alcohol abuse is perceived to be linked with town camps and public spaces such as the Todd River, for instance:

> Low security prisoners do cleaning up, Aboriginal people leave rubbish, Todd River a disgrace but areas I use in the CBD are fine, terrible smashed bottles are a problem, hard to keep up but they are trying! Need to educate the 'litterers' (*sic*) has improved from years ago, Need to clean Todd River up more often otherwise Ok, Doing their best under the circumstances, Whites do good job but Aboriginal people need to learn about picking up litter, there is different standards. They have their own community standards they have to respect our standards when they are in town. Noticed prisoners cleaning up [which is good they learn something]. Good if you like standing in the ocean and bailing, good considering the ways of some people, a lot around the town camps generally not too bad. Todd River needs more attention and many small supermarket areas. Riverside is bad, smashed bottles Young kids need another place to be. Too many wine casks around town. The river is a disgrace. Those who throw it down should pick it up! Informants (excluding the residents of town camps) were on the whole satisfied with the quality of garbage collection, except for the need for more awareness of recycling.

The major issues pertaining to public toilets were access to toilets that are first insufficient in number and second those located in some shopping areas are only accessible if payment is made or a key is given. Non-Indigenous informants were divided about the issue of access and the right of all people to dignified access. The ramifications of non-access exacerbate public health and social issues. The issue of maintenance of hygiene was stressed as a reason for non-access. 'No point in providing facilities unless staffed to prevent filth and vandalism.'

Responses to perception about the quality of emergency services varied. Informants considered the ambulance service prompt except when short staffing causes delays or when there are traffic delays. The most vehement concern was the central control of the service in Darwin: 'When you ring 000 you go to Darwin takes too long need to connect direct to local service.' Residents in town camps found it difficult to access services because of the lack of functional telephones on town camps.

According to participants the emergency services are largely unknown. There was much concern expressed about contingency plans for flooding. Many of the respondents stressed that they were newcomers and did not know much about the disaster preparations. Long-term residents expressed their exasperation as follows:

> Listen to—a long standing resident, this could be a disaster area, I listen to the old people, old Mrs. H talks of big floods we have done nothing to prevent it ... it could come anytime ... .

### 6.10.1 *Indicators of Accessible Services, Infrastructure and Community Life*

The logic of including a discussion of cemeteries and the perceived need for a crematorium in this section is that it is relevant to environmental health of residents, some of whom are transient but others see cemeteries as places for recording history and a sense of place. Many of the residents had not been to the three cemeteries and have no long-term connections in town. The older cemeteries are a concern to the long-term residents, particularly some Indigenous people who are unable to find the graves of their families. In one of the old cemeteries the only marked grave at the time was Namatyjara. The plan for the graves is said to be difficult to follow by an Indigenous informant. She experienced a sense of frustration. The landscaping of the cemetery, maintenance and cleaning up and identifying the walk paths were identified as concerns by rate payers.

The newer cemetery was regarded as acceptable by most, albeit some raised the problem of vandalism. Elderly people in particular complained that they felt unable to visit as regularly as they would like to, because of the distance factor. The new cemetery was considered to be too far out of town. The lack of a crematorium was raised as relevant to those who do not live long-term in Alice Springs and would prefer not to leave a grave or to travel elsewhere for a funeral.

Access to public transport, private transport and related transport concerns for drivers, cyclists, pedestrians, frail users of public transport and users of wheelchairs have an impact on quality of life. The case can be made that if the transport needs are assessed purely in terms of expressed need (measured in terms of current usage statistics), it could be concluded that the current system is under-utilised by the public and that if anything, the service could best be maintained to current levels. A closer examination of the situation, however, makes it clear that the current *user statistics* are limited by the current *limited routes and times* travelled by the public bus.

The data indicate that there is a case to be made in terms of felt need and normative need for the provision of a generic bus service with associated sheltered bus stops, which meet the needs of office workers, school children, people with a disability or frailty, parents with children in prams, residents of town camps, residents living outside the Gap, after hours recreational needs of a wide range of people and age groups but particularly young people who live locally and tourists. This would help to alleviate: (i) The number of people needing long-term parking in the CBD area, (ii) The needs of school children in all residential areas which may assist in cutting down on absenteeism in town camps with fewer bus runs, (iii) The isolation of people who are frail or elderly or who have young children in prams, (iv) The difficulties in accessing a range of services by people living currently in under-serviced, outlying areas, (v) The provision of safe, affordable public transport would help to ensure that people do not walk long distances at night and could thus play a role in crime prevention, (vi) Young people's access to transport as a means to ensure community participation and prevent problems associated with walking at night and (vii) Elderly people who currently limit their activities because of the cost of taxis.

At least 20% of the people in Alice Springs are frail or have a disability. Overall 16.3% of residents in Alice Springs are unemployed ($N$ = 711 unemployed and 4,999 not in the labour force). The rate of employment amongst 15–19-year-olds is 19% and amongst 20–25 year olds is 13%. Public transport is essential for ensuring that all people are able to access community services, employment, recreational opportunities and public spaces effectively. One of the ways to address social marginalisation (due to unemployment, age, ill health and/or immobility) is to ensure that public spaces are accessible via a generic transport system which is accessible in terms of times, routes (geography), cost and physical infrastructure. In particular the design needs to accommodate people using walking aids or wheel chairs and prams. A parent tending toddlers and pushing a pram would benefit from an accessible service as much as a frail, elderly person or a person with a disability. Access to public transport is central to addressing poverty through ensuring that barriers to education, employment and a wide range of community activities are minimised. In order to make it possible for people to access services and to participate in the workforce, mobility is a vital ingredient. The transport data need to be considered in relation to the demographic data and life chances of the citizens of Alice Springs. For some adults on incomes, which are, higher on average than elsewhere in Australia, access to public transport is unnecessary because they own multiple cars. Many young parents living in Alice do not have extended family in Alice Springs on whom they can rely. This makes it essential that parents of young children are able to access services and do not remain housebound. Furthermore, the number of single parents at home looking after children is increasing; it is essential that they do not become isolated. The same applies to the elderly living on their own. A case in point is an elderly man with a disability living close to the centre of town, who could not make use of the library because he did not know where or how to find a bus. His world is limited to grocery shopping and the television! He needed regular kidney dialysis and was too frail to walk any distance. Affordable, accessible public transport that is well advertised in many places (supermarkets, post office and pharmacies) would help to ensure that people are not constrained to merely meeting basic needs and are able to participate more fully in a range of activities.

The unemployed and Indigenous residents in town camps need access to regular public transport to increase life chances. By reducing the amount of money spent on taxis and ensuring that a wide range of education and health services are more accessible, life chances can be enhanced. The public bus service is currently underused.

It was stressed that access to a regular public bus service would make a considerable difference to school attendance and the use of a range of services (health, recreational and shopping). The cost of a shopping trip for groceries is so much higher for a Town Camp resident who has to pay for groceries and then for a taxi.

Users of public transport stressed the need for a regular and extended service accessible to people on wheels and crutches.

To sum up, a generic system would meet the needs of all citizens because many people have mobility needs which could be better served by means of an accessible transport system, for the following reasons: the level of diseases such as diabetes, cardiovascular diseases and injuries in the NT is well known. High percentage of parents using push chairs requires a generic service which is accessible for the frail on frames and crutches and the users of wheelchairs. The population is ageing and this is relevant in Alice Springs, where the population aged above 50 comprises about 16% of the population. The Indigenous community suffers illnesses associated with ageing at an earlier age and the number of physically disabled have increased nationally.

# 7

# Systemic Approach to Address the Process of Commodification

## Rights, Reconciliation and Reality: Creating Opportunities for Participation and Spiritual Well-being

The old woman sits bowed in a puddle
Two security guards stand over her.
One observes, arms folded.
The other radios for an ambulance.
The white, wooden table outside the public lavatory
is presided over by a woman who crochets.
In front of her is a rose china plate
filled with silver 50 cent pieces,
the price of entry

This chapter discusses creating opportunities for participation through developing social capital,[1] whilst mindful of the barriers caused by power imbalances and cultural differences. The central argument is that social

[1] This 'is the basis for trust in the community as a result of having, a sense that social opportunities do exist for all citizens irrespective of language, culture, gender or class'. It is summed up as '... features of social organisation such as networks and norms that facilitate co-ordination and co-operation of mutual benefit for the community. It builds the capacity to trust, and have a sense of security, social cohesion, stability and belonging'. Cox, E. (1995) *A Truly Civil Society*. NSW ABC books, Sydney. Boyer lectures.

marginalisation needs to be redressed systemically through improving health, education and employment opportunities that are responsive to a community, national and international audit of needs, opportunities and rights. Self-determination can also be seen as a reaction associated with what Pixley (1993) calls being a 'second class' citizen because of unemployment or because of a feeling of 'being shamed' in public places. A visit to the bank, a shop or just being on a pavement outside a shop can lead to being asked to 'move along' by police or security guards. The rationale from those in authority was linked with themes of pollution (noisiness, drunkenness and being untidy or unclean). The fact that many Indigenous people come to Alice for sorry business, to visit family and access essential services, means that people often camp with relatives or on the perimeter of town camps or in the Todd River. There are no public showers and Indigenous people have limited access to public amenities, many of which are available only during business hours and require payment to an attendant. At worst, it can cause situations like the one described above. The unnecessary response is the tail end of a series of misdirected systems. The link between current feelings about citizenship status and movements for self-determination is complex, but there is a link between the two. I am not however trying to reduce the land rights movement to an access issue, but a feeling of not being welcome 'in one's own land' and 'being shamed' does play a role in the nation within a nation movement that can restore a sense of pride in being Indigenous and a sense of well-being.

Harding (1992) reveals the implications of working with all the issues, including assumptions and values (emotions and beliefs should be included as well) because of their implications for the way reality is framed and knowledge is presented. Thus the tools for thinking and caring addressed in chapters 2–4 are valuable if they are put to use to help systemic analysis and systemic policy solutions that are developed by, with and for people within their environment.

When we put all the pieces together, the whole picture looks different. The issue of working in disciplines or within sectors is problematic not only because it leads to dishonest representations of reality, but also because it is not very helpful for problem-solving. The movements for connected thinking and social responsibility are discussed in this chapter and then some policy suggestions are made on the basis of this analysis. Social movements and integrated development are relevant in Alice Springs. The aim of this section is to describe a framework for policy and practice that is widely used nationally and internationally to guide best practice at the local government level. The Healthy Cities approach refers to built and natural environments as linked systems. The basic assumption is that the environmental and ecological context is integrated with the social and health context. It provides benchmarks for socio-cultural, political and economic development. These benchmarks are guidelines and local initiatives are encouraged to adapt the guidelines to suit local contexts. According to this approach all environments are understood to be arenas in which citizens need to participate in creating sustainable solutions that meet both social justice and long-term environmental concerns. The five key action areas for developing healthy settings within a healthy town, city or region are: '(a) healthy public policy, (b) supportive environments, (c) community action, (d) personal skills and (e) reorienting services'.[2] The approach has been implemented in Canada, America, Europe, parts of Africa and Australia (For instance, Noarlunga, Canberra and Illawarra) and its development is regarded as a social movement by participants in Australia and the Western Pacific Region. The challenge is to achieve harmony and balance. Life chances of citizens can be enhanced in multiple ways through inclusive, integrated policy and planning *by, with and for* people and a sustainable environment. The value of this integrated model is that it brings together research into citizenship (rights and responsibility) with public health research through the central concept of social capital, that is, a sense of well-being and trust based on having a real stake in the community. A real stake means participation and working together. In the civil arena[3] as defined by the World Health Organisation (WHO), a healthy city promotes:

1. A strong, mutually supportive and non-exploitative community.
2. A high degree of participation and control by the public over the decisions affecting their lives, health and well being.
3. Access to a wide variety of experiences and resources, with the chance for a wide variety of contacts, interaction and communication.

Such a role for citizens is active not passive, not merely as welfare recipients but as active shapers of their community. Building on the 1972 Stockholm Conference on the Human Environment, followed by the watershed Brundtlandt Commission Report in 1987 'Our common future', sustainable, systemic development is linked with social and environmental justice. The action occurs both outside organisations through social movements and through working with and within organisations. Ethical

[2] The original document stressed health services, but working across sectors and disciplines requires re-orienting all services. This has been stressed by Pat Mowbray in *Healthy Cities Illawara: Ten years on. A History of Healthy Cities Illawarra from 1987–1997*.

[3] Duhl, L. (2001) Systems and Service ISSS 45th Conference, Asilomar.

and sustainable justice concerns drive social movements that span sectors, so the limitations on activity usually caused by professional and organisational factors such as specialisation do not play a role. Social movements for self-determination and movements across space are part of an initiative to achieve greater self-determination (Coulehan 1997).

## *7.1 Social and Geographical Movement: Time, Space and Commodity—Exclusion as a Motivation for Land Rights*

Social movements are a means for bringing about social and environmental change based on wide ranging communication within and beyond formal organisations in the public, private and non-government sector. They too have the potential for both positive and negative change based on participatory design and governance versus the politics of social rights and reparation that is not balanced by personal responsibility. International movements for human and environmental rights influence social health movements for self-determination in Alice Springs. The local social movements for self-determination are defined in many ways by informants but with the common factors of striving to achieve an improvement in life chances and social health outcomes. Health, land and treaty (Pamphlet on National Treaty Support Group 2001) are part of a much wider Indigenous rights movement internationally:

> Land will continue to be an important priority, because land is crucial to our cultural and economic well-being.[4] Our land is our life. Since time began we have cared for our country and it has provided for us. We belong to the land. Our land is the source of our identity ...[5]

The mobile population of aboriginal people who move from remote areas to Alice to make use of its health services and to visit family members have needs that are currently not being adequately met. One of the areas of debate is the notion of definitions of citizenship as perceived by First Nations. Citizenship rights and responsibilities in these terms are re-defined to include links *across social and environmental goals*. Ties with the land are closely linked with a sense of health and well-being. This cultural value has been used as a political vehicle for resistance and social justice by Aboriginal people internationally. It is clear from discussions with

[4] 'Geoff Clark, the first fully elected chairperson of ATSIC' Koori Mail 12 January 2000.

[5] Central Land Council and Northern Land Council (1995) *Our Land, our Life. Aboriginal Land Rights in the Northern Territory.*

different interest groups amongst Aboriginal peoples that for some, rights within a Community Council Area are as important as recognition that they are first and foremost Indigenous Australians. The notion of a detribalised, deracialized citizenship (Rowse 1998: 210) based on liberal, capitalist notions of rights and responsibilities is perceived differently in the Indigenous land rights movements internationally.[6] Whilst for some Indigenous people self-determination is defined in narrow terms, for many citizenship can only be achieved if a Canadian type model is achieved[7] as follows: In democratic nations such as the United States, Canada and Australia, the rights to land continue to be debated in the form of First Nations versus colonisers. For example, the Sioux, the Inuit and Australian Aboriginal people emphasise that they are the caretakers of the land for their children and that their health is linked to the health of the land. The idea of a nation within a nation expressed in terms of land rights is not without precedent. In April 1999 the Inuit, Canada's Indigenous people obtained the right to self-government. The Inuit argued that they had achieved lower social health outcomes and that they would prefer their own place. In Nunavut they manage all their own affairs as a separate indigenous nation. This model could be of value for some Australian Aboriginal people who seek a separate identity of a nation within a nation based on communal rather than capitalist values. This provides a different option for citizenship, one that is equal but distinctive.[8] The New Zealand Waitangi Treaty is also under consideration as a potential local model. Self-determination[9] is defined very differently by different stakeholders ranging from government interpretations of

[6] The Meek shall inherit the earth. *Weekend Australian Review*, 3–4 April 1999.

[7] In April 1999 the Inuit, Canada's Indigenous people obtained the right to self-government after decades of negotiations. The Inuit suffered high rates of unemployment, high rates of youth suicide, high rates of alcohol and other drugs such as solvents, high rates of homelessness and faced life chances not dissimilar to those of Indigenous Australians. In Nunavut Territory they have a flag and a public government representative of local people who are 855 Inuit. They have a key say in the governance of the land and the people living there. A sense of ownership and responsibility is thus achieved as a separate indigenous nation. This model is understood by Australian Aboriginal people who seek a separate identity of 'a nation within a nation' based on communal rather than capitalist values. This provides a different option for citizenship, one that is equal but distinctive.

[8] *Weekend Australian Review*, 3 April 1999.

[9] 'Struggles for social justice and cultural autonomy by Indigenous Australians have constituted some of the most far-reaching challenges to the Australian

user pays (for services and rent) to rights spelled out in the Kalkaringi statement in August 1998 by the combined Aboriginal Nations of Central Australia at a constitutional convention, which details the concerns the Indigenous people have with the governance of the Northern Territory (NT) and specifically the move towards statehood. Striving for self-determination and a perceived sense of real citizenship can range from the struggle to become consumers in the user pays system to involvement in the landrights movement.[10]

> 'Voluntary regional agreements are starting to provide some indication of how the concerns of Aboriginal people to achieve better outcomes in terms of caring for people, caring for country and building sustainable regional Aboriginal economies can be pursued'. Such approaches could go some way to address

*Continued*

State. In the last twenty years Aborigines have gained official recognition as a people and support for self management and self determination policies. These apparent successes have resulted in an incorporation of indigenous communities and their politics into mainstream institutions in ways which can actually increase state supervision and threaten cultural independence. Partly this contradiction arises from the need to create peak bodies able to represent Aboriginal issues at the highest levels of government which run counter to the localised and land-based social networks which have enabled indigenous values to be maintained under welfare colonialism ...' This quotation exemplifies the notion of separateness through culture as a form of both resistance and persistence to use Keefe's (1988) terminology. The history of Aboriginality needs to be interpreted in terms of invasion, slaughter and inability to control labour, exclusion from citizenship, segregation and exclusion from property rights (Hollinsworth 1996).

[10] For some Indigenous people self-determination means first gaining further control over their lives in their own town camp area. For example, Mpwarnte (Abott's camp) would prefer to be declared 'a dry community' in addition to using a range of other means such as wardens, night patrol, truancy regulations and even the use of restricted legislation. This is a choice of action by a community; namely a group of people with shared interests is necessary in order to provide for a range of options. A community is based on self identification of shared interests, which may not necessarily be shared by others (even if they are extended family members). This point was argued at the Liquor Commission hearing for the extension of controls so that bringing in alcohol would become a criminal offence. The rights and responsibilities of the group to decide for itself were stressed. The decisions came from the camp leaders because they had incurred injury in attempting to keep the peace. It was argued by the police that if the area was declared to be a dry community then it would mean further criminalisation of behaviour and this could lead to more incarceration of Indigenous people.

the issue of the erosion of the Native Title Act through amendments under the current government post Mabo. Such steps could have particular relevance in addressing the concerns raised by Native Title.[11]

Indigenous social, cultural, political and economic concerns have been expressed through Native Title[12] Claims, The Alice Springs (Arrernte) Native Title Claim has successfully lodged The Land Act 1992, Amendments to the Pastoral Land Act 1992, Aboriginal Land Rights Act 1976 (NT), Sacred Sites Act 1989 and Aboriginal Heritage Act 1984 that is being amended. According to local informants in Alice Springs, the Reeves Report is received very differently by different interest groups. For some decentralisation of decision-making could empower local communities because there are more possibilities for self-determination in decision-making in local areas (e.g. town camps). For others, decentralisation is seen as a move towards disempowering the Indigenous nation-building process. There appears to be a need for both local control for health and development and also social movements for building a sense of national Indigenous identity by working across organisations.

National identity, citizenship and the Territory (statehood) do not always overlap. Nationality may be used to undermine or build states and some national identities may be recognised as full citizens and others not (Castells 1997) in many ways as indicated by a wide range of indicators of health, education, and incarceration rates, for example. The claim to crown land in and around Alice Springs has resulted in an acknowledgement by the NT Government to the continuing Arrernte connection to Alice Springs. To sum up, self-determination can be expressed in local and more general terms. The two are not mutually exclusive. The reality

[11] Howitt, R. (1998) Recognition, respect and reconciliation: steps towards decolonisation? *Australian Aboriginal Studies* 1 (29).

[12] Native title; an opportunity for understanding: proceedings of an induction course conducted by National Native Title Tribunal Areas to increase awareness of Aboriginal and Torres Strait islander and other cultural perspectives in the native title process. University of Western Australia, Nedlands 1–3 December 1994, edited by Frank McKeown and the Research Staff of the national Native Title Tribunal. According to Wilhelm (1999) '... many of the Reeves recommendations breach the Australian Constitution, international law and natural justice ... . A number of Reeves recommendations would result in an acquisition of property, which under the Australian constitution requires the payment of just term's compensation, he said. ... He questioned whether the Reeves report complied with the Racial Discrimination Act and the international Convention for the Elimination of All forms of Racial Discrimination'.

of Aboriginal politics (like all politics) is that interest groups (defined by family, language, access to land or some other criterion) compete for scarce resources and for using criteria to include and exclude. This means in practical terms that one language or descent group or a group with whom reciprocal links have been set up, or groups with a particular level of education are able to use some Aboriginal facilities and organisations but not others. Public space and facilities are divided in Alice Springs in terms of facilities that can be used by some groups and not others. Gender also plays a central role in the use and non-use of areas of significance.

### *7.1.1 The Context of Land Rights*

Closer links within and across all levels of government are required to improve service delivery. Coordination of organisations by the Central Land Council (CLC) is not without challenges too. For example, the CLC Annual Report (1997)[13] pointed out the problem associated with the implementation of the Aboriginal Councils and Associations Act 1976 currently reviewed by Aboriginal and Torres Strait Islander Commission (ATSIC). The CLC is concerned about the substance, implementation and administration of the act, and pointed out that it had raised these concerns ... over a number of years (CLC Annual Report 1998: 58).

Every activity of public and private sector organisations is constructed within the context of different stakeholders at local, Territory Level, Commonwealth level and international level who attribute different layers of meanings and values to the same issues and interventions. The Alice Springs (Arrernte) Native Title Claim has been lodged and judgement has been handed down.

From the point of view of the CLC,[14] the Northern Territory Land Corporation and NT Conservation Commission were said to be set up by a conservative NT Government to

> "combat land claims ... so land granted to them becomes alienated from the crown. However they are a statutory body and as such are not required to lodge company returns ... I see this as a dangerous institution of Government given that the NT is the only Gov. in Australia not to have any freedom of legislation information ... . The bipartisan push for statehood is a matter of deep concern to Aboriginal people. If the Land Rights Act came under the jurisdiction of Territory Government it could well spell disaster for Aboriginal Land owners." (CLC Annual Report 1997: 13). The Central Land Council 1998 cites Mckeon as follows.

[13] CLC (1997) Annual Report 1996–1997.

[14] Community Land Council Annual Report (1997).

"As a result of the common law recognition of native title and its protection and recognition through the Native Title Act, Aboriginal and Torres Strait Islander people are able to assert and express their relationship to country in the context of existing legal rights which are dependent upon the grace and favour of governments. And so in negotiating about native title, which is the primary function of the national Native Title Tribunal, we are doing business together in a way that places Aboriginal and Torres Strait islander people on a more equal footing than ever before … . Native title, in the legal sense, is an element of a broader range of issues between Aboriginal and Torres Strait Islander people and wider Australia. In recognising the importance of native title we must realise this. Native title is a legal concept which is declared and recognised by the common law of Australia, but it does not define the full relationship between Aboriginal and Torres Strait islander people and their country. It can be lost or extinguished … . Secondly, native title issues cannot be isolated from heritage issues, which may arise independently of the existence of native title, and of questions of legal extinguishment. Nor can they be isolated from questions relating to the use and protection of traditional country. In that context, we have already found in the course of mediation and discussion with Aboriginal and other people that the recognition of the traditional relationship to country independent of the existence or non existence of native title rights, is very important to meaningful negotiation—the sense in which one can say of a people, or of a group, that this is their country, according to traditional laws and customs.

If that recognition is made at an early stage in the negotiation process, it can be done without prejudice to questions of the precise boundaries and existence of native title. It can pave the way to a much improved relationship in terms of consultation between Aboriginal people and others, and recognition of their relationship to the land. Indeed we have already found in experiences, particularly for some local government authorities, that there is a readiness amongst Australians to embark upon that kind of recognition as a first step in the negotiation process … the fact that native title is an element of a wider range of issues, and cannot really be extracted neatly form them, points to the need to consider and develop intellectual framework for regional and local agreements and more comprehensive settlements of questions about land use and management. That is well developed in Canada, and it is possibly and probably … the direction in which Australia will move in the medium to longer term … . The legal concept of native title, the fact that it describes a legal right, gives rise to problems of communication and understanding. In summary the legal idea of native title is one which involves a right to communal origin or nature which can nevertheless give rise to individual rights or interests. Its content is determined according to the laws and customs of indigenous people with the relevant connection with the land. It does not matter that those laws or customs may have changes since European settlement, provided that the general nature of connection with the land remains. The kind of occupancy that supports native title does not necessarily import exclusive possession of the land. Native title is not precluded merely because more than one group may use the land … . There are unresolved legal issues concerning extinguishment, particularly the situation of pastoral leases, both with and without reservations, other forms of leasehold, and many forms of parks and reserves and so forth. Do they extinguish native title or to what extent does native title co-exist with these forms of grant? The content of native title at common law requires understanding of Aboriginal and Torres Strait Islander Law and traditions, and the extinguishment bring into sharp focus issues of non-Aboriginal property law." (McKeon 1998: 4–5).

The CLC report cites the role of the tribunal as a first step in negotiations; it is not a court.

"The concept of native title as it emerges from the common law declared by the High Court in Mabo requires exploration and fleshing out in the concrete context of specific applications. It also demands the application of new modes of dispute resolution in connection with the work of the national native Title Tribunal. It provides new opportunities for the development of an Australian experience and expertise in the understanding f native title and in the resolution of complex public disputes. It provides also an opportunity, through the process of negotiation, for more mature discourse between Aboriginal and non-Aboriginal Australians and associated with that a greater sense of our completeness as a nation ... (McKeon 1998: 243).

For the Indigenous people of Central Australia social movements and geographical movements are linked (Coulehan 1997: 6).[15] Rowse (in CLC 1998: 29–35) discusses the way in which Coombs criticized Stehlow's contention that culture is static and that in fact Aboriginality

[15] He describes the process as follows: 'From the 1960's and with increased momentum in the 1970's, Aboriginal people in the Northern Territory were taking steps to assert their right to choose where and how they lived. Aborigines began to "walk off" pastoral properties and to leave mission and government settlements to return to country and establish small disperses and more autonomous communities. By the 1970s, the exodus of Aboriginal people from the settlement sites to which they had been attracted or compelled by the agencies of mission and government intervention and mining and pastoral industries, had assumed the momentum of an Aboriginal social movement. The return of Aboriginal people to lands of social, cultural and economic significance has come to be called the "outstation" or "homelands" movement. This social movement was an Aboriginal initiative that preceded the advent of Aboriginal Land Rights legislation, which dates from the introduction of the federal Aboriginal Land Rights (Northern Territory) Act of 1976. ... While the outstation movement demonstrated Aboriginal affiliations to country and autonomy in patterns of movement and settlement, it ought not to be interpreted as an Aboriginal retreat from change nor as an intention to remain isolated in "country" Aboriginal groups have always insisted on their need for modern ... services ... in their remote homeland centres and outstations. Contemporary Aboriginal patterns of mobility and settlement reflect the desire of Aboriginal people to be more self-determining, whether in "country" or in urban centres such as Alice Springs'.

was adaptable as demonstrated by its survival. Land is one basis for political life and the other is self-determination. They are not mutually exclusive.[16]

The United Nations Commission on Human Rights Working Group on Indigenous Populations has sent delegations to consider indigenous

[16] Indigenous social, cultural, political and economic concerns have been expressed through Native Title Claims, The Alice Springs (Arrente) Native Title Claim that has been successfully lodged and a judgement has been made on native title. Native Title decisions are of particular relevance to Alice Springs. According to the Local Government Association (1999: 47–48):

*"Native title is the term used by the common law to recognize the pre-existing and continuing connection that Aboriginal and Torres Strait Islander peoples have in relation to land and waters.*

*Native title rights and interests are possessed under traditional rights of access, use, possession or occupation of land and water. Essentially native title rights and interests exist because of the traditional laws acknowledged and the customs observed by Aboriginal peoples and Torres Strait Islanders, and their connection with land or waters.*

*As the Justices of the High Court stated in their joint judgement in Fejo v the Northern Territory of Australia (1998) 156ALR 721:*

*Native Title has its origin in the traditional laws acknowledged and the customs observed by the indigenous people who possess the native title. Native title is neither an institution of the common law nor a form of common law tenure but it is recognized by the common law. The underlying existence of the traditional laws and customs is a necessary pre-requisite for native title but their existence is not a sufficient basis for recognizing native title.*

*The concept that Aboriginal and Torres Strait islander peoples' property rights pre-exist and survive the establishment of sovereignty in colonized land has existed in British common law for well over two centuries. Other former British colonies, such as New Zealand, Canada and the United States of America, have long recognized that two land tenure systems exist in their countries:*

- *the system introduced on colonization-from which freehold, leasehold and other titles arise; and*
- *a pre-existing indigenous system- from which indigenous property rights and interests derive.*

*In Australia, Aboriginal and Torres Strait Islander people's rights and intrests in land were not recognized until 1992 when the High Court delivered its historic judgement in the case of Mabo v the State of Queensland (no 2) (1992) 175 CLR.... The High Court recognised that Aboriginal and Torres Strait Islander peoples' native title had survived and that native title must be treated fairly before the law with other titles.*

*The common law recognises the existence of native title, and the native Title Act 1993(cth) protects and provides processes for dealing with native title rights and interests. ...."*

Justice Olney gave his reasons for judgement in the Alice Springs native title case (Hayes v Northern Territory) on 9th of September 1999. According to socio, political, cultural and economic issues in Alice Springs. According to Vadiveloo (1998) 'Mandatory sentencing has led "the Territory into the national and world-wide spotlight of organisations including Amnesty International and the Human Rights Commission"'.[17]

International attention has also been focused on changes to the Native Title Act by the UN Committee on the Elimination of Racial discrimination and found that the Commonwealth Government has discriminated against Indigenous peoples.[18] Stakeholders locally and internationally have interpreted land rights differently. One of the areas of debate is the notion of definitions of citizenship as perceived by First Nations. Citizenship rights and responsibilities in these terms are re-defined to include links across social and environmental goals. Ties with the land are closely linked with a sense of health and well-being. This cultural value has been used as a political vehicle for resistance by Aboriginal people nationally and internationally.[19]

*Continued*

Anderson (1999), to quote and summarise:

*"The application was brought by three Arrernte estate groups to 166 parcels of land and water within the municipal boundary of Alice Springs in central Australia.... The applicants sought a determination that they held native title rights and interests in the claim area which entitled them to possession, occupation, use and enjoyment of the lands and waters to the exclusion of all others. The Northern territory argued that native title within the claim area had been wholly extinguished by the grant of pastoral and other leases, reservations, public works and the grant of other inconsistent interests affecting the claim area.*

*Olney J did not accept fully the case put by either party. His Honour found that the applicants had native title with respect to the whole or part of approximately 113 of the 166 areas claimed. However, he concluded that the native title rights and interests of the applicants do not give them the right to possession, occupation, use and enjoyment of the claim area to the exclusion of all others. Although Olney J found that native title had been extinguished by some leases, reservations, and public works he rejected the Northern Territory's submission that pastoral leases granted both befor and adter the surrender of the territory to the Commonwealth n 1911 extinguished native title rights and interests....."*

The claim to crown land in and around Alice Springs has resulted in an acknowledgement by the NT Government to the continuing Arrernte connection to Alice Springs.

[17] *Alice Springs News*, 25 November 1998, 5(43).

[18] *Land Rights News*, March 1999, page 14.

[19] The notion of a detribalised, deracialized citizenship (Rowse 1998a: 210) based on liberal, capitalist notions of rights and responsibilities is perceived differently in the Indigenous land rights movements internationally. Whilst for some

The use of the social movement approach for change through networking locally, nationally and internationally is a vital tool for learning and for self-determination (that does not lapse into a limiting form of nationalism), but poverty has excluded most Indigenous people from communication tools. Indigenous citizens have overall minimal literacy and numeracy levels not to mention little opportunity to access technology. People living on the town camps in Alice Springs struggle even to find a functional telephone and have to walk to access most of the services because transport both private and public is not readily accessible. The role of schools, libraries and youth centres in creating a culture of learning is vital to enable young people to learn and to use communication technology such as the Internet for development, if they are not to fall further behind in the development stakes (due to lack of access to public space and cyberspace). Negroponte (1995) stressed that the digital age is one that will bring changes in all areas of life and recently he has addressed the gap between the haves and have-nots by setting up computer spaces in Cambodia to provide educational opportunities for young people with limited life chances. Libraries, schools, youth clubs and other suitable public spaces could be adapted into knowledge centres as free public spaces that could be used as a means of enhancing literacy, learning and recreation in an environment that meets the cultural needs of the users. These could become starting points or portals for acknowledging common denominators and bridging the differences by creating 'transcultural webs of meaning' (McIntyre-Mills 2000) across diverse paradigms of knowing.

Recognizing the personal knowledge of people is important in education, health and employment, in fact all areas of governance. Taking ideas without acknowledging them and calling them by a 'fancy name' and then imposing the rhetoric of the idea whilst filleting out the guts is just another form of colonization, another form of exerting power over the other. It is perceived as a betrayal and a reason to distrust intimacy that

*Continued*

Indigenous people self-determination is defined in broad social justice terms, for others citizenship rights are linked with land rights. In democratic nations such as the United States, Canada and Australia, the rights to land continue to be debated (Weekend Australian Review 3 April 1999). For example, the Sioux, the Inuit and Australian Aboriginal people emphasise that they are the caretakers of the land for their children and that their health is linked to the health of the land. Parts of Canada, such as a municipal area outside Montreal is run in ways determined by Canada's Indigenous people according to a guided tour whilst participating in the International Sociological Association Conference in Montreal in 1998. A sense of ownership and responsibility is thus achieved to manage resources wisely as a nation within a nation.

can be used to control and devalue another. The anger that people feel is either directed outwards as violence or inwards as depression. The statistics in this study present a systemic picture to complement the qualitative data. It is clear that the answer to the problem of poverty lies in the way it is defined and who defines it and in whose interests.

## 7.2 *The Potential of Social Capital for Inclusive Governance*

It is debatable whether social capital is a useful term; perhaps it is not as useful as wellbeing, a much broader term that resonates with Indigenous ideas about spirituality and oneness with the land. Well-being is also more distinct from economic terminology and it is broader, which brings both advantages and disadvantages from a researcher's point of view. Nevertheless the term social capital is not necessarily very meaningful to participants who have particular attitudes to capital and commodity. Also, according to White (2002), the sense in which social capital has been used has differed quite widely from one social analyst to another. Some have paid scant attention to power and have placed emphasis on trust (Cox 1995) and access to networks without much attention to the issue of power and that increased participation needs to consider the power dynamics of the participants. Pierre Bourdieu discusses that cultural capital and social capital in relation to power can open the dialogue.[20]

In this section it is argued that the concept of social capital is relevant to thinking, practice and policy in sofar as it can contribute to:

(1) understanding poverty in socio-cultural, political and economic terms, that is systemically;
(2) building citizenship rights and responsibilities contextually;
(3) addressing reconciliation and human dignity;
(4) improving access in terms of information, communication, attitude and infrastructure;
(5) extending crime prevention opportunities; and
(6) enhancing social health and well-being.

### *7.2.1 Understanding Poverty in Individual, Social, Political and Economic Terms*

By thinking about poverty not only in *individual* but also in *social, political and economic* terms the focus is on what both society and individual stakeholders can do to address challenges in sectors such as housing, health,

[20] See Bourdieu, P. and Wacquant, L.J.D. 1992, White 2002.

education, employment, and policing. The individual exercise of self-control and personal liability needs to be considered alongside a sense of social responsibility. Both individual and social responses are necessary for problem-solving. A sense of reciprocal rights and responsibility is the basis for developing a shared future for all Australians.[21] The value of critical systems theory is that it strives to counterbalance individualisation policies towards issues such as poor health, unemployment and crime (associated with poverty) with policies that focus on both individual and co-created social responsibility.

The following addresses some aspects of the policy environment within which development takes place. An important NT social responsibility approach is the Public Health Policy that advocates primary health care and prevention, rather than a curative approach. It also stresses a cross-organisational and integrated approach by encouraging participation in working across sectors (education, health, housing etc.) and disciplines (engineering, teaching, nursing, sociology etc.). This approach can more effectively address interrelated social issues, thus enhancing life chances and quality of life. The impact of devolution of services and privatisation could however impact on achieving this goal; for instance, the following policies make the implementation of the public health policy challenging. Privatisation involves shifting the responsibility of government for aspects of service delivery to non-government organisations and the business sector. The NT Government's strategy 'Planning for Growth was translated in terms of some proposed privatisation of public services. The following policies reflect the individualisation of responsibility'. In the NT the impacts of cut backs in the welfare state and the tendency to shift responsibility from government back to the individual has particular ramifications in a population with diverse life chances. Contractualism is the legal contract between service providers and service users. Contracting out means that organisations award contracts on tender to individuals and other organisations to provide specific services. The organisation manages the contract and the contractee carries out the contracted service. Privatisation has an impact on human services, community development and community governance. Michael Jones[22] described the language and meaning of development changing because 'privatisation', 'market testing' and 'contracting out' are commercial management

[21] Positive and negative welfare and Australia's Indigenous communities *Australian Institute of Family Studies in Family Matters* (1999) 54 (Springs/Summer).

[22] One of the keynote speakers at the Winds of Change Conference for Local Government (hosted and run by the local government Community Services Association of South Australia) held in 1995 in Adelaide.

methods, which are applied at local government level. He warned that in pursuing efficiencies, effectiveness should not be lost. Jeff Tate warned that if contractualism were taken too far it would 'change the relationship between citizens and their local government to a relationship of service provider and consumers'. This limits the relationship merely to a legal contract and diminishes the sense of wider responsibility to democratic values and to the building of social capital. The shift in thinking about citizenship to thinking about consumerism or customership needs to be redressed if governance is not to become merely a business contract.

In this context of the shrinking welfare state, contracts or legal arrangements can begin to replace rights. In the move away from universal welfare rights, residual, limited or at best, qualified rights have been emphasised. This is driven by economic rationalism in response to the need to become competitive in global markets (created by removing tariff barriers and by the information and Internet age). This means shifting spending from welfare to spending on ways to increase economic competition through increasing an emphasis on user pays, through an emphasis on the economic, rather than the social. The social and the economic aspects of society are however linked. Social, cultural, economic and environmental capital need to be considered, because the long-term impact on life chances could also impact on economic well-being. Ensuring, for example, that school retention rates are improved, employment rates are high and crimes rates are low ensures social and economic well-being in the long term, because democratic and stable societies are attractive to investors. The user pays notion of citizenship responsibility needs to be seen within the context of the shifts from universal welfare to residual welfare policy and the shift to conceptualising citizens as consumers first and citizens second. The withdrawal of ATSIC funding a few years ago has not made a great impact on camp residents until recently, because Tangentyere has paid for maintenance and repairs. But because at the time of the research some Housing Association Committees are in debt as a result of the high level of damage and poor maintenance, there is difficulty in moving to the desired financial independence expected by the Territory Government.

In some instances residents of town camps appear to have little understanding of its implications for their monthly budget. This is an example of the dissonance across a *capitalist culture*, which stresses individual responsibility and a *transitional culture* that believes that responsibility to family ought to be limited to a certain extent and a *traditional culture*[23] that stresses family and community responsibility.

[23] According to Rowse (1998b: 210–211): 'The liberal notion of "the citizen"—encumbered neither by Indigenous tradition nor by colonial forms of patronage and subjection—was associated with the emergence of labour power as

### *7.2.2 Building Citizenship Rights and Responsibilities Through Addressing Governmentality*

Social capital can be measured in a geographical community by means of demographic, social, political and economic indicators.[24] Qualitative constructs of what social capital means and if in fact it is meaningful need to be considered by all stakeholders. At the end of this research process, some Indigenous informants stressed that social well-being and spirituality would be better concepts as far as they were concerned. Nevertheless in crude terms it can be argued that in a geographical community where there are high levels of social well-being the quality of life is indicated in terms of positive health status outcomes, high levels of employment, high educational outcomes, representation of the diversity of the population in key organisations and low crime rates. Public spaces (parks, pavements and facilities) are regarded as places where citizens are welcome and safe (irrespective of age, gender, cultural background or level of physical and mental ability). As detailed in previous chapters, where citizens feel marginalised the opposite is the case.

Bearing in mind this human rights environment, the National General Assembly of Local Government stressed at the 1996 General Assembly the role of local and community government as fostering 'harmonious relations amongst all Australians' in accordance with the Statement on Community Tolerance:

> The National General Assembly of Local Government: Reaffirms its commitment to the right of all Australians to enjoy equal rights and be treated

*Continued*

a commodity whose sale guarantees the seller's economic autonomy from welfare dependency. This project of emancipation consigned to the past (or some secondary psychological realm) the individual's affections for place for kin beyond the nuclear family/household. This vision of Indigenous people's progress to modernity therefore induced in practitioners of assimilation a tendency to overlook or understate the continuing bonds of kin and country'. *Through Tangentyere Council* 'A space was starting to open for indigenous people to articulate the distinctive forms of their urbanity and modernity … . The work of Tangentyere Council since 1978—developing custom-designed houses and town camps—has made it possible to write in more positive terms about cultural differences in ways of living in a town such as Alice Springs' (Rowse 1998b: 203).

[24] WHO (1988a) Promoting health in the urban context. *Healthy Cities Project Paper No. 1*, WHO Healthy Cities Project Office, Copenhagen. WHO (1988b) A guide to assessing healthy cities. *Healthy Cities Project Paper No. 3*, WHO Healthy Cities Project Office, Copenhagen. Davies, J. and Kelly, M. (1993) *Healthy Cities: Research and Practice*. Routledge, London.

> with equal respect, regardless of race, colour, creed or origin. Reaffirms its commitment to maintaining an immigration policy wholly non-discriminatory on ground of race, colour, creed or origin. Reaffirms its commitment to the processes of reconciliation with Aboriginal and Torres Strait Islander people, in the context of redressing their profound social and economic disadvantage. Reaffirms its commitment to maintaining Australia as a culturally diverse, tolerant and open society, united by an overriding commitment to our nation, and its democratic institutions and values. Denounces racial intolerance in any form as incompatible with the kind of society we are and want to be. Further, this National Assembly calls upon Councils throughout Australia to give practical effect to the above commitment by actively promoting the benefits of a cohesive, multicultural society. Supporting the Council of Aboriginal Reconciliation's Vision for a united Australia, and local declarations of Reconciliation[25] with our (*sic*) Indigenous peoples. Promoting access and equity in service provision for all members of their communities. Addressing wherever possible the special needs of disadvantaged groups.[26]

At the 103rd Annual Conference of the Local Government Association of Queensland, Raynor said that Australia needed to resist becoming

> 'laid back about the value of democracy which we tend to take for granted' and pointed out the valuable role which local and community government can play in preserving grass roots democracy ... . Local government is the best evidence we have that we are committed to democracy. She said its closeness to the daily lives of people gives it responsibility for upholding democratic ideals. She urged delegates to remember that they are a government not just an efficient provider of services. Local Government is about much more than good management. 'It is not just a matter of providing value for service and not just about consultation. It is about encouraging participation ... . Professor Donald Horne said the theme of the conference got down to one of the great cornerstones of a modern, liberal, democratic society ... he urged Local Government to work to enrich the social and cultural life of citizens and to speak up on national issues particularly where the national government seems to fall down'.[27]

Despite the ambivalence or cynicism expressed by survey respondents and informants in Alice Springs, about the extent to which local and

[25] The way in which reconciliation will address the right to self-determination through councils has not been spelt out in any detail because there is a sense that local solutions need to be worked out not only regionally with an acceptance of the need to balance the needs of local councils for decentralised self-determination, but also with the Central Land Council as a centralised organisation to represent their concerns.

[26] Australian Local Government—Association: promoting access and equity in local government—services for all March 1999.

[27] Strong local government makes democracy work. (October 1999) *Local Government Focus* 4(9).

community government could make a difference, it was stressed that more could be achieved if more people and organisations were prepared to work together across sectors for more integrated or holistic interventions.

### *7.2.3 Addressing Reconciliation and Human Dignity*

Informants stressed that the meaning of social rights[28] and social responsibility is tied to a historical and cultural context. The differences in life chances across the population in Alice Springs was made evident by the survey population, Australian Bureau of Statistics (ABS) and the Australian Institute of Health and Welfare Reports. One of the key indications of the lack of self-respect and dignity are the mortality and morbidity rates associated with alcohol, violence and social marginalisation. The issue of alcohol in town camps[29] needs to be considered in a discussion of building social capital.[30] One of the ways of building social capital

[28] Rights are spelt out in commonwealth and state legislation, for example. The Commonwealth Disability Discrimination Act of 1992 that has been drawn up within a wider legal and policy context, for instance, The Equal Opportunity Legislation of 1984 that addresses the different gender issues faced by men and women as workers, The Racial Discrimination Act 1975, The NT Anti-discrimination Act 1993, The UN Committee on the Elimination of Racial Discrimination (to which Australia is a signatory nation and required to report every two years), The Commonwealth Regulation Impact Statement (RIS) on Draft Disability Standards for Accessible Public Transport Standards of the Anti Discrimination Act and Accessible Public Infrastructure introduced in 1996 and endorsed in 1999.

[29] For instance the request by the members of Mpwarnte Camp to 'remain dry' with the assistance of the police.

[30] The fencing of the camp and its award as a tidy camp were cited as ways in which the community attempted to delineate itself as a safe community of people with shared interests. Despite having extended family who believed they had a right to enter the camp whenever they pleased, which is culturally appropriate. It was requested that when people (including extended family) are violent and disorderly as a result of alcohol, they obtain police assistance to maintain the area as a dry community. The Housing Association with the assistance of Tangentyere Council staff gave witness to the fact that alcohol disrupts the way of life of the community and makes it difficult for children to sleep and to study. It contributes to family violence and the modelling of violent behaviour to young children, which has a negative impact on mental and physical well-being of the community. Whilst it is appreciated that the police do not want to extend their powers, the argument was that the measures that they already have in place, such as wardens and night patrol, require yet another level of intervention.

is by building citizenship rights; the other is through building a *sense of responsibility* through giving people a stake in society. This is achieved through greater participation in decision-making across all the sectors. Both can be considered as vehicles for development. Communities are created out of separate interest groups as a result of a sense of shared meanings and shared life chances. Where there is a high level of social capital across the community, the life chances of different interest group are less likely to differ dramatically and the life chances can be assessed in terms of freely available data. The interest groups are more likely to share a sense of place and share a sense of being a community.

Although some initiatives have been undertaken to progress reconciliation, informants stressed that reconciliation would have more meaning locally when some of the pressing life chance concerns are addressed. Indigenous people have had the vote since 1967 but sections of traditional and transitional population perceive that they are still second class citizens as far as the wider population of Alice Springs is concerned. Also successful Indigenous capitalists in Alice Springs are regarded with some ambivalence as evident in conversations about the need for this group to develop social responsibility towards those who are less successful.

The 'cash work nexus' (Pixley 1993) is central to definitions of citizenship. The majority of Aboriginal people have difficulty in obtaining work opportunities. This was particularly so after obtaining citizenship and the notion of minimum wage associated with citizenship led to many pastoral workers becoming unemployed. Employment for young people and the disabled (physical and intellectual) needs to be given attention. Historically those who received welfare were excluded from voting, which also effectively included Aboriginal people. The current circumstances of people reflect this historical legacy.

### 7.2.4 *Improving Access in Terms of Information, Communication, Attitude and Infrastructure*

The Guide to the DDA Act of 1992 stresses that:

> The Act uses a very broad definition of disability and covers disabilities that are physical, intellectual, psychiatric, sensory and neurological. It also covers physical disfigurement and the presence in the body of an organism capable of causing disease, such as HIV ... . People with disabilities from Aboriginal, Torres Strait Islander and non-English speaking backgrounds often encounter additional barriers in attempting to access services. Access is impeded not only because of the potential client's disability, but also because services are not offered in a way which is culturally and/or linguistically appropriate ... . The act follows the broad definitions of disability mentioned above i.e.: intellectual, psychiatric sensory and neurological disabilities. It covers physical

> disfigurement and the presence of organisms, which can cause disease such as HIV that can in most cases lead to AIDS … . The Act includes people with current, future and past disabilities and it renders discrimination against a person whose associates have a disability. Both direct and indirect discrimination is covered by the 1992 Act.

According to the Disability Discrimination Act of 1992 access is defined in terms of physical aspects, communication aspects, and information and attitudinal aspects.

> Australia is a party to a number of international instruments that require it to uphold the basic human rights of all Australians. Many of these human rights have particular relevance to people with a disability. They include the right to respect for human dignity and freedom, equality before law, privacy, and protection against discrimination and equal opportunity in employment.[31] Australia can discharge its international obligations and responsibilities by legislative reform or by Executive action.[32]

The specific cultural and historical context of colonisation and immigration need to be born in mind as particularly relevant to addressing communication, information and attitudinal aspects of access.

### *7.2.5 Extending Pathways for Prevention*

Building *crime prevention* measures in Alice Springs in terms of the pathways to prevention approach could enhance the life chances of young people in Alice Springs, who are our future. This approach is a positive and proactive form of social intervention, which will not necessarily be more costly to implement than the current system of mandatory sentencing and Zero Tolerance policing[33] given the cost of mandatory sentencing to the community is in excess of $54,000 per prisoner annually: the social

[31] International Covenant on Civil and Political rights; United Nations Universal Declaration of Human Rights; International Convention on the elimination of all Forms of Racial Discrimination, Convention on the Elimination of all Forms of discrimination Against Women, International labour Organisation No 111 Discrimination (Employment and Occupation) Convention.

[32] According to the Australian Law Reform Commission No. 79 (1996) cited in *Making Rights Count: Services for People with a Disability: New Disability Services Legislation for Commonwealth*. Australian Government Publishing Service.

[33] Stone, S. (1998) Ministerial statement reclaiming the streets: Zero Tolerance policing and the Northern Territory, August. The policy of Zero Tolerance policing introduced in August 1998 can be summarised as being introduced to target specific 'problem areas'. It follows the 'broken window' philosophy; namely that if there is an opening people will take it. Once people realise that there are no openings the crime statistics will decrease.

and economic impacts of mandatory sentencing. Mandatory sentencing was perceived to have limited benefit to the community, because the costs in social terms are too high.

According to Malcolm Fraser:

> ... It is contrary to the principles of natural justice; it is contrary to the long established principles of the Common Law whose purpose has been to allow courts full discretion to take into account all the circumstances of a case. To deny the courts that discretion is to deny justice. It denies the need for a sentence to be proportional to the crime. The Common Law does not sanction arbitrary detention; neither does it sanction preventative detention. It does not accept excessive periods of detention for the sole purpose of protecting the community from repeat offenders ... . Mandatory sentencing of the kinds we have seen, represents a reversion to old practices which had been largely put aside since the last decades of the 18th century.[34]

Specifically mandatory sentencing was criticised by young informants with a wide range of life chances and age groups and by a wide range of service providers and members of the legal profession, because it traumatises young people and erodes their faith in society and in themselves at an impressionable stage in life when young people the world over are at their most rebellious. It was stressed that it limited the professional autonomy of the legal profession, that it breached international human rights legislation and that it has little impact on crime prevention.[35] According to

[34] Sunday, 5 March 2000.

[35] According to the Northern Territories Correctional Services Annual Report 1997/1998, 24: 'Aboriginal people continue to be over-represented in custody. On 30th June 1998, 72.6% of adult prisoners in the NT were Aboriginal. The rate of imprisonment of Aboriginal people is 1460 per 100,000 adult Aboriginal population. When comparing the percentage of adult Aboriginal (17 years and over) in custody to those in the wider community, the NT compares favourably to other jurisdictions. For instance, adult Aboriginal People comprise 22.5% of the NT adult population and 72.6% of the prison population; this is then 3.2 times their proportion in the wider community. By way of comparison, adult Aboriginal comprise 1.2% of the total NSW population yet comprise 14.5% of those imprisoned. This is almost 12 times their proportion in the wider community. If each jurisdiction is examined in terms of the number of Aboriginal people in custody compared to the numbers residing in each population, the NT actually has the lowest rate of Aboriginal incarceration'. It needs to be noted that the rate at which Aboriginal people are held in 'safe custody' could contribute to this figure; also the tendency to organise home detentions and alternatives for juveniles. According to the Northern Territories Correctional Services Annual Report 1997/1998, 21: 'The design capacity of the Alice Springs Correctional Services Centre is 400 (including low security cottages) and the average number

human service providers working with young people in Alice Springs, the Mandatory sentencing law[36] in its current form has received considerable attention from human rights, equal opportunity and community groups because of its impact on the life chances of young citizens. Informants stressed that crime preventative measures could address *the social* as well as *the individual* context of crime through building more programs to address family violence, youth unemployment, truancy, poor school retention rates, a sense of self-esteem, self-control and optimism about the future.

### *7.2.6 Enhancing Social Health and Well-being*

Higher mortality and morbidity rates in Alice Springs and Northern Territory discussed above are outcomes associated with violence and road deaths as well as diseases directly and indirectly linked with substance abuse and the associated poor nutrition (as a result of spending money on alcohol and other drugs as well as the unavailability and cost of food in remote communities). Substance misuse is an effect and a cause of social ills. A major environmental hazard is the lack of safety as a result of extensive violence associated with alcohol abuse. The damage to housing stock and the overall quality of an environment is linked with alcohol abuse.

*Continued*

of prisoners held was 294 (minimum 253 and maximum 349). This compares with a daily yearly average of 283 for the 1996/97 financial year …'.

Alice Springs Correctional Centre has become the Territory's major maximum security facility for long term prisoner's. Probation according to CS refers to the court's order for supervision as part of an 'Order for Release on Bond' or 'Order Suspending Sentence'. Probation orders for Juvenile offenders for the Southern region show that from 1994 to 30th June 1998 the trend has been downwards from over 60 to less than 30. Community Service Orders as on the 30th of June show that there has been a move from 40 in 1995, 45 in 1996, to less than 15 in 1997 and less than 10 in 1998. Juveniles in detention in the Southern region as on 30th of June 1998 1 in remand and 1 in detention compared with 3 in remand and 3 in detention in 1996. According to "The National Aboriginal and Torres Strait Islander Survey" conducted in 1994 10.2% (3200) of Indigenous persons aged 13 years and over in the Northern territory had been physically attacked or verbally threatened in the 12 months preceding the interview in Alice Springs. The old maximum-security goal has been decommissioned and another built 23 kilometres from Alice Springs. The design was to meet the Royal Commission into Aboriginal Deaths in Custody. Aboriginal people continue to be over-represented in NT's prisons, accounting for 72.6% in 1994–1995.

[36] The concept of mandatory sentencing was introduced in the NT in November 1996 and on 8 March 1997 the amendments to the NT Sentencing Act came into effect.

The social movement to address social ills associated with alcohol has existed in Alice Springs and the region for many years. An awareness of the problem and its social impact led to a protest march led by Aboriginal women to mourn the effects of alcohol and to demand an end to the soaring rate of rape and murder'[37] For a decade people including Indigenous leaders have lobbied for a range of measures to address the problems associated with alcohol misuse. This is indicative that social capital exists across the community.

According to a resident:

> In late 1989, some senior Aboriginal men from Central Australia asked the then Health and Community Services Minister, Steve Hatton that his female ministerial officer become an advocate for Aboriginal women to voice their needs/ concerns to government. As a result of this request, I travelled throughout Central Australia, listening to these women. Overwhelmingly, the main concern was the effect alcohol was having on Aboriginal life; including violence, child rearing, education, etc. In April 1990, a group of women from Ntaria asked that NT ministers attend a march—the women wanted to protest to government about the effect alcohol was having on their lives. This happened on 5th May 1990. At least 500 people from remote communities marched, over 200 in ceremonial paint, including the most senior men. As a result of this, the then Chief Minister established the Living With Alcohol Program, with a levy put on alcohol ... to raise money for alcohol reform. Even before this time it has been Aboriginal organisations that have led the debate on this issue. Any ongoing reforms that can assist groups retrieving some quality of life needs to be supported fully by government.

We need to be interactive not merely proactive, to use Banathy's (1991) concept that distinguishes between looking to the future in order to predict and control the way we manage our organisations. We should try to co-create our futures. Our assumptions and values play a role in the way we see time and space, the way we perceive knowledge and its role and the way we perceive that it should be applied in management, governance, decision, policy and action across sectors.

The notion of self-determination is defined differently by different stakeholders: (i) standing on one's own feet and getting off welfare is a goal that is defined in terms of a particular world view. To understand why it is necessary or desirable requires a certain frame of mind, a certain capitalistic understanding of economics and finances. (ii) Self determination from the point of view of some indigenous groups in Alice Springs is the right to declare a locality a dry zone and to concentrate on creating

[37] Alice a town without pity. *The Sunday Age* 3 June 1990 highlights the concerns expressed a decade ago.

a safe space for family whilst keeping out some family who do not share their vision. (iii) For some the issue and the vision is broader and hinges on obtaining land rights as a vehicle for socio-cultural and political progress. The three areas are not mutually exclusive. For example, Mpwarnte (Abott's camp) would prefer to be declared 'a dry community' in addition to using a range of other means such as wardens, night patrol, truancy regulations and even the use of restricted legislation. This choice of action by a community (namely a group of people with family ties and shared interests) was considered necessary for the well-being of the residents. Their application to the Liquor Commission was not upheld. Two years after the request was rejected, the community has spiralled from a motivated community that won the 'Tidy Camp Award' to a dysfunctional community. The decisions came from the camp leaders because they had incurred injury in attempting to keep the peace. It was argued by the police that if the area was declared to be a dry community then it would mean further criminalisation of behaviour and this could lead to more incarceration of Indigenous people. Tangentyere is the peak organisation for a group of housing associations that represent the town camps. Each housing association has its own Incorporation Act and can be used as a means for determining group norms and camp rules. A community is based on self-identification of shared interests, which may not necessarily be shared by others (even if they are extended family members). This point was argued at the Liquor Commission hearing for the extension of controls so that bringing in alcohol would become a criminal offence. The rights and responsibilities of the group to decide for itself were stressed.

The balance of power between local councils and overarching land councils needs to be maintained despite the suggestions in the Reeves Report that the central councils should be divided to decentralise power. Centralised power through the land councils as expressed in the Kalkaringi Report is considered essential for self-determination as a group of people. This quest for landrights is linked with the limited extent to which Indigenous people achieved equitable citizenship rights in terms of socio-political and -economic indicators such as language and cultural rights, educational opportunities, health status, crime and freedom of movement.

The role of dependency on the social wage is one that defines the current situation of many Aboriginal people. Rowse (1998)[38] outlines the way in which dependency was created; rations were at one level a means

[38] Rowse, T. *White Flour: White Power: From Rations to Citizenship*. Cambridge University press.

to 'civilise' at another 'a means to control' and 'manage aboriginal people'.[39] At both levels they were a top-down approach to development, the agenda was certainly not of their own writing. If one sees 'the gift'[40] as a means to obtain reciprocal gains, namely compliance or access to land, then it implies that the power relationship was equal. Rowse (1998b: 40) emphasises that 'Effort and reward' are however well understood in traditional societies and this was effectively undermined by the system of rations without work. Taking away the right to self-respect by taking away their responsibility to address their needs and not to be dependent on others therefore creates dependency—working in harmony with the environment, rather than conquering the environment (such as Cromwell Drive along the Golf Course, with watered lawns, despite the aridity of the environment).

The major problem is that Indigenous citizenship rights and responsibility have been at the level of rhetoric and the real life of many Indigenous people is still one based on survival. A culture of poverty limits expectations and limits the belief that it is possible to bring about change in the world. Horizons for many are day to day, hand to mouth existence, rather than planning ahead and believing that the rhetoric of rights applies to individuals in their own private lives and public lives. Similarly the rhetoric of responsibility means that it needs to be applied in their private and public lives.

Citizenship and governance are flipsides of the same coin. But these concepts are meaningless to colonised people who have been welfare

[39] The movement of people to sources of food is traditional and the movement to missions and pastoral stations in pursuit of food made sense, albeit the acceptance of rations was prompted by ambivalence: desire for commodities and a fear of the new forms of power that they confronted which undermined relations between kin and led to a destruction of a way of life.

[40] Marcel Mauss 'The gift' White flour, beads, sugar was the trade for white power. This is the exchange that underpinned much of the 'trade' by colonists. But moving along to this day, the relevance of the exchange remains, the suggestion in the Alice in Ten that aboriginal people should collect their social wage away from the town centre in their remote communities, once again echoes the notion that the rights as citizens are not so great for those who do not earn a cash wage themselves. The dependency relationship on commodities such as alcohol is another aspect of the issue. In a communal society the notion of gift is quite different. Possessions are owned communally and it is a right and a responsibility to receive and share commodities. The notion of gratitude is not the same. The notion of receiving money as a social wage or the dole is perhaps seen as simply part of the natural scheme of things. The concept of property is also quite different in traditional communities.

subjects and wards of the state, whose movement has been controlled and whose very family structure was often not of their own choosing. Generations have experienced the impact of state power and shifting towards another way of thinking and practising requires an act of faith in oneself and in the potential of governance structures to represent the interests of Indigenous people. Mistrust is based on a legacy of colonisation and remaking the future through participatory design and governance has never been possible. Leaders (Indigenous and Non-Indigenous) are regarded with suspicion, because of the discourses on the legacy of co-option and corruption

Social rights need to be balanced with social responsibility at personal and at a public level. It requires owning issues and not projecting all the blame outwards (even though it is historically warranted). Self-determination[41] needs to take an additional step to create workable self-governance at a community level and to limit the current over-servicing and over-dependency. This does little to advance the cause of Aboriginal people. Hollinsworth quotes Tatz as saying: 'Without doubt, Aborigines are per capita the most over-administered minority anywhere in the world'(1972: 101 in Hollinsworth 1996: 119). The challenge is to move away from being administered to being able to apply self-governance principles at community level, based on dignity and respect irrespective of age, gender or language.

## *7.3 Development Approaches to Enhance the Life Chances of Young People and their Families*

The discourses about youth and crime could be addressed through approaches that stress citizenship rights and responsibilities irrespective of age, gender, culture or level of ability. For this to occur, a sense of respect from adults is required plus an understanding of the difficulties faced by young people attempting to come to terms not only with adolescence but also with a non-Indigenous capitalist culture that requires standards of behaviour that are quite different from traditional ways of life. Young people grow up fast. Risk-taking behaviour without worrying about the consequences (such as binge drinking and using inhalants)

[41] 'Self-determination, land rights and cultural autonomy emerged as the central platforms of this more strident Aboriginal politics. In street demonstrations and student organisations there was widespread non-Aboriginal support as indigenous activists aligned themselves with international liberation movements ...' (Hollinsworth 1996: 117).

is prevalent amongst all young people and Indigenous youth are no exception. According to Bowden (1994: 27) the life chances of Indigenous people can only be understood in terms of understanding:

> Aboriginal cultural survival strategies of passive resistance, separate identity and varied levels of integration to balanda culture. The choices are difficult to make. As young people have to choose between alienation from their peers if they become too integrated in western, capitalistic culture. Young people in particular have a very difficult stage in their lives when they rebel not only against their elders and their systems of Aboriginal law but also against the authority of a wider culture of which they have limited understanding. Young people from outlying camps who decide to rebel and come to town meet another culture and legal system head on. Aboriginal adolescents growing up required to learn to negotiate the rules of two very different cultural systems that in many ways have very different values, because one is capitalistic and the other egalitarian ...

At the time of the research, mandatory sentencing was linked with the high rates of imprisonment in the NT. According to the Criminal Law Editorial: 'Before the legislation came into effect, the NT had the highest rate of imprisonment of any jurisdiction in Australia, some three times more, than the State with the next highest rate of imprisonment, Western Australia'. As on March 1997, in the NT the average daily prisoner population per 100,000 population was 305.7.[42] A review of the literature by Schetzet in 1998 indicates that some studies locally and overseas demonstrate that tougher sentencing systems do not have a major impact on crime rates.[43, 44] According to Schetzet (1998):

> ... Whilst there are no available figures on the total number of people sentenced to mandatory periods of imprisonment to date, the rate of imprisonment in the NT has clearly increased since mandatory sentencing laws were introduced. By the end of 1997, the NT prisoner population per 100 000 population had increased to 435.2—an increase in excess of 42% since mandatory sentencing was introduced.[45]

The NT has the highest ratio of police per population and the highest rate of imprisonment in Australia. It also has the highest suicide rates

[42] Darwin Community Legal Service, Mandatory sentencing: an information kit, August 1997, page 7.

[43] Harding R. (1993) Opportunity costs: alternative strategies for the prevention and control of juvenile crime, in Harding, R., Repeat Juvenile Offenders: The Failure of Selective Incapacitation in Western Australia. Research Report No. 10, University of Western Australia, Crime Research Centre, page 141.

[44] Cain, M. (1996) Recidivism of juvenile offenders in NSW, Department of Juvenile Justice quoted in Darwin Community Legal Service, page 6.

[45] Australian Bureau of Statistics, National figures on crime and punishment, *Year Book Australia 1998.*

nationally. The statistics show that youth unemployment is linked with youth crime.

The challenges faced by service providers, according to the Youth Justice Coalition (1999–2000) include the need for addressing community development as a means of crime prevention instead of resorting to mandatory sentencing. They stressed the lack of recreational options (other than sports) alongside the lack of accessible public transport at night as issues. Others also raised the need for more opportunities for involvement in decision making, such as The Chief Ministers' Round Table of Young Territorians[46] and youth service providers confirmed this. Service providers at the Central Australian Aboriginal Congress (31 October 1998) in a wide range of interviews stressed the gap in the delivery of services to young people less than 15 years of age who are intoxicated. They are not catered for by the alcohol services or youth services (which provide a service for those over 15 years of age). The need is far greater than we can cope with'. 'Homelessness as a result of dysfunctional families is an issue that needs to be addressed by capacity building'. For instance: 'just learning how to shop so that you are not being ripped off at the corner store', because of innumeracy and illiteracy are basic life skills that many need to learn. Building the self-esteem of young people to prevent their dropping out of school because of a sense of hopelessness is another priority. Lack of resources in remote Indigenous locations impacts on access to information and life skills such as information literacy and computer skills. In remote areas computing skills could deal with 'the tyranny of distance' and ensure that they are not cut off from life chances. Keeping track or 'Tracking children' who move into the city and are 'lost' was an issue raised and identified as being of particular importance to families in remote communities. Negotiating two systems: a capitalist and non-capitalist system, with two very different sets of values, remains an ongoing challenge that could be alleviated through achieving access to the Internet. This could help to address the divides and service links between towns and the Aboriginal outstations needs to be built up. A combination of all these factors contribute to a sense of being marginalised and the associated depression leading sometimes to suicide and to copycat behaviour. A multi-disciplinary response to mental health is required. No one program could address all the needs. An interagency response is essential to youth at risk.

The challenge is to deliver services to young people and their families in the developed, transitional and less developed sections of the population. Family dysfunction and domestic violence occurs across

[46] Peter Davis; Letter to the Director of Office of Youth Affairs.

all sections of the population but is concentrated in sections of the population that face the challenges of unemployment, low levels of education and limited hope for the future.

One of the key pillars of social policy that creates opportunities for people in the future is education; it is a way out of welfare. Promotion of life chances through creating a culture of learning beyond the school walls is discussed as one of the best examples of this approach in Alice Springs. Policy and practice need to support the Irrekerlantye Learning Centre for young people and their families (where appropriate) and the Alice Outcomes program needs to support learning beyond the school walls by adding to the resources of mainstream educational organisations: Young people and Indigenous people experience the highest levels of unemployment in Alice Springs. Poor education outcomes, high levels of youth suicide, high levels of youth incarceration, high levels of domestic violence, high injury rates along with high morbidity and mortality rates, associated with alcohol and other drugs and poor nutrition, spell out the different life chances. These statistics when considered alongside the youth suicide rate of the NT and the poor retention rates in schools are indicative that young people and their families face both private troubles and public issues.

Two-way education is required for young people to learn their rights and responsibilities. For this to occur, a sense of respect from adults is required plus an understanding of the difficulties faced by young people attempting to come to terms not only with adolescence but also with a non-Indigenous capitalist culture that requires standards of behaviour that are in some respects quite different from traditional ways of life.

A social health promotion approach builds life skills initiatives such as a youth precinct safe for young people and patrolled by youth workers and night patrol officers was raised by a consultant to a government department. In the opinion of some professionals this could be used to assert that some spaces are more suited to young people than others and could undermine their common-law rights of young people to move throughout the town, even though the intention is to give further rights to young people. Nevertheless there is a need for young citizens to have access to safe spaces.

Accommodation for those less than 15 years of age is scarce and gaol is not an acceptable option for safe custody for this age group. The nexus of a gap in services for the under 15 who are in need of detoxification from binge drinking and the lack of safe alternative emergency accommodation for those awaiting trial means that the full extent of mandatory sentencing weighs heavily upon young people who commit property crimes. Preventative measures could address *the social* as well as *the individual* context of crime. The question needs to be asked, how did the preventative measures and the commonwealth pathways to prevention policies at that time link logically with the state and territory policies in Western Australia and the Territory? Mandatory sentencing was introduced in the NT in November 1996 and on 8 March 1997 the amendments to the NT Sentencing Act came into effect. The policy of Zero Tolerance policing was introduced in August 1998, which can be summarised as being introduced to target specific 'problem areas'. Zero Tolerance follows what has been described as the 'broken window' philosophy; namely that if there is an opening, people will take it. However, more not fewer people were being incarcerated. In the words of a group of young people:

> There are too many take away places ... . People need places where they can stop and drink ... . There should be more places for adults and young people to dry out. There should be more education and adverts on TV about drinking. Also there is very little for young people to do. Just fighting and drinking ... . Movies are not free. Youth Centres should be open longer. They should be open all night. We would sleep right here on the floor. That would be better. Young people walk home at night ... this can lead to problems. There should be more youth workers, youth patrols ... I would do it if they paid me ...
>
> The responsibility is with people: 'If you drink you take the consequences. If our parents drink they can go to jail ... they can also go to DASA ...'. 'If we make trouble we can go to jail ... those are the rules ...'. Some talked of "the young person from the island who went to jail for a small thing ... for bigger things its fair to go to jail ... we know the rules ... If we could do just one thing it would be to have a place for young people, not run by people like the police but by youth workers. They should give young people coffee and feed them and counselling ... have follow up work."

The rule of law under mandatory sentencing becomes subject to parliament, not to courts. It becomes political rather than judicial. Amendments were made in March 2000 for diversionary programmes to be set up for those aged 15 and 16 who have committed minor offences. The punishment for repeat property crimes is thus in some instances on par with punishments for violent crimes and fraud. Diversionary programs whilst preferable to incarceration for young aged 16 and 17 were criticised for only dealing with individual circumstances rather than also addressing the social context of poverty in which much of the crime occurs. By August 2002, mandatory sentencing was revoked in the NT by the incoming Labour Government after amendments had been made by the Liberal National Government[47] in the wake of widespread civilian protests.

[47] Service providers welcomed the review of mandatory sentencing but believed that the changes would have little real impact because the diversionary options that have been introduced, whilst welcome in principle had not addressed the

According to a spokesperson for Alice Springs Youth Accommodation and Support Services (ASYASS), 'There is no accommodation for people under 15, except at XXX House or through Family, Youth and Community Services, if they are referred. The need is far greater than we can cope with'.

Fragmentation in services was interpreted as partly due to some professional's deliberate decisions to operate only with the agencies

*Continued*

major concerns of the Law Council of Australia, namely restoring the decision-making authority to the legal fraternity. A Legal Officer from the Central Australian Aboriginal Legal Aid Service questioned 'whether Mr Burke (and previously Stone) understands sentencing principles—what courts try to achieve through sentencing'. Punishment is the primary objective of mandatory sentencing, rather than lowering the crime rate that was Stone's objective. The mandate for sentencing was a community mandate rather than a mandate based on the legal principle that the punishment must match the crime, called proportionality. The responses in March 2000 to the criticism was to keep in place mandatory sentencing per se, but to provide more diversionary services. Tom Stodulka, in his capacity as Director Policy, Attorney General's Department and Helmy Bakermans, Policy Officer, Attorney General's Department discussed the New Amendments to the Juvenile Justice Act. Diversionary programs for first offenders were described as being attempts to prevent further offending and so to prevent young people from being incarcerated. It was stressed that a wide range of community organisations (both government and non-government organisations) have been invited to offer assistance and to tender for these diversionary programs. Another strategy of the Attorney General's Department is to build more victim–offender conferencing programs because it teaches a sense of responsibility for actions, makes people confront what they have done and teaches how it effects the victim. It was stressed that it is a very strong punishment as well, not an easy option for the offender. Further, the victims are able to let go some of their anger about the crime, when they understand why the act occurred and some of the circumstances. They stressed the importance of programmes for those who commit minor offences. The property offences, which led at that time to mandatory sentencing of those aged 18 years and above David Bambers pers comm and (see Schetzet 1998 page 118 and Flynn 1997) were 'Theft regardless of the value of the property, (excluding shop lifting and theft when the offender was lawfully on the premises), criminal damage, unlawful entry to buildings, unlawful use of vessel, motor vehicle, caravan, or trailer (irrespective of whether as a passenger or a driver), Receiving stolen property (regardless of value), Receiving after change of ownership, taking reward for recovery of property obtained by means of a crime, assault with intent to steal, robbery. White-collar crimes involving fraud and deception, however are exempted.'

with which service providers feel confident to work. Differences based on values and assumptions underpinning models of service delivery were also an issue at the time of the research. A different understanding of the nature of development, for instance, as proactive, preventative or reactive, participatory or top-down. Risk behaviours begin early in Alice Springs, for instance, according to human service providers, the use of drugs by young Indigenous and non-Indigenous people (including immigrants) can occur as young as 13 years. Drugs, such as marijuana are readily available on the street. Petrol sniffing, smoking and abuse of alcohol are also a concern. Alcohol, its availability and usage plays a role in shaping the life chances of citizens by rendering individuals, families and sections of the community dysfunctional. Young people and their families face the

*Continued*
"The following sections is extracted from the ATSIC Report (1999, pages 11–13)

| | no | rate | no | rate | over-representation |
|---|---|---|---|---|---|
| New South Wales | 132[a] | 915[b] | 225[c] | 34[d] | 27.2 |
| Victoria | 8 | 244 | 63 | 13 | 19.1 |
| Queensland | 71 | 475 | 55 | 15 | 32.8 |
| Western Australia | 70 | 783 | 41 | 20 | 38.3 |
| South Australia | 17 | 541 | 60 | 38 | 14.2 |
| Tasmania | 7 | 343 | 15 | 27 | 12.7 |
| Northern Territory | 20 | 227 | 1 | 7 | 32.2 |
| ACT | 3 | 811 | 13 | 38 | 21.5 |
| Australia | 328 | 583 | 473 | 24 | 24.7 |

*Notes*: [a] For the purposes of standardisation, these numbers do not include young people over the age of 17 years who are held in detention. Some jurisdictions such as New South Wales have significant numbers of Indigenous young people in this age category and could add as much as 20% to the figures cited above.
[b] Rate per 100,000 of the population.
[c] Ratio of Indigenous rate to non-Indigenous rate.
*Source*: ATSIC 1997: 90–91.

Because national data identifying Aboriginality has only been collected for a relatively short period, it is difficult to identify national trends. However, available data show upward trends in the number of Indigenous youth incarcerated, the rate of incarceration and the level of over-representation since 1993. The number of non-Indigenous youth in detention centres has remained stable between 1993 and 1997. During the same period the number of Indigenous youth incarcerated increased by 55% ATSIC 1997: 91–93). The highest rate of over-representation occurred in Western Australia. In that State Indigenous young people were 38 times more likely to be incarcerated than non-Indigenous youth. The Australian Institute of Criminology has released more recent data on juvenile incarceration. Table 7.1 shows the number and rates for all young people in Australia during June 1997 and 1998."

ramifications of domestic violence leading to lack of sleep, lack of ability to concentrate, lack of home learning supports, poor nutrition and concentration levels, and physical and mental illness.

Demographically young people are the fastest growing section of the population. The demographic analysis indicated that this section of the population is significant (Table 7.2).

The National Health Committee New Zealand, Ministry of Maori Development Strengthening Youth well being, New Zealand Youth Suicide

*Continued*
"This is also extracted from the ATSIC Report (1999, 11–13)

*TABLE*
*"Young People (10–17 Years) in Detention, Australia, 30 June 1997 and 30 June 1998*

| State | 1997 | | 1998 | |
|---|---|---|---|---|
| | Number | Rate | Number | Rate |
| New South Wales | 357 | 51.4 | 336 | 48.0 |
| Victoria | 71 | 14.1 | 67 | 13.2 |
| Queensland | 100 | 24.9 | 136 | 33.6 |
| Western Australia | 111 | 52.0 | 136 | 62.7 |
| South Australia | 77 | 47.8 | 50 | 30.9 |
| Tasmania | 23 | 40.2 | 19 | 33.5 |
| Northern Territory | 21 | 89.4 | 25 | 103.5 |
| ACT | 16 | 43.7 | 11 | 30.4 |
| Australia | 776 | 37.1 | 780 | 37.0 |

*Source*: Carcach and Muscat 1999. Australian Institute of Criminology cited in ATSIC 1999. Many jurisdictions saw a drop in the number and rate of juveniles incarcerated between 1997 and 1998, including New South Wales, Victoria, South Australia, Tasmania and the ACT. The two mandatory sentencing jurisdictions Western Australia and NT saw an increase, along with Queensland."

The following sections is extracted from the ATSIC Report (1999, pages 11–13)

*TABLE*
*"Aboriginal and Non-Aboriginal Percentages of the Population in Different Age Groups in the Northern Territory Southern Region, 1997*

| | Aboriginal | Non-Aboriginal |
|---|---|---|
| Under 5 yrs | 50.9 | 49.1 |
| Age 5–19 | 49.7 | 50.3 |
| All ages | 37.7 | 62.3 |

*Source*: Boughton (1999: 16) cites unpublished data."

The following sections is extracted from the ATSIC Report (1999, pages 11–13)

Prevention Strategy (1997) has drawn the connection between promoting general well-being and reducing the risk of suicide: 'Young people who attempt suicide, often but not invariably, come from family backgrounds characterised by multiple problems, difficulties and stress ... ' (Beautrais et al. 1997:10). It stresses that programmes for suicide prevention should promote well-being in a wide range of ways and that suicide should not be normalised by discussing it at school or in other contexts as if it were a normal every day event in society. The New Zealand Ministry of Health

*Continued*

"Those States with the highest rates of incarceration of young people are NT and Western Australia. Both these rates also increased between 1997 and 1998. Western Australia has nearly double the national rate and the NT nearly three times the national rate. In relation to the NT, the Australian Institute of Criminology noted that the tendency in the Territory had been 'to lower imprisonment rates until 1997 when it recovered its upward trend. The rate of juvenile incarceration observed during June 1998 (103.5 per 100,000) was almost double that observed the same month in 1996, which no doubt is the direct result of the three-strike legislation introduced by the territory's government in 1997' (Carcach and Muscat 1999).

The following table shows the changes between 1997 and 1998 for Indigenous young people aged 10–17 years for Western Australia, the NT and Australia overall ATSIC 1999, pages 11–13."

***TABLE***

*Indigenous Young People (10–17 Years) in Detention, Western Australia, Northern Territory (NT) and Australia, 30 June 1997 and 30 June 1998*

| State | 1997 | | 1998 | |
|---|---|---|---|---|
| | No | Rate | No | Rate |
| Western Australia | 70 | 649.2 | 87 | 758.8 |
| Northern Territory | 20 | 209.5 | 21 | 216.1 |
| Australia | 312 | 419.2 | 326 | 411.4 |

*Source*: Carcach and Muscat 1999 in ATSIC 1999.

In both the NT and Western Australia the rates of Indigenous incarceration have increased between 1997 and 1998. The increase has been particularly pronounced in Western Australia. This is against the national trend that saw the Indigenous rate fall slightly between 1997 and 1998. Western Australia and the NT were the only jurisdictions besides Queensland that registered an increase in the rate of Indigenous juvenile incarceration. In all other jurisdictions the rates declined. It should also be noted that the above information refers to prison census data and not flow-through data. It is well established in criminological literature that census data underestimates prisoners serving short-term sentences. The importance of this is that in the NT it is significantly likely to underestimate the increased level of over-representation of Indigenous children and young people serving shorter 28-day mandatory sentences".

(1997) stressed that prevention of suicide amongst Maori in New Zealand is by means of creating a society where Maori identity and culture is valued, nurtured and strengthened by enhancing safety, opportunities for decision-making, partnership and a sense of identity. Thus a wide range of underlying causes are addressed. The wide range of preventative measures range from ensuring safe accommodation, food, educational and recreational activities.

Lack of sleep because of a noisy environment linked with alcohol misuse does not enhance school performance. Lack of regular public transport to school is an added concern; if a child oversleeps she will miss the only school bus run from town camps. If a child is hungry then she will not have the energy to get up early. A lack of food security to use the current term is as much a result of poor budgeting skills as it is a result of the higher cost of food in remote areas and the fact that parents who are intoxicated do not look after their children's best interests. If the child arrives at school and does not concentrate she will soon fall behind and lose interest in the school system. If she manages to reach the higher grades she faces the challenges of peer pressure to use alcohol and other drugs. Intoxicated binge drinkers commit foolish acts that can lead to property crimes and the consequences of breaking the law in the NT. Also, the high rates of unemployment amongst Indigenous young people do not provide role models or an incentive. Thus to address something as apparently simple as improving attendance at school can require addressing the issue by means of teams cooperating across sectors to promote a healthy environment for young people and their families, which maximises the life chances of citizens). The Youth Justice Coalition (1999–2000) stresses the need for addressing community development as a means of crime prevention instead of resorting to mandatory sentencing. Young informants from a range of backgrounds stressed that alternatives to mandatory sentencing would be appreciated, because incarceration could lead to young people losing faith in society and becoming hardened to a life of crime. According to ATSIC (1999: 11) nationally Indigenous young people are 24.7 times more likely to be detained in a juvenile detention centre than non-Indigenous youth and this report details the extent of the incarceration.

A systemic and integrated approach to preventing social ills means addressing health, education, employment and crime prevention through coordinated policies and programmes of action. Integrated budget lines are needed to make this a reality. The following raft of measures together could make an impact on crime control. Controlling alcohol availability could assist directly in promoting functional families by reducing domestic violence and creating a context that supports school attendance and learning after school hours. The provision of a sobering up and counselling facility for young people under 15 years of age who are not assisted by current adult oriented programs is urgently required in Alice Springs. All young people as citizens need to contribute to youth policy decisions on a regular basis through school, local and territory government organisations. Young people do not want to sit in meetings; they want to be able to ring a call number or post an email or speak to an accessible representative.

The mission is to build functional, literate and numerate families who are able to access information with confidence, through broad-based community development and education in life skills.

Young people and Indigenous people experience the highest levels of unemployment in Alice Springs. Poor education outcomes, high levels of youth suicide, high levels of youth incarceration, high levels of domestic violence, high injury rates along with high morbidity and mortality rates, associated with alcohol and other drugs and poor nutrition, spell out the different life chances. These statistics when considered alongside the youth suicide rate of the NT and the poor retention rates in schools are indicative that young people and their families face both private troubles and public issues.

## *7.4 Promotion of Life Chances Through Enabling a Generative Learning Community Beyond the School Walls for Young People and their Families*

> Open societies will emphasize a questioning and involving education system, the easy movement of people, and the rapid transmission of ideas. And you can't do that half-heartedly or only in part … . Despite the hopes of some … it is not possible to take the economically advantageous bits of the information revolution and expect to suppress the rest of it …
>
> The information age and its abacus the computer, has spawned a new way of thinking. What will underpin growth in the future is the capacity of countries to develop a milieu that sustains creativity. Provided the creativity is there, new industries will emerge from the flux, many of them having little to do with information itself. In this new environment, knowledge workers will be in shorter supply than capital … Education must be the core of any government's response to the challenge of the new age. But a new sort of education and literacy is required. … (Keating 2000: 287)

The challenge is to ensure that young people have a stake in the future. Children who are starting primary school need to have a wide range of supports in place to ensure that by the time they are adolescents they will feel that they have a stake in society and faith in authority. Unemployment of young people is directly linked with crime rates in other parts of the world and there is no reason to consider that the NT is an exception.[48]

Poor education and employment outcomes are key indicators of youth disadvantage and they contribute to undermining a sense of well-being. A local Alice Springs model such as the Irrekelentya School strives to provide education for both young people and their families instead of trying to teach within inappropriate structures that are recognised as problematic by teachers. This model is based on intergenerational learning so that young people can receive the support they need to become successful learners, for instance, encouragement just to make it to school and emphasis on recognising the social context of learning. Achieving job readiness however is not only about numeracy and literacy, it is about engaging in two-way learning across cultures, in order to learn from one another and thus to enrich knowledge and experience. Thus learning about Western social customs, generic business and management practices is as important as learning a respect for Indigenous knowledge such as a holistic understanding of the world.

Educational challenges need to be addressed systemically to ensure job readiness through computer skills to obtain information literacy. Education according to Banathy (1991) needs to respond to the requirements of the post-industrial age that requires innovation and flexibility as key learning outcomes, rather than merely the ability to become literate in predetermined subject-related content or predetermined skills. We have moved from requiring machine technology to requiring intellectual technology (Banathy 1991: 24). The pace of change has accelerated and as we move from print to cybernetic technology and from nation states to potentially global networks of association, we need to open our minds from deterministic to systemic modes of thought. Unfortunately many social institutions model closed and deterministic thinking and practice. It is essential that in order to maintain the pillars of citizenship (associated in the past with the nation state and the linked rights and responsibilities) that are essential for maintaining human dignity, we model systemic designs in key social institutions. If the focus is on learning outcomes then it is possible to create alternative design structures that focus on the needs of the learning, so that she will be able to operate effectively in the future.

This is a major requirement for the type of work that will characterise the economy of the NT. The Territory needs appropriate skills in order for it to compete in the global economy. Because of its geographical

[48] Local statistics on crime by demographic characteristics were not available at the time of the research from the NT Government.

isolation Alice Springs needs to rely heavily on being able to access the Internet. Tangentyere Council, the agency for Indigenous people living in town camps, has been aware for a long time of the large numbers of young people not attending school. As a result innovative education measures have been implemented to support intergenerational learning. Education, communication and social movements have systemic relevance to the vision for a learning community. The use of the social movement approach for change through networking locally, nationally and internationally by using the Internet to communicate with relevant organisations is a vital tool for learning and self-determination. As yet it is of little help for those who could benefit most from it. The Indigenous citizens of Central Australia have overall minimal literacy and numeracy levels and little opportunity to access technology.[49]

We need to work across sectors and adopt the healthy city approach to health and development (that takes into account the Indigenous perspectives of health and land), which would be of value to promote, foster and facilitate social health, in particular education and employment outcomes that would enhance the life chances of all citizens. We need open communication styles and tools that are needed to address education to promote the management of diversity. Fostering close links across organisations is important.

The development program could help to prevent crime not only by teaching literacy and language skills that do not undermine identity, but also by providing more opportunities and developing numeracy and budgeting skills. Limited budgets that are poorly managed leads to a lack of money that in turn prevents attending school or alternative education programs daily and obtaining a stake in society. Working with families by means of an intergenerational approach, piloted by the alternative programs mentioned below, builds capacity within families.

The public, private and volunteer sectors could extend its contribution to this initiative through the provision of library programs to assist users to access the Internet, CDROMs and other sources of information. Students are frequently undernourished and best educational results can

[49] Dewey, J. (1997) *Democracy and Education*. Simon and Schuster.
Fals-Borda, O. and Rathman, M.A. (1991) *Action and Knowledge: Breaking the Monopoly with Participatory Action Research*. Intermediate Technology, London.
Friedman, J. (1992) *Empowerment: The Politics of Alternative Development*. Blackwell, Oxford. Negroponte (1995) *Being Digital*. Vintage, New York Negroponte has set up computer laboratories in Cambodia to provide educational opportunities for young people with limited life chances. Children as young as 8 years are using the computers to obtain essential life skills.

only be achieved when the primary needs have been met, thus provision of food within home environments need to be addressed. This educational approach builds upon the WHO healthy settings approach to development that is multi-disciplinary and intersectoral and attempts to involve as many organisations as possible in Alice Springs. The educational principles are humanistic, participatory and transcultural. The starting premise is that unless educational processes suit the cultural framework of young people at risk, they are unlikely to make use of education opportunities. The programme would encourage learning for problem-solving and learning to obtain life skills. The curriculum would be open and intergenerational and has already been successfully piloted through the Detour Program over a period of 2 years.

The lack of space (both public and private) for young people to engage in productive activities has been well documented as one of the factors leading to policing responses.[50] Alternative forms of community education have been successfully trialled and implemented in order to overcome the inappropriate nature of formal main stream schooling to Aboriginal people who have limited school readiness skills and are without the home resources to undertake homework (Bowden 1994)[51]. The Irrkerlantye Learning Centre has found that intergenerational classes work well for Aboriginal people. Men, women and children can attend some classes together, particularly those on vocational training and talking about cultural heritage, pride and identity. The Detour Program makes use of a fund to give Aboriginal people a chance to travel and to obtain a first hand experience of living in a different environment. 'For example just seeing the numbers on a post box in a city far away from Alice Springs can demonstrate the value of being numerate and literate, so as to write a letter to a member of the family.' (Totham 1999, personal communication).

[50] Cannon, C. and McDonald, D. (1996) *Keeping Aboriginal and Torres Strait Islander People out of Custody: An Evaluation of the Implementation of the Recommendations of the Royal Commission in Aboriginal Deaths in Custody*. Australian Institute of Criminology, Canberra.
White, R. and Alder, C. (1994) *The Police and Young People in Australia*. Cambridge University Press.

[51] Durnan, D. and Boughton, B. 1998. Detour *Project Consultancy: 1997 Evaluation and 1998 Planning. A Report to Tangentyere Council*. Tangentyere Council, Alice Springs.
Traves, N. (1998) The Detour Project: An intergenerational place of learning for Aboriginal youth. Background Paper 1998, Centralian College, Alice Springs.
Wright, K. (1998) *ASYASS: The Alternative Family Model of Working with Young People*. ASYASS, Alice Springs.

ASYASS has operated for over a decade in Alice Springs to provide not only accommodation, but also advocacy and guidance to young people at risk. The philosophy is based on mutual respect and caring. The high crime rate and high incarceration rate in Alice Springs requires a crime prevention approach, not merely a crime control approach.[52] Pathways for marginalised people need to be created back into the mainstream.[53] Educational philosophy draws widely on the sociology of education and development. 'Learning by doing' can be enhanced by linking educational

[52] Australian Youth Policy Action Coalition (AYPAC) Newsletter at http://alice.topend.com.au.
Flynn, M. (1997) One Strike and you are out. *Alternative Law Journal*, 22(2), pages 72–76.
Flynn, M. (1998) Editorial. *Criminal Law Journal*, 22(August), page 201.
Gearn, S. (1997) Locking up the children. Balance Northern Territory Women Lawyers' Association.
Short, M. (1997) Mandatory sentencing. Balance Northern Territory Women Lawyers' Association.
Tippet, J. (1997) Mandatory sentencing and other matters. Balance Criminal Lawyers' Association.
Harding, R.W. (1995) Repeat juvenile offenders: The failure of selective incapacitation in Western Australia. Report No. 10, 2nd edition. The University of Western Australia, The Crime Research Centre. *Alice Springs News*, 5(46), 16 Dec 1998.
Schetzet, L. (1998) A year of bad policy. *Alternative Law Journal*, 23(3).
Podesta, L. and Jones, P. (1993) Delinquency and homelessness: towards an integrated response. National Conference on Juvenile Justice: Conference Proceedings No. 22, Australian Institute of Criminology, edited by Atkinson, L. and Gerull, S.A. Cain, M. (1996) Recidivism of juvenile offenders in NSW, Department of Juvenile Justice in Darwin Community Legal Service, page 6.
Australian Bureau of Statistics (1998) National figures on crime and punishment. *Year Book Australia*.

[53] Braithwaite, J. (1998) *Crime, Shame and Reintegration*. Cambridge Press, Melbourne.
Cochrane, J. (1999) Technology's shining light for orphans' bleak life. *Australian* 17 January 1999.
Harding, R. (1993) Opportunity costs: alternative strategies for the prevention and control of juvenile crime, in Harding, R. *Repeat Juvenile Offenders: The Failure of Selective Incapacitation in Western Australia*. Research Report No. 10, University of Western Australia, Crime Research Centre, page 141.
Jamrozik, A. and Nocella, L. (1998) *The Sociology of Social Problems: Theoretical Perspectives and Methods of Intervention*. Cambridge University Press, Cambridge.
McIntyre, J. (1996) *Tools for Ethical Thinking and Caring*. Community Quarterly, Melbourne.

tasks with familiar daily tasks and using computer technology as a tool to achieve this end.[54] Computers are a key vehicle for education. They could help to build upon young people's enthusiasm for learning by tapping into their enthusiasm for computer technology and games. Feeling computer literate is a good starting point for building self-esteem, which can then lead to further enthusiasm for learning. Children as young as 8 years can use computers to obtain essential life skills. This project could have significance for a number of government agencies. It could contribute to health, education and employment crime prevention policies.

### *7.4.1 Discourses on Life-Long Community Learning and Information Literacy*

Libraries along with youth clubs and the woman's centre are much appreciated as places where computer literacy can be learned.

> More library displays would be welcomed on 'how the library works'. Perhaps this could be a permanent fixture or display. Better clearer signage throughout the library 'like in a bookshop'. More work on literacy promotion through the volunteer's network, Friends of the Library would be useful in this regard. More outreach work for volunteers needs to be undertaken to draw in the users of alternative education projects and CDEP workers.

Residents from town camps commented that low levels of literacy and numeracy means that the library is considered to be a place for young people to learn. Young pre-school and primary level children use the library mostly. Most young people who are in their mid to late teens tend to be boarders at Yirara College and so make use of their own school library facilities. Once young adults return to live on town camps their opportunities to pursue further education are not promoted in an environment of crowding, domestic violence and alcohol abuse. The lack of accessible and readily available transport limits the use of the public library by all age groups. The lack of electricity, desks, books and computers and limited disposable income means that a very limited home study environment is created in many homes. Overcrowding and a high level of domestic violence is not conducive to home study.

Respondents who work with CDEP workers and the elderly made the following suggestions:

> Their bus could drop off books to readers and collect books from the Library. We need to enrol their staff and users through the library. The CDEP workers

[54] Dewey, J. (1997) *Democracy and Education.* Simon and Schuster.
Fals-Borda, O. and Rathman, M.A. (1991) *Action and Knowledge: Breaking the Monopoly with Participatory Action Research.* Intermediate technology, London.

could be invited to attend library discovery tours. The library should be a friendly place where you can gain knowledge without feeling threatened. Five years ago the library was perceived to be a place for white middle class people.

Because of a bad experience an informant said that he had not used the library, but conceded that he had heard some positive things about the library and that he would probably join with his children.

According to Greg Snowden (2000, personal communication):

"Indigenous people are very concerned if they perceive something to be sacred—even if it is implicit. Indigenous people are adept at seeing the sacred in the world. The library needs to be demystified by explaining how it works and what it is aiming to do. In other ways we need to also understand that the library is a sacred space, and is used by people in that way. The question is do we want to support its meditative and sacred elements or open it up as a more effective and efficient learning institution. If the library is to be opened up more to Aboriginal people it may need to change its quiet and contemplative culture—which may not be what other users want. Perhaps there needs to be a second section of the library that is more interactive—group discussions for younger people and groups It was suggested that there should be some 'hands on scientific models that give practical and interactive demonstrations of scientific principles.

- Aboriginal learning tends to be in groups, with a lot of verbal explanation and interaction. Learning tends to be experiential. Thus a discussion/interaction room would be helpful. Important knowledge is kept until people are ready, until they have sufficient life experience and wisdom to understand and respect what they have been given. Knowledge is relational in that it is passed on depending upon who the student is, what country they come from and what knowledge they need to know. By contrast, libraries tend to be fairly solitary institutions with individuals pursuing their own interests separated from other people. There is almost no sound whatsoever—talking to others is frowned upon. Knowledge is available in books, no matter your life experience or psychological state or where you come from or your position in the social group. While there are some limitations in what is available for someone to read, ultimately if you want to know something you can find a book to explain it for you. This appears to be a very different approach to learning from traditional Aboriginal cultural approaches. Aboriginal culture is very much an oral culture rather than a literacy based culture. Information is passed on in a social, relational and experiential context. The giver of the information is as important as the information itself. This is very different from the world of books where we don't personally know the authors, let alone feel a need to be related to them, we are not listening to their voice or necessarily even care what their experiences have been. From an Aboriginal perspective, the library appears to be almost a 'whitefella sacred site'. When people go inside it is as if they are going into a church. There is a 'hushed' culture, where everyone is very quiet. It has a spiritual atmosphere, a place of meditative contemplation.

There are a lot of 'secrets' which some people know about (the catalogue systems, the ways of placing books, how people take notes), and others from outside have great difficulty understanding and entering this sacred cosmology.

- Video material is produced in the NT and SA particularly by local communities and regional Aboriginal organisations. These depict various aspects of Aboriginal culture and life. They could be advertised and shown at libranes and schools, for example.[55]
- More use of the outdoor area to teach library users about arid plants. More life long learning through hands on activities, such as plant identification and closer links with environmental libraries."

## *7.5 Creation of Employment Pathways*

The respondents and informants prioritized the need to address education and unemployment needs of young people in Alice Springs. In the discussion on employment and unemployment indicators it was clear that despite the overall high levels of employment there are pockets of high unemployment associated with low skill levels that need to be borne in mind when planning responsive development initiatives. Local models that successfully meet the specific education and employment needs of Indigenous residents are discussed below.

Participants in the Employ Alice planning committee stressed that successful employment in tourism, entertainment, construction and mining sectors and others (education, health and retail) could pave the way for further opportunities to be made available. Participants from Centrelink stressed that the challenge is to meet the needs of their funding bodies and to achieve outcomes.

At the time of the research the Chamber of Commerce in Alice Springs planned to play a role in assisting industry, social welfare services through Centrelink and a number of Indigenous organisations to place job-ready applicants. This requires coordination and liaison in order to establish: (i) the size of the pool, (ii) the level of available skills, (iii) the extent of employment opportunities available, (iv) barriers to obtaining employment and (v) barriers to remaining in employment. Developing a database that covers the opportunities that could be provided by a wide range of organisations was considered to be a useful goal.

Barriers to education and employment are socio-cultural, political and economic, but specifically alcohol and the associated family dysfunction play a role as an affect of wider issues. According to a psychologist

[55] For example, Walpiri Media, Institute for Aboriginal Development and ATSIC.

(based on her assessments for community support over the previous year and a half) young people highlighted the following concerns: 'a sense of boredom, the desire to be able to read and write, the desire to be a valued member of the community, the realization that people who are employed are more valued than those who are unemployed'.

A continuum of opportunities could create pathways from Community Development Employment Program (CDEP) and on-the-job training with sponsored mentors to full-time employment to address the current unemployment figures. The continuum could include programs to build self-confidence, personal presentation, communication and literacy, interpersonal skills, numeracy, ongoing assessment of skill levels and tailored training to achieve basic skills, matching interests and abilities with jobs and setting up placement programs with mentors within a range of industry sectors. At the time of the study this was an area that was being addressed by staff of Tangentyere Council with the support of a government grant to set up a Job Network at Tangentyere Council.

The creation of employment pathways entails enhancing access to information, decision-making, resources and markets. The critical socio-economic and environmental constraints are associated with developing participatory planning to enhance the ability of the poorest households[56] to access resources, and microcredit plus services to improve ability of the poorest to generate income. The focus could be on (i) enhancing access to information, resources and decision-making, (ii) fostering links to improve governance (accountability through elections, finances, communication and information to the poorest)[57] and (iii) to maximise resources and to create multiplier effects. The challenge is to facilitate links with local organisations across the sectors and building communication channels with Indigenous committees and forums representing the most marginalised interest groups. Phasing in the educational components relating to environmental and social sustainability of enterprise at the outset is important to ensure that the self-generated ideas by the participants incorporate these ideas. Group formation to generate their own responses could then logically follow on from the educational programs. The project components are to be implemented flexibly to ensure that local priorities can be taken into account to enhance a sense of community ownership. The program needs to be open to the suggestions and contributions of participants.

[56] AusAid (2001) *Reducing Poverty—The Central Integrating Factor of Australia's Aid Program*. Commonwealth of Australia.

[57] United Nations Development Program Poverty Report (2000).

The links between Aboriginality and poverty are evident in the national and local data. Those who make up the poorest households have limited resources: Poverty overall is linked with a complex range of factors, such as age, ethnicity, gender, level of education and geographic location. The links between gender and poverty also need to be considered in planning. One of the greatest strengths of a microfinance project is the potential to develop diversified cash incomes, and also to develop the interface between the marginalised poor, particularly women and local government through developing the capacity of members of poor households to participate as representatives. Like all other aspects of development, microfinance can cause more harm than good, if not carefully implemented. The sustainability of microfinance along a modified Grameen Bank Model (GBM).[58] The challenge is to incorporate indigenous elements and to be responsive to the local indigenous context and gender issues (opportunities for women and men), but in line with the cultural norm (of men's business and women's business being separate) and ensure that repayment of loans is not achieved through further indebtedness. When women are given opportunities to enhance their capacity through literacy, numeracy and microenterprise, the overall health and well-being of women and their children improves.

### *7.5.1 Systemic Initiatives to Promote Employment*

The job shop and bank located on the site of the housing association provides integrated service delivery and an opportunity to learn about employment, commodities, housing and employment in a safe environment.

> It works because the approach is process and people oriented. If a person needs to be supported for the first few days of work by being taken each morning, then this is provided. Little things can make a big difference … . Job Matching is available to all unemployed people. It is being run as a proper business, so it has to at least break even. … We aim to have a friendly, personalised approach for people looking for work. This means making the Shop a place where Aboriginal people will feel comfortable and welcome. This may well include a cup of tea and a yarn.

Systemic digital mapping for job creation opportunities across all sectors could help to support the job shop. A social environment mapping project could provide a local interactive Internet site and add information to existing websites; so as to enhance the social, cultural, economic and

[58] The most important aspects of the Grameen Bank Model for microfinance is group solidarity and social capital (defined as the basis for trust in society).

environmental profile of Alice Springs and to help to create employment pathways in the public and the private sector. The value of digital mapping is the use of hyperlinks to make connections across categories and levels in directories on the Internet. A digital mapping project could also provide educational opportunities together with primary, secondary and tertiary institutions to promote integration within the community by giving citizens an opportunity to contribute their ideas. It could stimulate participation and enhance the opportunities for discussion for creative problem-solving. The potential value of the mapping process and Internet site lies in: (i) adding information and knowledge content, (ii) creating employment pathways by working with public and private sector, (iii) breaking down barriers to communication and understanding, (iv) empowering people from all walks of life irrespective of age, gender, culture or occupation to contribute through directly accessing the Internet site or (v) by making written or recorded contributions. For example, young people could gather oral histories from older citizens through the CDEP Program and young people will be trained in interview skills and the use of recording equipment. History has been documented through oral history and existing work could form the basis for further contributions from the public. Stories could be augmented by means of photographs and drawings from existing collections. (vi) Providing opportunities for recording and documenting oral histories, material culture of the social and natural environment.

The outcomes could be an increase in transcultural understanding as well as the creation of business opportunities. This could be a tool for accessing information and enhancing information literacy within a wide range of learning forums (formal schools, alternative schools, youth groups, colleges, universities and libraries). This project could be a useful two—way educational opportunity for alternative learning. The need for creative approaches to education using computer technology creatively has been documented (Negroponte 1995). Pathways for marginalized people into the mainstream can be created if social issues are approached not merely with information literacy but with a realisation of the opportunity costs of not involving this section of the community.

# 8

# *Health, Education and Employment*

## *Articulating Axial Themes Through Participatory Design Processes*

> ... other serving—is ... not about helping, fixing, or changing the 'other'. This form of service is based on a systemic relationship of mutuality and dignity. This type of service is distinct from helping, which by its very name creates a unilateral relationship .... In a relationship where there is an exchange of value of equivalencies there is no inequity, inferiority, domination, debt incurred, or unilateral control ...
>
> *Nelson 2001*[1]

### *8.1 Resisting Commodification Across the Sectors of Health, Education and Employment*

Participatory design is fraught with many difficulties and it is vital to consider complexity at every stage of the intervention and the very process of intervention itself. Who should be involved, why and how and in whose interests? The main challenge is that no so-called community is homogeneous and so we always have to consider the complexity of the community in which we are working. The mental maps vary by age, gender, language, level of income, kinship ties, political position and experience in town or in country, to mention just a few aspects of diversity. When we work with the stakeholders as outsiders we can be unaware of the power dynamics at work. We may think we are doing one thing, but in fact may be pawns in a chess game of which we are quite unaware (or if we are lucky we may have some insights into the multiple and cross

[1] Nelson Harold (2001) Letter from the President of ISSS for the 45th conference.

cutting issues at stake). Also it is a good idea to look explicitly at the issues of ontological maps or maps of knowledge and the way they differ across the different participants and of course the way our own maps intersect with the diverse interest groups in the community. We may be told that 'everyone agrees', but in fact it may only be those with the most power. Our participatory planning may help to entrench a particular interest group without our even realising it.

Indigenous citizens need to be the subjects, not the objects of social planning and design. To create a viable space in the civil arena is the challenge. How can marginalised people who have been the subject of over-bureaucratisation become involved in decision-making? Housing associations to administer public housing, gardens and shared public spaces can provide an entry point for developing civil governance. This chapter discusses ways to prevent the process of commodification by encouraging active involvement in designing their communities.

When people are part of the process of decision-making and the management of knowledge narratives, they are less likely to be controlled as 'the other' or as 'those people'; and will be actively responsible for their own destinies as citizens with rights and responsibilities. By resisting commodification at the local level, Indigenous people are likely to achieve self-determination. As stressed throughout, achieving better life chances for citizens could be achieved through better access to health, education and employment as well as control of their own governance. We need to address the social problems systemically and go beyond the limits of our blinkers (created by not seeing the connection across self-determination and social well-being). People have the answers to their own problems. They need the space to own and address the issues. We need to see the whole picture and consider the ramifications of working narrowly within only one sector. As stressed in chapter 1, Alice has almost twice the number of human services as the national average and yet the outcomes in terms of social health are dramatically below the national average. Because the life chances of Indigenous citizens in Alice Springs have remained significantly lower than for non-Indigenous citizens, in terms of health, education and employment indicators, a different approach, not more of the same is needed. The community is already overserviced by professionals. The barriers to achieving health, education and employment outcomes for some Indigenous citizens living in Alice Springs are both personal and social due to systemic socio-cultural, historical, demographic, geographic, economic and political factors that have increased dependency. Planning needs to address conceptual and practical issues such as the lack of continuity in service delivery, because of the high turnover in staff in human service organisations. Added to this any plan needs to take into account the historical and global context and the regional service nature of Alice Springs and the way in which the health services are accessed. Treatment services in town appear to be used by Indigenous citizens once they are already at a chronic stage of their illness. Thus there is a need for more preventative and promotive services in remote areas. Accessibility to services is always a problem in terms of cost, distance and sense of appropriateness when vast geographical and social distances need to be covered by poor, remote users with a different cultural sense of need.

Unemployment and marginalisation are directly linked with crime rates in other parts of the world and there is no reason to consider that the Northern Territory (NT) is an exception.[2] Low skilled jobs are disappearing which means that the unemployment rate amongst Indigenous young people will only be addressed by better preparation for a skilled service industry and by creating specific opportunities. The introduction of changes in the conditions of work (relations of production), types of commodities and changes to the very fabric of cultural identity (as a result of social change) have far ranging ramifications for those effected locally. The types of economic activity include (i) knowledge production for sustainable living, (ii) localised, traditional approaches to traditional industries in primary, manufacturing and service work without information literacy and a local, rather than a global approach, (iii) welfare dependency on the social wage and those struggling to find employment and to shrug off the legacy of what Rowse (1998b) called the dependency created by rationing: 'white flour, white power' in the history of colonisation, (iv) traditional Indigenous communal lifestyle. The 'cash work nexus' (Pixley 1993) is central to definitions of non-Indigenous citizenship. The majority of Aboriginal people have difficulty in obtaining work opportunities. The dynamics of the town are spelled out in terms of (i) the social indicators of quality of life and (ii) the social health outcomes of citizens. Class and culture continue to define life chances today that can best be addressed through systemic measures. The following participatory governance approach is adapted from the process of setting up healthy cities/settings, which maximises the multiplier effects of working across sectors and disciplines.[3]

This approach can assist in (i) shifting thinking from compartmentalisation to thinking systemically, (ii) informing interventions so as to avoid making changes that do not consider the wider implications,

[2] Local statistics on crime by demographic characteristics were not available at the time of the research from NT Government.

[3] Adapted from Mowbray, P. (2000) *Healthy Cities Illawara: Ten Years on A History of Healthy Cities from 1987–1997.*

(iii) enhancing an understanding of the linked nature of social and environmental concerns, (iv) developing problem-solving strategies.

The promotion of a healthy environment needs to be placed as a central assumption of planning. The challenge is to find ways to encourage existing organisations (that in the past have tended to work separately) to work together. Sustainable integrated development assumes the short-term benefits of financial profit and the long-term benefits of economic sustainability to be of equal importance. The costs to society and the environment are built into all decisions and rational decision-making needs to address social, political, economic and environmental costs. This approach is concerned about the future of economic initiatives and the implications for all members of the community.

In Alice Springs creating the conceptual and practical links between improving health, development and social well-being will be facilitated through working regionally across public, private and volunteer sectors. A sustainable integrated approach to health is vital to address both the physical and mental health of the Indigenous residents of Alice Springs and the Central Region.

Local organisations are ideally placed to model healthy settings development drawing on the United Nations (UN) and the World Health Organisation (WHO) charters that underline that health and development are integral to one another.[4] Local grass roots governance could

[4] In the past institutional thinking has been bound to disciplinary knowledge bases and these have been used for defining the frameworks for policy and practice and the competencies of practitioners. The end result is that much development work has been segmented and compartmentalized. Hence the exhortation made in the Ottawa Health Charter (1986) to see the links between health and development and to work in a 'multidisciplinary and intersectoral manner'. This approach is suitable to address the challenges in Alice Springs. A range of linked social issues could be addressed such as the social marginalisation of Indigenous people, high unemployment particularly amongst Indigenous young people, high suicide rates (particularly amongst males [aged 14–24]) and incarceration rates, high rates of domestic violence and one of the highest levels of alcohol consumption nationally. The cycle of hopelessness can only be addressed through a multifaceted approach that addresses the systemic nature of the problem and takes an action research approach to addressing integrated development. A health department such as Territory Health Services cannot address the systemic problems alone. It requires a sustainable and integrated approach that draws on the Indigenous knowledge of people to achieve self-determination. This traditional Indigenous knowledge has been recognised in the Ottawa Health Charter (1986) that echoes what Indigenous people the world over have known: that the web of life needs to be respected and people are part of this integral whole rather than no masters as global social and environmental challenges testify.

contribute to the implementation of the National Mental Health Strategy (1999) by adopting a partnership approach across sectors and disciplines. Such an approach would help to promote health outcomes and enable people with a mental illness to be assisted in many ways within the community, not merely through institutional treatment and clinical intervention.[5] Introducing change through modelling integrated sustainable solutions to address health and environmental issues through Indigenous spirituality at the grass roots could be an excellent programme focus. Learning by doing (through Participatory Action Research [PAR] for integrated development together with a wide range of participants across sectors and disciplines) could lead to making a real change to the quality of life of the citizens of Alice Springs. To make an impact in Alice Springs on alcohol consumption, domestic violence, poor physical and mental health outcomes, poor retention rates at schools, low levels of Indigenous, particularly Indigenous youth employment, high suicide rates and to address the high incarceration rates, it is necessary to adopt a new approach.

The focus of all governance initiatives needs to be context specific but basic qualitative and quantitative indicators (socio-cultural, political, economic and environmental) provide guidelines for assessing the extent to which goals have been reached in terms of each one.

The premise on which the systemic approach is based is that systems are linked (Stafford Beer 1974, Churchman 1979a,b, Banathy 2001, Bausch 2001) and that management and policy to address health and development need to engage multiple disciplines and sectors, but they also need to be sensitive to the definitions and meanings of interest groups: men and women of all ages. The social, cultural, political and economic contextual issues faced by age groups and by gender groups would be a focus for action learning to address governance as it pertains to citizenship rights and responsibilities on town camps. This is in line with the Ottawa Health Charter of 1986 and UN Agenda 21 policy that stresses the need to understand the links across health and development, but the focus will be on how to make the rhetoric workable on the ground through translating policy into practice by means of PAR.

Learning to use PAR tools can help to enhance policy and practice that work with, rather than within boundaries to achieve integrated solutions. This requires matrix team approaches to design, plan and implement policy to address the complex, interrelated social challenges we faced currently and in the future. The community of practice

[5] O'Kane, A. and Tsey, K. (1999) *Shifting the Balance—Services for People with Mental Illness in Central Australia*. Menzies School of Health Research, Northern Territory.

management approach is suited to this challenge.[6] Facilitation would draw on contemporary systems thinking and practice as per Flood,[7] Romm,[8] Jackson[9] (current president of International Systems Sciences Organisation), Banathy[10] and Duhl[11] (founder member of WHO Healthy Cities). The five key action areas (as per the Ottawa Health Charter of 1986) are promoting 'healthy public policy, supportive environments, community action, personal skills and reorienting organisations'.

With higher levels of turnover in staff in the public sector in Australia, currently and in the future (as per Case 2001)[12] and higher levels expected in the future, it is vital to find alternative means to document, manage and retain the explicit and implicit knowledge within and across sectors at all levels of government (including local and community government). The argument is applicable in Alice Springs where the turnover of population is in the region of 20–30% every 2 years (as per General Practitioners Survey 1998). A community of practice (COP) using digital records could assist the process of managing knowledge (personal knowledge that is both explicit

[6] Nichols, F. (2000) *Communities of Practice: Definitions, Indicators and Identifying Characteristics*. Wenger, E. (1998) *Communities of Practice: Learning, Meaning and Identity*. Cambridge University Press.

[7] Flood, R. and Romm, N. (1996) *Diversity Management: Triple Loop Learning*. Wiley, Chichester.

[8] Romm, N. (1998) The process of validity checking through paradigm dialogues, 14th World Congress of Sociology, Montreal.
Romm, N. (2001) Our responsibilities as Systemic Thinkers. 45th International Conference of International Society for the Systems Sciences, Asilomar, USA.

[9] Jackson, M. (1995) *Systems Methodology for the Management Sciences*. Plenum, New York.

[10] Banathy, B. (2000) *Guided Evolution of Society: A systems View* Kluwer, London.
Banathy, B. (1996) *Designing Social Systems in a Changing World*. Plenum, New York.
Banathy, B. (2001a) Self-guided evolution of society: the challenge: the self-guided evolution of ISSS, 45th International Conference International Society for the Systems Sciences, 8–13 July.
Banathy, B. (2001b) The Agora project: self guided social and societal evolution: the new agoras of the 21st century, 45th International Conference International Society for the Systems Sciences, 8–13 July, 2001.
Banathy, B. (2001c) We enter the 21st century with schooling designed in the nineteenth. *Systems Research and Behavioural Science*, 18(4), 287–290.

[11] Duhl, L. (2001) Systems and service at School of Public Health, Berkley. *International Systems Sciences*.

[12] Speech by Paul Case, Commissioner for Public Employment PSA Conference, 14 August 2001.

and implicit) and to limit the reinvention of the wheel. Decision-making needs to be owned by Indigenous people. Existing conceptual skills are demonstrated through elaborate traditional mapping skills that have been usefully applied in public health contexts.[13]

## 8.2 *A Community of Practice*

The COP[14] could enable Indigenous staff members (who support 18 housing associations under an incorporated organisation) to develop diversity management approaches (systemic design, planning, facilitation, monitoring and evaluation) with the housing association members on town camps (Table 8.1). The process could create value by building connections, relationships, common ground and documentation of processes and thus enhance effective health and development practice (Institute of Knowledge Management 2001).[15] Since 1986 there has been progress in translating the vision of the Ottawa Health Charter through specific projects internationally[16]), specific agendas such as Agenda 21[17] that flowed from the UNCED summit (14 June 1992) recognise the links across disciplines for problem-solving.

The goal is to increase the viability of preventative and promotive approaches by facilitating an Indigenous COP to act as a hub for implementing change to improve governance. This approach would build on the local organisational structure and realities, rather than re-inventing another network and then trying to get Indigenous people to join. It is

[13] Conceptual maps depicted in the form of dot paintings showing systemic collaboration are evident at the Gap Youth Centre, while Catherine Abbot, a health worker with many years' experience has developed systems diagrams showing how health care could be delivered more effectively by working widely across sectors.

[14] This could be the means to address safer water, sanitation, refuse removal, drainage, electricity/power. It can also open the way for discussions on safe spaces for sleeping, recreation, learning, cooking and storage of goods. Monitoring housing functions and developing a collaborative approach to developing an environment with stakeholders, irrespective of age or gender can help to address power differences that lead to conflict.

[15] Lesser, E. Communities of practice and organisational performance, IBM Institute for Knowledge management, Boston University, Unpublished White paper.

[16] Davies, J. and Kelly, M. (1993) *Healthy Cities: Research and Practice*. Routledge, London.

[17] International Council for environmental Initiatives (1998) *Our Community Our Future: A Guide to Local Agenda 21.*

*TABLE 8.1 Considerations for Governance with an Indigenous Housing Association*[a]

| | |
|---|---|
| Socio-demographic | 16–20% of the population on any one day in Alice Springs are Indigenous.<br>Characteristics are: transient; fastest growing section of the population are young people.<br>High levels of morbidity and mortality, higher incarceration rates for Indigenous population than for the non-Indigenous population and higher levels of injury and domestic violence.[b]<br>Only 30% of Indigenous housing in Central Australia, according to Runcie and Bailie (2002), meets the most basic functional prerequisites, namely cooking, washing, storage, rubbish removal and drainage. Sense of safety, sense of well-being in the home, sense of being able to fulfil more than basic functions, such as studying were not included in the study.<br>Overall in Alice Springs the Indigenous population covers the full spectrum from owners of property to those relying on the social wage, but the majority are unemployed or on Community Development Employment Programs (CDEP). |
| Political considerations | History of colonisation, dispossession, missionary settlements, control of movement, Pastoral Wage Act and consequent loss of jobs, land rights and mandatory sentencing.<br>Over-bureaucratisation with little active civil governance to keep the elected members on track. |
| Legal considerations | The impact of legal considerations such as mandatory sentencing (repealed 23 August 2001) and the impact of the 2-km law.<br>Diverse quality of housing on town camps, but overall the impact of alcohol-related damage by residents and those escaping the 2-km law is relevant to safety and housing management. This has a direct impact on the quality of life and life chances of residents of housing association.<br>The stresses associated with young people being charged under mandatory sentencing caused grief within the extended families that comprise the housing associations. |
| Economic considerations | Reliance on the social wage.<br>Budgeting impacted by fluctuating numbers of people in the household.<br>Cost of food in remote communities is high.<br>Contingencies associated with illness and violence make budgeting difficult. |
| Policy considerations | Systemic problem-solving to promote safe communities through participatory design and promote civil society.<br>Policy research aims to demonstrate a practical approach to enable Indigenous citizens to improve their management skills through understanding rights and responsibilities (Weeks 1996, Rowse (1993a,b, 1998), as they pertain to housing[c] management to address safe water, drainage, sanitation, electricity, refuse removal, storage space, sleeping space, safe learning and recreational space, in other words their quality of life. The residents of town camps have the lowest health, education and |

*TABLE 8.1 (Continued)*

| | |
|---|---|
| | employment outcomes. The participatory action research will develop governance skills using action learning with specific age and gender interest groups (Cox 1991, Winter 2000) on town camps with a view of preventing conflict and injury. The study will focus on goal setting with selected town camps to improve housing, gardens for food and recreation and shared public meeting space on town camps, monitoring the functions of housing, gardens and living spaces and problem-solving (through appropriate communication with stakeholders). The strategy is to develop the governance skills of town camp residents by increasing access to resources, information, decision-making and enhancing relationships within town camps and with the wider community. |

[a] Social, cultural, political and economic context shape management, policy and planning at the local level.

[b] Memmot, P., Stacey R., Chambers, C. and Keys, C. (2001) Report to Crime Prevention Branch of the Attorney General's Department, Violence in Indigenous Communities, Commonwealth Government.

[c] Runcie and Bailie (2002) *Evaluation of Environmental Health Survey* on Housing Menzies School of Health.

vital that local people who have been developing innovative changes own and run the COP initiative. For example, the COP could support action learning or learning by doing so the participants could identify key issues on which they wish to work in age and gender groups.

## 8.3 *A Design for Participatory Governance*

The following are the incremental steps for setting up a transformative COP that maximises the multiplier effects of working across sectors and disciplines[18]:

1. Identify the issues by means of working with representative interest groups. The key point of PAR is that Housing Association members are the researchers and the actors. Non-Indigenous participants can act as facilitators of Indigenous facilitators. The vision of the housing participants would be a starting point. What would good town camp governance look like in terms of health, housing, education, employment? What would spaces for recreation look like? What spaces should be shared areas, Where should there be boundaries? What do different age

[18] See footnote 3.

and gender groups think and why? Developing a co-created construction or picture of what people want the project to entail through working with age and gender groups on selected town camps. Focus group discussion will be facilitated using diagrams: *analogue representations* of public areas/decorations/garden/spaces and *conceptual representations* to describe the relationships across ideas or processes. Diagrams are culturally appropriate to define the project domains for age and gender groups, analyse the systemic issues raised, communicate ideas and set up an action plan with associated tasks. It is possible that the vision for different age and gender groups will be similar in some ways and different in others. Areas of convergence and divergence could be important areas for development. The first step entailed establishing whether they think the PAR process (participation for keeping strong and keeping people healthy and happy on well-run town camps) was worthwhile as an option. This process was accepted as 'where Indigenous staff are at'. PAR is about helping participants to meet their own vision for their town camps, their own goals and identified needs (information-legal, housing management, material and human resources, relationship skills for management within and beyond the town camps and decision-making/communication/conflict resolution skills, processes to implement the goals for housing management. Issues of visitors, rental payments, notions of property, rights, responsibilities may be relevant.

2. Deciding what is needed and how the groups will work to meet realistic goals. Deciding on what success will look like. This will form the basis for evaluation. Identifying the resources that people on town camps can bring to the project and how the enthusiasm and motivation can be maintained.

3. Establishing ways that the town camps can work to assist one another and how other organisations can assist their COP (network to enable and build Housing Association capacity).

4. Establishing a meaningful work program that could follow the ideas of the potential plan and access resources.

5. Involve the members of Housing Associations in understanding that 'health is where people live, work and play' (Mike Sparks, 1999, Healthy Cities Coordinator, personal communication). It is created by people in their interactions with one another and with the physical environment. Achieving health in terms of physical, mental and spiritual well-being is about working in partnership to address complex challenges.

6. Set up a community of practice with Tangentyere Council by providing a practical understanding of systemic thinking and practice as applied to management and governance.

7. Identify an achievable project that gains widespread support across interest groups on a town camp, because it addresses a key issue for many people in the community.

8. Build by means of incremental steps, starting with one setting at a time. As each project achieves a successful outcome on each town camp, it needs to be documented in the media and celebrated as a step towards achieving the goal of Indigenous citizens creating successful governance systems and determining their own futures through their own designs. This design proposes supporting local governance via Indigenous housing associations (for town camps) to set up a COP. The proposal is to use PAR to assess its value and impact on improving governance and to guide and design the future development of Indigenous living choices. It is envisaged that this proposal to improve governance and management would support existing initiatives and priorities. The pilot project needs to be located at two town camps, selected in terms of their accessibility to alcohol outlets and to 'outsiders' who use the camp as a temporary base. Maximum and minimum accessibility to alcohol and outsiders would be the basis for selection. The central design is to surface assumptions and values about task, process and outcomes and to learn by doing small incremental tasks. Rethinking and remaking boundaries is vital for problem-definition and problem-solving. It is about understanding the big and the small picture and reworking design and management through interactive task-oriented processes. Nothing succeeds like learning from failures and making the next initiative better (Churchman 1982). Designing alternatives requires making opportunities to rethink and envisage alternatives, beyond the existing boundaries. The process involves enabling participants to engage in learning to meet organisational challenges. It involves conversation and learning from doing. To sum up PAR:

(a) applies action learning on problem construction and systemic approaches governance;
(b) teaches contemporary systems thinking in such a way that it builds on the live experience and skills of the Indigenous participants and resonates with their experiences and values;
(c) shares action learning tools to enable residents to take ownership of issues that are of importance to them via PAR[19] with Tangentyere staff and associates in the field.

[19] PAR is based on 'learning by doing'. It involves the participants at all stages of the research process and aims to transfer skills to the participants.

The process of participatory action learning (using systemic approaches) is one of learning through dialogue, observation and action with age and gender groups (Fonow and Cook 1991, Stanley and Wise 1993), with the specific goal of improving social dynamics and empowering the participants to address their sense of well-being. This is particularly relevant to empowering research that strives to ensure that participants determine the constructs and the direction of the interventions that form the basis for learning.

It is based on thinking skills called 'unfolding' meanings and their implications for stakeholders (as per Ulrich 2001) and 'sweeping in' social, cultural, political and economic considerations (as per Churchman 1979, 1982, Banathy 1996). These skills can be taught through the use of graphic, conceptual drawings and could build on painting and dramatic and narrative skills (story telling) as a starting point for identifying an axial or central issue on which the Housing Association members could work (in age and gender specific groups and as integrated, intergenerational groups) to improve the quality of life. Quality is a concept that is perceived as an aesthetic, cultural, moral, political, spiritual/religious concept, as per Churchman (1979a and b), but it has physical implications. Physical, mental and spiritual well-being can only be achieved through integrated approaches.

The workshops address:

1. Accountability as a focus of citizenship rights and responsibility.
2. Participatory governance and participatory design (as per Banathy 2000).
3. Facilitating democratic Indigenous governance. The lessons would demonstrate the belief in the ability of ordinary citizens to make a contribution to conceptualizing, planning, monitoring and evaluating development. The underpinning assumption is that the process of involvement is as important as the so-called tangible outputs of the project. The intangible outcomes are building the capacity and the confidence of people to engage in designs that 'leap out' (Banathy 2000) of existing boundaries through working creatively, confidently and respectfully with one another.
4. Action learning uses the process to build confidence and to provide an accessible overview of tools. For instance, tools for surfacing values, drama and narrative, heartstorming for participatory design (Banathy 2000) and linking these with the notion of what citizenship and governance for self-determination means (see Rowse 1998a, b).
5. Mapping systems using Checkland and Scholes' (1990) soft systems perceptual holons versus hard systems technical flow diagrams

(Jackson 1991: 260), pictograms showing the priorities of the participants and symbol scales for people with limited literacy and numeracy. In order to facilitate understanding, conceptual diagrams would be created and used by participants. One of the most useful is 'The Cone' (Estrella 2000: 45) that illustrates the participant's vision for a project (at an individual, interest group level by age and gender, household, organisational and town camp level) in terms of both tangible and so-called 'intangible' benefits (Estrella 2000). This conceptual tool provides a basis for developing indicators that could address both and thus meet the potential criticisms of those who require rigor in research. Strategies for combining different methods at different levels would be discussed, as is the importance of both qualitative and quantitative measure to fulfil different tasks. Rigor is particularly important when using transformative methodology because it must stand up to scrutiny and criticism from a wide range of stakeholders.

6. Systemic Management (as per Jackson 1991, current president of International Systems Sciences Association).

7. The process of participatory approaches for co-creative management.

8. Addressing citizenship rights and responsibility through systemic approaches to governance.

9. Supporting existing initiatives to address systemic life skills and capacity at a personal, household, town camp level and within the wider social, political and economic context.

10. Livelihood strategies, based on an understanding that not all poor households are the same and that understanding why (using story typologies) can make a considerable difference to life choices and chances.

11. Story telling for data gathering to improve governance on town camps. This approach will be used to focus questions and recording relevant points that pertain to 'checking out' how specific actions are unfolding. The participants would use this skill to monitor the progress of governance initiatives that they have identified for action.

12. Systemic maps showing, for example, the implications of including one extra investment, on the overall sustainability of a household.

13. Action learning on measurement and meaning. An introduction on ranking and scoring or giving a value to an issue.

The process is facilitated through the COP approach. This approach uses matrix, fluid strategies to learn by doing. It favours an incremental approach, not blueprints of direction from the outset. The members of the COP guide the process. The Housing Association members as part of the COP would direct all stages of the project, because they will own them.

The five steps involve:

1. Finding a meaningful focus for PAR.
2. Funding a part-time or full-time project officer with administrative back up to implement the project in partnership with relevant organisations.
3. 'Giving credit to everyone for his or her efforts', 'look for the positives and be prepared to work with everyone' (see Mowbray 2000).
4. 'Being prepared to help others and be helped by others' (Mowbray, Illawara Healthy Cities)
5. Developing a COP website and training for knowledge management.

The design uses conversation as a key learning tool. The distinction between emic (or insider knowledge) and etic (or outsider knowledge) is the key to understanding the role that identity, values and mindfulness play in framing a participant's personal understanding of issues. Shared Indigenous and non-Indigenous reality can only be co-created through respectful conversation that aims at mutual understanding and draws on the particular skills of the participants. The challenge is that even the most basic ideas (like 'management', 'rights', 'responsibility' or 'learning') are perceived and used differently. Differences stem from a host of systemically linked factors and are the result of historical, socio-cultural, political, economic and environmental variables.

In the words of Wenger (1998: 220) in reference to the COP:

> It … seek [s] the reconfigurations necessary to make its leaning empowering—locally and in other relevant contexts … . A learning community is therefore fundamentally involved in social reconfiguration: its own internally as well as its position within broader configurations … . This means that COPs need to consider the social, political and economic contexts in which it operates.

To further quote Wenger (1998):

> Issues such as the acquisition of specific subject matters, involvement in civic concerns, and people's relations to their jobs are actually implicated in the structures of meaning, even though they are often cast in terms of personal choices and abilities … . Of course availability of information is important in supporting learning. But information by itself, removed from forms of participation, is not knowledge, it can actually be disempowering, overwhelming, and alienating … what makes information knowledge—what makes it empowering—is the way in which it can be integrated within an identity of participation.

In other words, the etic (or outsider knowledge) needs to be understood through engaging in conversation to enhance mutual understanding.

Conversations for co-creating meaning can address what 'competence' and 'management' mean to the different parties. Greater mutual understanding can lead to creative solutions across sectors and disciplines.

At a praxis level (this means linking thinking and doing) the workshops and learning materials could provide a resource for those practitioners and citizens who wish to make a difference by working across sectors and in an interdisciplinary manner. It is both practical and idealistic in its message. This is not a contradiction. Unless we can empower people to voice their vision through respectful dialogue (as per Habermas 1974), creativity is lost and democracy is stunted (Banathy 1996, 2001 a, b, McIntyre 2000, Romm 2001). Scaling up participatory approaches from the local, to the national and international context is increasingly important as a means to promote mutual understanding.

At a specifically theoretical level, the seminars and conversations that make up the fabric of the COP could contribute to critical and systemic thinking and will demonstrate the inherent value in open discussions that can improve design and 'diversity management'. In this way the Indigenous COP could contribute to the field of management and organisational learning (e.g. see Flood and Romm 1996, Romm 2001). The importance of building trust is stressed throughout. Without trust the process of evaluation and organisational learning can be seen (at best) as rhetoric or (at worst) as just another means to control the participants.

History, language, religion, politics and the environment would be 'swept into' the discussions, rather than framed out by rigid approaches to research, management and evaluation (see Ulrich 2001). This systemic approach is vital to ensure that multiple variables are held in mind. Ignoring 'just one variable' can make all the difference: a mistranslation of a term, ignoring cultural nuances, token gender considerations, forgetting the importance of social dynamics and their political/historical context could undermine the viability of a project.

Similarly, the systemic approach allows for the process to consider both the intended and the unintended results of interventions. The pros and cons of each are considered valuable lessons from which to learn, so that the next phase of the cycle can benefit from what was thought, said and done previously.

Through PAR the participants could obtain a greater understanding of the mechanics of enhancing Indigenous governance skills for specific age and gender groups. The process could provide:

1. An opportunity to introduce action learning on governance issues, that is, age, gender and culturally sensitive. Improving governance skills with a specific focus on promoting healthy

household dynamics through the housing associations. The starting point would be working with interest groups and intergenerational groups to establish specific ways to promote their health, well-being and a safe environment. It is quite likely that the Indigenous COP may be re-named to become culturally appropriate, but it needs to be an ongoing daily arrangement on town camps for thinking and doing in separate age and gender groups and intergenerational groups, not a one-off talk festival.

2. Training Indigenous staff and residents who participate in the COP to focus on citizenship rights and responsibility as they pertain to gender and age.
3. Facilitating and documenting action learning to promote effective practice and to ensure that learned skills are retained by the residents of town camps.
4. Evaluating the progress in governance of town camps via relevant town camp based organisations, such as the Housing Associations as a means to learn more about effective change management praxis (as per Fuerstein 1986, Fetterman 1989, Friedman 1992, McIntyre 1996).

Values are at the heart of the definition of well-being and are at the heart of all development initiatives. Unless the initial definitions are owned by specific interest groups (age and gender specific) and shared to develop a co-created sense of citizenship rights and responsibilities (McIntyre-Mills 2000, Romm 2001), then the process of development is meaningless. Definitions that are owned and that reflect the needs across interest groups, that reflect the meaning of rights and responsibility can form the basis of conversations and practice that 'have radiance' (Churchman 1979a,b) and power to transform. Radiance is the difference between meanings that flow from self-confidence and a sense of dignity and identity, to meanings that are imposed.

The techniques for action learning stem from the belief in the ability of people to change their worlds through thought and action. The playing out of options to address issues can act as 'mental walk throughs' and a means to address practical concerns in the future. The conceptual skills can be taught in simple and direct ways using action learning techniques that build on conversation (as design and practice tools). This approach can be usefully applied by staff and housing association members (Table 8.2).

To sum up a COP could enhance (i) governance skills and well-being and (ii) professional learning about systemic approaches to managing complexity. In practice it could lead to enhancing the skills, well-being and life chances of the citizens who participate.

The value of this approach to management and action learning is that it is systemic and it appeals equally to those who are used to westernised ways of operating and those with a more traditional, egalitarian way of life. It is particularly useful for working with and between Indigenous and non-Indigenous groups. Its potential for problem-solving and co-creating solutions where there are differences is worth pursuing (provided generative communication can be achieved) because COPs are based on trust and a sense of resonance. Participation enables having a stake in

*TABLE 8.2*

*Systemic Learning, Design and Action to Enhance Governance and Manage Change*

| | Level 1: Indigenous community of practice (COP) | Level 2: Facilitation of Participatory Action Research (PAR) |
|---|---|---|
| Focus | Pilot of a potential healthy setting hub for Indigenous people with a focus on health/well-being, education and employment | Facilitate PAR and provide learning support for COP members on diversity management. |
| Task | Institutional strengthening through PAR to facilitate learning by doing. | Staff member appointed to facilitate the COP. |
| Process | Coordination of initiatives | Systemic approaches to achieving health and development:<br>• application of systemic practice tools to undertake action learning, research and evaluation;<br>• the development of diversity management and knowledge management.<br>COP as a vehicle for learning and applying multi-disciplinary and intersectoral approaches. |
| Potential outcomes | A COP for sustainable, healthy living, learning and re-creating that is supported by organisations such as universities, and the Gloria Lee Wellbeing Centre (that addresses health and development in an exceptional natural setting with eco-tourism/alternative retreat/conference centre potential). It has the environmental potential to be commercially viable Aboriginal enterprise that could be of benefit to | Contribute to setting up and replicating healthy settings on town camps and in the wider community through the COP. A potential outcome is the creation of a Healthy Settings Co-operative for existing collaborative projects, for example. |

*TABLE 8.2* (*Continued*)

| | Level 1: Indigenous community of practice (COP) | Level 2: Facilitation of Participatory Action Research (PAR) |
|---|---|---|
| | members of an Indigenous COP. Commercial enterprises could subsidise non-commercial activities to enhance social capital. | |
| Potential outputs | Workshops to demonstrate the existing innovative infrastructure:<br>• Solar power technology.<br>• Mud brick building techniques.<br>• Construction and maintenance of organic latrine.<br>• Recycling materials for building, gardening, crafts.<br>• Food security through growing healthy, affordable, shade-giving plants as food and through budgeting.<br>• Environmental protection, for example, soil quality and removal of invasive plants, building a fence to protect crops from feral horses and camels and to protect indigenous animals.<br>• Piloting a community exchange system.<br>• Planning a microfinance program with GO and NGO support. Marketing Gloria Lee as a well-being and conference center to support community development activities. Therapeutic activities and workshops such as recycling as art, bush walks to appreciate Indigenous plants and animals, narrative therapy for recovery from alcohol and other drugs for men and women of all age groups, facilitated appropriately with professional advice.<br>• The building of self-esteem for men, women and young people from town and from remote community in specific workshops. | • Seminars, workshops and training material for the transfer of research and management skills to local professionals and the COP.<br>• Website for COP to document and manage knowledge.<br>• Increased resources for hub demonstration projects. |

*TABLE 8.2 (Continued)*

| | Level 1: Indigenous community of practice (COP) | Level 2: Facilitation of Participatory Action Research (PAR) |
|---|---|---|
| | • Communication as therapy through narrative, drama, dance, music and poetry.<br>• Recreational craft activities and therapeutic conversation for young people.<br>• Camping and working day trips and longer stay holidays for people from Alice, interstate and overseas. | |

ones own future. Remaining on the sidelines through passive resistance or apathy can lead to being the object of another's future, rather than an author of one's own future.

# 9

# *Conclusion: Addressing Complex Reality*

## *Systems, Barriers and Portals: Identity, Nationalism and Globalisation*

> In order to be able to close the gap between technological and socio-cultural intelligence, a major shift toward attaining more understanding and human wisdom is required. We should create learning that enhances critical thinking, the understanding of the self, the systems and the environments in which we live, and the situations we experience. We can nurture wisdom by creating learning resources and arrangements by which to relate knowledge acquired and understanding gained to pragmatic, moral, ethical, and affective issues. We can then apply wisdom in making judgements, managing problem situations, making decisions, and living our lives enriched by our growing wisdom.
>
> (*Banathy 1991: 77*)

### *9.1 Summing up the Challenges for CSP*

As I write this chapter the news of a change in community dynamics came about with the new Chief Minister Claire Martin, declaring that she would end mandatory sentencing (23rd August 2001) and the Yeperenye (Butterfly) Centenary of Federation was celebrated in Alice Springs, led by bands such as Yothu Yindi celebrating 60,000 years of Aboriginality. It highlighted survival of the spirit in Alice Springs (known as Mparntwe or 'caterpillar dreaming') through colonisation, separation through the stolen generation, the nuclear testing at Maralinga and the early land rights appeals to current day appeals for a treaty. The voices retain a strident sense that land is an issue and that reparation remains an unaddressed concern. It is in this context that participation needs to be addressed in planning for the future if a nation within a nation is to become a nation of respected citizens who participate in practical solutions for development, rather than a nation based on separation and reparation. A new potential for generating a shared sense of community is possible and the future is ripe for implementing participatory designs that could make a real difference for the future. Banathy (1996) stresses this as the distinction between 'generative' and 'strategic dialogue'. The one must precede the other for trust to be developed (Laszlow 2001).

> "We need a new mentality and a new attitude towards our fellow human beings, everywhere on the globe, if we are not merely to survive physically, but to save for future generations the treasures of the spirit, the rules of conduct, the ideals, the faiths that are our finest heritage from the past. More than mere material progress is needed in out times: true civilisation demands civilised minds" (Strehlow 1956: 12).

Respect for the other is the prerequisite for the next stage of cocreated evolution. Self directed designs are a reality of the future. Technology and (unfortunately) not as yet our ability to think ethically, systemically and wisely has allowed us to move closer to thinking that we can bridge the divide between mortality and immortality, between the sacred and the profane, between humanity and divinity. Let us be mindful of the responsibility to self, other and the environment and remember that these links are the web of life.

We need to respond interactively to address the challenges that we face. They are not restricted to any one discipline of knowledge or the public sector. As the world faces the challenge of new forms of conflict fought in civilian contexts and as the competition is expressed in specific political, social, cultural, economic and religious narratives, it becomes ever more important to establish ways to engage in dialogue that can open closed mindsets to grand narratives, that can be glossed very understandably as evil by the populations of people who suffer under the yoke of oppression, defined albeit in very different terms. All human beings and in fact all sentient beings feel pain and we share in common our biology and our planet. If lasting peace is to be achieved we need to work together to achieve solutions that are systemic and that go beyond the political, the economic and the religious (that can become distorted proxies). We need to nurture the web of life and remember that this web is sacred. Humility and wisdom are needed now more than ever to co-create solutions. This is the hope for the future.

Constraints need to be addressed through human agency. CSP is based on a belief in the potential of human beings to construct and reconstruct their futures. In the new millennium the challenge is to ensure that biodiversity is maintained and that the powerful do not silence those currently with limited access to communicate their knowledge narratives. This is not an argument for universalising language or knowledge narratives, because the cognitive meaning maps associated with specific language, specific place and specific time[1] and specific discourses can be of vital use for future planning, because of the insights they provide into understanding not only ourselves, but also the environment. Examples of 'bush knowledge' are commonly cited but the spiritual relationships we have across self, other and the environment are often less cited nor is the most important consideration that it helps human beings to better understand themselves. The lived experience stored within language is based on empirical testing to provide a means of understanding and engaging within the world in a particular way. The structure of language (as per Levy Strauss 1987, Winch 1958) tells us about the structure of thinking associated with particular ways of life and systems of meaning, albeit extreme forms of relativism does not help the process of co-creation for enlightenment. But if we assume the rational potential of human agents (drawing on Habermas 1974) then we need to strive through open, generative dialogue (as per Banathy 1996), towards drawing on the liberative potential (as per Gouldner 1971, 1980) in order to shape our world responsibly, but with due regard to people's constructs of truth and that power and knowledge are indeed linked (Foucault and Gordon 1980). The quality of our communication is all-important in creating a shared future, if we are to avoid retrogressive moves into separatism and isolation that are the product as much of fear as they are of aggression. 'Our entire history is connected to space and place, geometry and geography ...' (Negroponte 1995: 238) Despite Negroponte's opinion that the national state is no longer a relevant political entity it needs to be balanced by the reality: nationalism as closure born of resistance. However his belief that cyberspace is important is irrefutable. To participate in cyberspace is the challenge. The barriers are those of poverty, the same barriers to geographical space and recognition of conceptual space (the old power and knowledge connection as per Foucault). Governance can and is likely to be achieved through

[1] Schwandt, T.A. (1994) Constructivist, interpretivist approaches to human inquiry. In Denzin, N and Lincoln, Y. *Handbook of qualitative research.* Sage, London.

cyberspace; there are no

> physical limits or boundaries other than the contour of the planet itself. In the same way that media have gotten bigger and smaller at the same time, so must world governance ... This will not occur overnight, but there are signs of it happening faster in some communities than in others. I don't mean physical communities. I mean the likes of financial or academic communities ... . Some people find this threatening. I find this exhilarating (Negroponte 1995: 239).

Negroponte's science and technology narrative does not include a recognition of the value of cognitive maps that will not be represented because the digital portals remained barred by illiteracy, innumeracy and the lack of access to a computer. Democracy in education remains a challenge. People need to be empowered to build on their personal lived experiences (as per Freire 1982) to be actively involved in shaping their futures through becoming computer literate and mapping their discourses. These needed to be developed in Indigenous languages but also communicated in more universal languages, so the knowledge can be shared. Facilitation of computer literacy and access to computers and electricity is vital for participation in a digital age.

Ecological humanism argues that by virtue of our common humanity and shared planet we need to achieve a balance between our common denominators and the value that can be gained from preserving biodiversity and cultural diversity, as it is a jump lead of creativity and survival. Uniformity brings conformity and a lack of resources for responding to challenges. CSP strives to understand the dynamics of social interaction so as to empower those who are marginalised (conceptually and in terms of cyberspace) as a result of poverty.

Alice Springs faces some of the same challenges as other regional and non-regional areas in Australia (and elsewhere) such as the impact of privatisation on services and the reduction in the size of government. This impacts on human services in particular. Service providers frequently rush in to address the effects of social and environmental problems without a thorough understanding of the nature of the problems (as effects of historical, social, political and economic processes). Diversity management is particularly relevant to responding to the diverse social, political and economic needs of the population, some of whom are winners and others losers in the development stakes. Change agents need to map conceptually the values and assumptions of organisations and stakeholders to facilitate dialogue and problem-solving (McIntyre 1996). But transcultural thinking needs to honour the integrity of Indigenous narratives through developing documentation that reflects the diversity of thinking (Fiona Walsh 2002, Tangentyere Planning Day).

One of the greatest challenges to systemic co-operation is a real distrust of the (so-called and much vaunted) policy of building partnerships. The rhetoric is taken on board by most of the public sector locally, but the organisational structures and management style remain the same, even if the new concepts are mentioned like a mantra. Mergers and cuts are the feared result of working across organisational sectors and giving away 'trade secrets'. Jeopardising the funding opportunities of a section or unit is anathema, because in fact organisations are structured to work in competition and not to think collaboratively. It is simply not the way people are 'wired' to work. Resistance to change is due as much to a fear of cuts as it is to cynicism about the motives for change. Also for successful cooperation to occur there is a need to know and understand the landscape of organisations across the sectors and the contextual details. This sort of knowledge is built up over years. The nature of short-term contracts for sectors of the workforce (the knowledge workers) does not help to facilitate change, because so many industry workers in the human services have a short-term, rather than a long-term understanding of social issues.[2]

Competition for resources across the sectors is another barrier. A shift away from full-time government work to competitive tendering, contracting and part-time work as a result of 'downsizing' not only jeopardises collaborative work but reduces the professional human service worker's sense of career development (Territory Health Services Planning for Child Care at Community Services (February 12th 1999). The devolution of responsibility away from government to the private sector and non-government organisations has considerable implications for both staff and the citizens who use the services. The notion of user pays is a recurrent theme as governments move away from universal to residual health care.

> Direct government employment in the town has already reduced by 19% from 1986–1996 whilst the total employment has grown by 19%. Undoubtedly some of the direct employment cuts have been replaced with program funding and/or subcontracting of services. However, privatisation and subcontracting of services opens up a very competitive market where interest is expressed from outside Alice Springs and where price cutting is commonplace (Street Ryan and Assoc 1999).

The split between purchaser and provider encapsulates placing dollar value on (the commodification of) human services and emphasises efficiency and outputs, rather than effectiveness in the long term. A related concern is that mergers will result in more generic service delivery as a result of systemic thinking. Diversified human services are

[2] Tsey, K. (1999) Menzies School of Health. Alice Springs News Vol. 6, No. 1: 5.

considered vital to meet the specific health, housing, education and legal needs of Indigenous people, for instance. Another fear is the paradox that working too closely with the third or volunteer sector can help to fuel further cuts to paid work positions and not necessarily as a pathway to building more options for the unemployed.

A cynical attitude to systemic ideas that merely colonises the terms as empty rhetoric (to bolster narrow managerialism) remains as much of a challenge as a refusal to consider the value of the approach, or wrenching out the liberative potential through editing and controlling the framework of what constitutes knowledge. Even when holistic and sustainable approaches to development are taken onboard by bureaucratic organisations in name, bureaucrats that still think in terms of building power bases can eliminate the liberative potential. A related issue is bureaucratic control of research that can be described as follows: first appointing what is thought to be the right person, that is, in terms of meeting the specific requirements of the organisation, then the research brief and terms of reference can be shaped in terms sufficiently narrow to suit the needs of management.

Where there are competing values in terms of understanding the nature of development and the process of development it becomes particularly difficult to bring about change, because it requires allowing space for diverse narratives within one landscape of options, but also requires a shift in values to place social and environmental justice as central goals for all dialogue and action. Furthermore, instead of implementing organisational management for managing diversity, the traditional scientific management style (suited to controlling stakeholders, rather than drawing on the creative potential of 'human resources') continues to be the order of the day in old style bureaucracies. Even worse, the bureaucracies continue to focus on compartmentalised areas, rather than facilitating an integrated approach to development. Toffler (1990) identified 'cubbyhole' thinking more than a decade ago. It can be regarded as strait-jacketing thinking. This adds to the concern raised by Foucault and Gordon (1980) on the links between power and knowledge. Because of the close links across academia, the public and the private sector it tends to have a negative impact on stakeholder's ability to be creative. Diversity management theory (Flood and Romm 1996) is largely ignored or at best referred to rhetorically in organisations that function as closed, hierarchical systems that control thinking and practice. With the rationalisation of public service organisations, the notion of being objective researchers in tenured positions is no longer common. The control of research at all the stages of conceptualisation, design, process of data collection and the process of writing it up is increasingly the preserve of management. 'Who is controlling the research?' is a question asked quite openly by managers and politicians who shape the terms of reference. Parameters can ensure that some questions are asked, rather than others. Research needs to ensure that all stakeholders are co-researchers who develop the concepts and determine the frame of reference. This ensures that research is relevant and able to represent the range of opinions. Once the range has been mapped, then points of overlap can be found or created through respectful dialogue, to ensure that shared rights and responsibilities can be developed.

Policies at the global and local level impact on the conceptualisation of welfare and development. Internationally, globalisation has been translated by Western democracies into economic rationalism. In general terms, the rationalisation of the welfare state internationally is a response by governments in a bid to remain competitive in global markets. Social problems in terms of this approach are increasingly individualised and citizenship models emphasise the responsibility of individuals and families. Privatisation in the human services also leads to focusing on easy to process 'clients' rather than working with clients who are likely to be expensive to educate, treat, counsel or service in general. This impacts on the life chances of people. Some current development models that are non-systemic (psychological models, medical models, education models, crime prevention models) and economic models stress that individuals need to take responsibility for their problems (Foucault 1979) rather than both individuals and society. A response to a fast changing competitive world does not have to be economic rationalist in orientation; it can be systemic and sustainable. Strategies to address systemic and sustainable models stress that social, political, economic and environmental variables need to be considered. Social, economic and environmental capital needs to be addressed if sustainable and socially just local development is to be achieved. The gap between the ordinary citizens and elected representatives needs to be closed by an active civil society.

When people are excluded from participation in governance they cannot co-create the design of their own futures. For this reason the triple loop learning approach needs to include the fourth loop of 'who decides'. In order to communicate and participate in conceptual, geographical and cyberspace numeracy and literacy are vital. But so is respect for the personal knowledge that people have developed as a result of their lived experiences. This is relevant to individuals, organisations and the cultures of interest groups of all age groups. Decision-makers need to understand the legitimacy not only of gender, class, culture, race, level of ability, but also age as a category to be considered when diverse interests are considered for participatory planning. This involves planning with, not planning for young people! The citizenship rights of young people were particularly highlighted at the time of the research as they felt that mandatory sentencing and zero policing made young people fear to use public space. Coping with grief and sadness is a central concern if people are to participate meaningfully in designing for the future. Facilitating optimism and a sense of fun can help to generate communication in a range of media.[3]

## 9.2 *Policy Suggestions and Interactive Design*

We need to understand the new approach to development thinking. CSP entails understanding the links within and across sectors (e.g. education, health, housing, recreation, crime protection) and disciplines (e.g. primary, secondary and tertiary education, public health, engineering and architecture, recreational studies and criminology) in order to address major concerns. A model is of value if it is used for problem-solving to address identified linked problem areas. Improving health, development and social well-being is only possible through working across sectors and understanding that issues are linked. This requires that management of organisations be based on open, non-hierarchical communication.

Life chances of citizens can be enhanced in multiple ways through inclusive, integrated policy and planning *by, with* and *for* people and a sustainable environment. This requires a process of including *representatives of all interest groups,* irrespective of age, gender or socio-cultural background. Focusing on the links across apparently discrete problem areas can enhance the multiplier effects of programs.

We need to strive for participatory governance approaches to ensure that people are the subjects of their own futures, not the objects of design that may not reflect their needs. Participatory Action Research, that is 'learning by doing' can develop the civil arena and place the issue of power at the centre of discussions between the representatives and the ordinary citizens, by asking:

*Who has participated in the process of decision-making?*
*In whose interests are the decisions?*
*Do the participants represent a wide range of people across sectors and disciplines?*
*Have all the participants in the project had a say in setting the goals and objectives?*
*Have a wide range of interest groups been represented?*

[3] For instance at a time when NT government was moving into a rationalisation approach to services such as education, health, power and water, funds for art projects could be found in surprising places if some lateral thinking was used, for instance using paint maintenance budgets creatively for art projects.

*Have the social, cultural, political, economic and environmental aspects of the decision been considered in terms of a SWOT (Strengths, Weaknesses Opportunities and Threats) chart?*
*Who will participate in the process of implementation and why?*
*Are the goals and objectives in line with sustainable development?*
*Has there been an adequate opportunity to review these objectives in the light of the social, political, cultural, economic and environmental factors?*
*What is the organisational design of the project?*
*What are the formal and informal communication channels?*
*Where are the breakdowns in communication and why?*
*What is the structural context of the organisation/s doing the implementation?*
*What are the barriers to implementation and how are these being addressed?*
*What are the expected and unexpected outcomes of the project?*
Who or what will be positively impacted?
Who or what will be negatively impacted?
*What steps have been taken to maximise long term benefits to the environment and social justice in order to ensure a sustainable future?*
*What are the outcomes?*
*Has the environment benefited?*

An integrated model for governance spans sectors. The approach incorporates both a socio-cultural focus and environmental focus. It provides clear steps for action through a set of questions to guide ethical, systemic thinking that clearly links social, cultural, economic and environmental capital. Another advantage is that the process of evaluation begins when the stakeholders make decisions on what the long-term goals and objectives should be.

When we slip from co-creation to zealotry or cynicism, these are more likely emotions. When identities are defined in opposition and based on competition, encouraged by the market rules paradigm, there is little scope for co-created solutions. Barriers are distrust of the other based on fear, greed and contempt. The challenge is to build trust through selling the message that co-created solutions make sense and are pragmatic for our mutual survival. The process for achieving integrated development outcomes involves the following steps:

- Multiple meanings and their relevance to problem solving.
- The way structures are shaped by people's interactions.
- The way process (interaction) and structure are linked through management and governance.
- The way systemic or holistic thinking and practice can influence the situation.

I have tried to show the implications of current governance decisions on the life chances of all the citizens of Alice, how people could shape their futures when given an opportunity to do so. The implications of non-participation and reactions to non-participation are also addressed.

Open systems of communication are responsive to diversity. Closed, systems associated with multilayered hierarchy do not foster creativity and iterative communication that is necessary for problem-solving. The management styles associated with closed hierarchies are discussed in so far as they impact right across the social, political, economic and environmental system. The detailed case study demonstrates the impact that closed management systems have on governance and decision-making. The suggested approach to interrelated social, political, and economic and environmental challenge was to introduce a critical systems praxis of healthy settings that is as mindful of power as it is of empowerment. Intervention needs to be systemic, not mechanistic, because the understanding of societies is still reminiscent of the old approaches to science based on machines, rather than systems. It is essential to work with diversity and to co-create meaning, not to try to impose one outcome or one set of meanings. The reality is that the social world is complex and to impose one dominating approach restricts opportunity and does not allow for creative solutions to change. Development concepts need to be defined in culturally acceptable ways. For instance, self-actualisation will mean different things to different cultural groups. It could be defined in terms of land and landscape as being essential for health and well-being. Self-esteem is a Western concept and also needs to be associated with community well-being not individual well-being for some cultures. A sense of well-being is a prerequisite and pathway for the prevention of a number of social problems (Ministry of Maori Development and New Zealand Ministry for Health 1997). Models from New Zealand and Canada are useful in this regard. The emphasis is on a broad-based raft of preventative measures to build well-being. The starting point for development needs to be an acceptance that diverse cultural positions need to be accommodated in terms of their perceptions of need, if a sense of well-being associated with self-determination is to be achieved. When community well-being is lost it is difficult to co-create meaning, because confidence is required to share personal knowledge (as per Polanyi 1962) with the other.

## 9.3 *Building and Sustaining Life-Long Learning*

A participatory governance approach can sweep in the considerations of power and reduce the distance between the elected representatives and ordinary citizens. PAR processes can help to build up the arena of civil society as conceptualised by Gaventa (2001) and Gaventa and Valderrama (1999) can provide a means to implement a modified form of WHO healthy settings to address the needs of people marginalised in terms of social, conceptual, geographical, and cyberspace (with the attempted addition of power and empowerment and all the associated problems) stress tools for ethical thinking and caring to redress the impact of the market economy (commodification of body and mind). The focus on participatory design for the future is very important and the means to participate in information-sharing and information-creation via basic skills such as numeracy, literacy and confidence in one's own self-worth. Confidence is linked with respect for the value of one's own history and experience. Analogies from the natural sciences are useful insofar as we can develop our understanding of living social systems. In mathematical terms we need to find the common denominators through co-creating meaning despite the divides caused by academic controls, managerial controls and state control on what constitutes legitimate knowledge for social policy decision-making! The big challenge is to find ways to convince human beings that co-operative, rather than competitive solutions are best for the long-term benefit of everyone. The building of incentives for people to want to co-create is the challenge. Why should people want to trust one another, only for pragmatic reasons (the low road to morality, as Alinsky, [1972] a critical community operations researcher/practitioner called it). So systems thinking can play a role in showing the connections in the short- and long-term of co-creating rather than working in competition, but if people are so caught up in the benefits of economic rationalism for them in the short-term (as per the refusal of some industrial nations to sign the Kyoto agreement in 2002) then we are a long way off achieving this goal. Only when people realise the boomerang effect (to use Ulrich Beck's 1992 term) of pollution (social/political/economic) across boundaries of class and power will people realise that pragmatically it is better to co-create than to dominate. Class is alive and well again and with the end of the welfare state critical thinking has an important role to play.

Systems thinking and practice can engineer social changes that silence diversity (the zealotry option) in line with a belief in one grand and powerful narrative of the market rules, because it must (as it is a natural system). Or systems thinking can be used to co-create solutions in the interests of preserving both diversity per se and the conditions for diversity to flourish. I don't believe that we have multiple narratives at the moment because they are silenced. Diversity is the jump lead for creativity, which is necessary for evolution based on the contributions made by all lifeforms. Ecohumanism[4] is really just another version of 'dependent organisation', the Buddhist concept, (Koizumi 2001). The convergences

[4] See McIntyre Mills (2000) for a constructivist essay on this topic.

with aspects of Buddha's thought and systems science and also with aspects of Christianity (love thy neighbour as thy self and the notion that the word or information precedes all) is really exciting in that it addresses the issue of dependent organisation and the origins of life and the evolution of human beings and the planet.

The only problem is that of course if we are interested in issues such as illness, pain, and prolonging the life chances of some people, then we have to address the issues through political, economic and social co-creation in the human organisations in which we operate, as well as through our individual thinking, prayer, meditation and philosophy, important as it is. Troncale (2001)[5] demonstrated the awesome synergies across the sciences and intimated that the social sciences are still in the dark ages, which they probably are, but we still need to realise that people have to go a long way to realise that systemic thinking is once again a tool that can be used to engineer accountable change or simply to shore up the interests of the powerful. Accountability (as per Romm 2001) and critical thinking will be needed to ensure that systems thinking isn't just used to empower the powerful even more. Even if evil is about 'rewiring one's brain differently so as to see the implications of one's actions' (Troncale 2001, personal communication). Education is important but behavioural choices are still shaped by values. I argue that if people can see in pragmatic terms that they are all members of one species (at the moment) and we have one planet and thus we share the same ultimate needs we have the basis for co-creating meaning based on mutual advantage. Unfortunately if the divides between haves and have-nots continue then perhaps some people will be left far behind (and not realise it because they will be controlled through multiple systems). We need to ask ourselves how close we are to this stage of our evolution. The point I make about citizenship versus being passive consumers is important in this regard. What is worrying is when people can no longer understand why this is important.

Systems thinking and practice can either engineer social changes that silence diversity or multiple narratives can co-exist in the interests of preserving both diversity per se and the conditions for diversity to flourish. Indigenous egalitarian maps of commodities (as items for sharing and giving) are quite different from economic rationalist models. Co-creating meaning can create less extreme cognitive maps and garner respect for both storing up power through profit and through giving.

[5] Troncale, L. SYSML Towards a Widely-Available Internet Tool for systems design/research that promotes systems education and consensus.

Bits of data based on the Boolean choice of 'either or' approach is meaningless, unless sense is made of them in contextual terms. For this to occur 'both and' thinking also has a role in some contexts. What one does with information depends on one's ability to think systemically. Hooks or points of reference can be found to construct meaning that fits a category, but the challenge is to find ways of managing and making sense of multiple, diverse paradigms and worldviews. Only when information is processed with theoretical and methodological literacy can knowledge be created and used as a basis for decision-making that acknowledges diversity. Systemic thinking means that from the outset, conceptualisation will be multisemic. Development concepts need to reflect the diversity of the population. Instead of models such as Maslow's hierarchy of needs as a starting points (namely physical, safety, belonging, self-esteem, self-actualisation), Bradshaw's model (Kettner et al 1985) was considered more appropriate as it recognises the complexity in the definition of need and encompasses perceived need, expressed need, comparative need and normative need/right. Such a model is culturally appropriate because it recognises the diversity of citizens. This model is not hierarchical and lends itself to systemic development models.

The programmes for prevention to address interrelated indicators of social marginalisation and despair need to redress real and perceived social marginalisation through integrated development across sectors (and in so doing build social capital across all interest groups). It is suggested that integrated development programmes using a multidisciplinary and intersectoral approach advocated by the Ottawa Health Charter (1996) need to be implemented. Key tools useful for systemic development across the three tiers of government, the business and voluntary sector are: (i) Open communication for enhanced participation that stimulates the development of matrices and networks, as an antidote to rigid structures tied to organisational structures. (ii) Trying things out through projects using a Participatory Action Research process. This involves learning new ways of doing things through working with people as participants at every stage of the process and documenting the perceived strengths and weaknesses of the process. (iii) Diversity management that is responsive to diverse socio-demographic patterns is about 'multiple loop learning' (Flood and Romm 1996)[6] that is iterative and ongoing. It means that everyone shapes viewpoints and the process is inclusive and adds to developmental effectiveness. Inclusive decision-making processes

[6] This refers to the links across power, authority and what is thought to constitute legitimate knowledge.

replace hierarchical decision-making. 'Might right thinking' (Flood and Romm 1996) is replaced by reflections on the implications of assumptions. Improving development outcomes can however only be achieved in partnership across the public, private and volunteer sector. Diversity management involves citizens because their contributions as workers/service users to problem-solving are valued as a resource. This is motivating and good for production levels, but also good for management, because workers 'personal knowledge' (Polanyi 1962) helps enhance creative thinking and reduce risks in decision-making.

Integrated development strategies are required to promote health, well-being and development. The emphasis is on a broad-based raft of preventative measures to build well-being. Social movements for integrated development such as the healthy settings/cities and environments resonate with some of the Indigenous concerns about health, well-being and a sense of place. According to the International Healthy Cities Foundation[7]

> ... It began to highlight the interconnections among what seem to be diverse elements and problems in society. And finally, it suggested the solutions to both community and quality of life problems also may be interwoven.

This approach refers to built, natural and social environments as linked systems. The basic assumption is that the environmental and ecological context is integrated with the social and health context. It provides benchmarks for socio-cultural, political and economic development to suit local contexts. According to this approach all environments are understood to be arenas in which citizens need to participate in creating sustainable solutions that meet both social justice and long-term environmental concerns. The challenge is to achieve harmony and balance. Life chances of citizens can be enhanced in multiple ways through inclusive, integrated policy and planning by, with and for people within a sustainable environment. The field of community health, primary health care, community development, urban planning, sociology, public health and more recently environmental studies are the forerunners of the Healthy City/Environment Movement. These disciplines made the connections across social and environmental health and well-being and stressed the value of considering preventative rather than merely curative and reactive models.

A shift in thinking has occurred away from reacting to the problems of individuals to proactively create appropriate environments to promote health (physical, mental, spiritual) and social well-being. The value of integrated models is that it brings together research into citizenship (rights

[7] Website (webmaster@healthycities.org).

and responsibility) with public health research and environmental research through the central concepts of sustainability and social capital, that is, a sense of well-being and trust based on having a real stake in a sustainable community. A real stake means participation and working together.

The goal of the movement is to build a stake in a socially and environmentally just society. Such a role for citizens is one of active shapers outside organisations (through social movements) and through working with and within organisations. The seven principles for sustainable development[8] provide the framework for organising, implementing and evaluating (ICLEI 1998). The focus of all initiatives need to respond to the specific contexts, but basic qualitative and quantitative indicators (sociocultural, political, economic and environmental) provide guidelines for assessing the extent to which goals have been reached. The integration across disciplines and sectors is recognised as the basis for bringing about change and solving problems. New development thinking is about the origins of health and well-being and ways to promote health in a number of linked program approaches, rather than about working in a reactive and separate manner through separate departments and separate projects. Throughout Australia and the world, towns and cities have joined this global development movement for benchmarking their levels of development, with reference to a wider national and global context. The Healthy City/Environment Movement is not merely about experts finding technical solutions, it is about citizens working together to bring about sustainable health and development by, with and for people and their environment locally, nationally and internationally. By promoting health and development through integrated policies, much could be done to eliminate sickness and social problems

> Medicine is a social science and politics are nothing more than medicine on a larger scale (Virchow cited in: Kelly and Davies (1993: 3).
>
> The healthy cities program is … about a change in power relations … and a fundamental … shift in the conceptualisation of health itself … Healthy cities (and healthy environments) and the new public health are postmodern movements[9] … underpinned by a … moral view of health rather than a biological or physical definition (Kelly and Davies 1993: 7).

[8] Partnerships, Participation and Transparency, Systemic Approach, Concern for the Future, Accountability, Equity and Justice and Ecological Limits.

[9] This means that unlike the modernist approach which pigeon-holed or boxed topics, it strives to break down divisions and to blur boundaries which led to fragmentation in thinking and did not permit holistic, integrated problem solving … instead in an organisational sense it helped to build up separate bureaucratic empires. What are needed are structures that promote working in matrices and networking across disciplines and sectors.

Such a role for citizens is active not passive, not merely as welfare recipients but as active shapers of their community. Contributions to development can be made in many arenas. The concept of social capital is relevant to thinking, practice and policy insofar as: (i) understanding poverty in social, political and economic terms, that is systemically, (ii) building citizenship rights and responsibilities in the context of local government, (iii) addressing reconciliation and human dignity, (iv) improving access in terms of information, communication, attitude and infrastructure, (v) extending community safety, (vi) enhancing social health and well-being.

By thinking about poverty not only in individual but also in social, political and economic terms the focus is on what both society and individuals ought to do to address challenges in sectors such as: housing, health, education, employment, and policing, for example. Self-control and personal liability needs to be considered alongside a sense of social responsibility. Both individual and social responses are necessary for problem solving. Social capital[10] is measured in a geographical community by means of demographic, social, political and economic indicators of well-being. These indicators have been developed by the WHO[11] and were used as policy and planning indicators.

The old arenas for practicing development tended to be single focus and compartmentalised in their approach (for instance: groups, organisations, private and public sector bureaucracies, political parties). The new arenas tend to be multiple focus and systemic in their approach (for instance: networking across arenas for problem-solving including government, non-government, private sectors (management and labour), Matrix teams for flexible, creative solutions and social movements at local, national and international level to implement integrated, sustainable and socially just solutions.

[10] This 'is the basis for trust in the community as a result of having a sense that social opportunities do exist for all citizens irrespective of language, culture, gender or class'. It is summed up as '... features of social organisation such as networks and norms that facilitate coordination and co-operation of mutual benefit for the community. It builds the capacity to trust, and have a sense of security, social cohesion, stability and belonging'. Cox, E. (1995) *A truly civil society* Sydney, NSW ABC books. Boyer lectures.

[11] World Health Organisation's (WHO 1988) *Healthy Cities project paper* No 1. Promoting health in the urban context. WHO Healthy Cities Project Office: Copenhagen, World Health Organisation's (WHO 1988) *Healthy Cities project paper* No 3. A guide to assessing healthy cities WHO Healthy Cities Project Office: Copenhagen. Davies, J. and Kelly, M. (1993) *Healthy Cities: Research and practice*. Routledge, London.

Development contributions can be undertaken at an organisational level by addressing strategies to enhance diversity management through improved (open, reflexive, respectful) communication that is geared to rational outcomes, based on the assumption that the closest we can get to truth is through dialogue. By addressing the process, content and style of management people could work flexibly and creatively and at an inter-organisational level together with government, non-government, third sector (volunteering) and business, so that a concerted effort could be made to address the major community development challenges at project level in the community through (i) communication and working across sectors and disciplines to achieve greater outcomes from an integrated development approach. (ii) Building community information services to assist members of the public to access information and to facilitate the integration of service delivery. (iii) Modeling innovation and creativity to inspire policy change through specific projects.

## 9.4 *Building the Capacity to Make Sense of Data and Information for Decision-Making*

'Transcultural webs of meaning' (McIntyre -Mills 2000) could be created through informed analysis of data for creation of knowledge and for decision-making that can build shared understanding based on learning that values lived experience (as per Polanyi 1962, Wenger 1998, Fiona Walsh 2002, personal communication). This requires building lifelong learning capacity by means of accessible learning centres. Public learning centres could include public libraries, educational institutions or collectives[12] that emphasise that they are free public spaces that can be used as a means of enhancing literacy, learning and recreation, self-development and social and environmental justice in an environment that meets the cultural needs of the users. Learning centres can create opportunities for diverse worldviews to be represented. For this to occur there needs to be participation in the decision-making, so as to model principles of democratic decision-making and to draw in diverse groups of citizens to use community resources that are meaningful to them.

[12] Jacques Boulet for example has facilitated a suburban university in Melbourne Victoria that addresses the needs of a range of community users. A church building has been allocated for the initiative, but a website, newsletter and phone links expand the impact to other centres locally and internationally.

The relevance of learning centres to facilitate learning through building on areas of existing knowledge and then to use these knowledge narratives as portals for future learning is a vital role for the democratisation of education and knowledge. Learning centres need to include local knowledge in the form of oral and recorded stories, songs, art (conceptual and analogue) and artefacts in building social capital amongst citizens, irrespective of age, gender, culture, socio-economic background or level of ability could energise learning and creativity. Information technology could become starting points or portals for bridging the differences and creating 'transcultural webs of meaning'[13] in response to diverse paradigms of knowing.

Pathways for marginalised people need to be created by means of educational opportunities.[14] Computers are key vehicles for education. They help to build upon young people's enthusiasm for learning by tapping into their enthusiasm for computer technology and games. Feeling computer literate is a good starting point for building dignity, which can then lead to further enthusiasm for learning. The statistics show that youth unemployment is linked with youth crime. Lifelong education beyond the school helps to enhance literacy and language skills. Learning centres could provide a welcoming space for transcultural healing, well-being and co-creation of knowledge for the future.

Citizenship pertains to the rights of young Australians. Young people unlike those over 18 years do not currently have the right to vote but young people have the right to safe accommodation, access to public spaces and specific services to meet their particular needs, irrespective of their age.[15] The lack of space (both public and private) for young people to engage in productive activities has been well documented as one of the

[13] McIntyre-Mills, J. (2000) Global citizenship and social movements: Creating transcultural webs of *meaning for the new millennium*. Harwood.

[14] Braithwaite, J. (1998) *Crime, shame and reintegration*: Cambridge Press, Melbourne.
Cochrane, J. (1999). *Technology's shining light for orphans' bleak life*. Australian 17/1/99.
Harding, R. (1993) Opportunity costs: Alternative Strategies for the Prevention and Control of Juvenile Crime in Harding, R. (1992). *Repeat Juvenile Offenders: The failure of selective incapacitation in Western Australia*. Research Report No 10, University of Western Australia, Crime Research Centre page 141.
Jamrozik, A. and Nocella, L. (1998). *The sociology of social problems: Theoretical perspectives and methods of intervention*. Cambridge University Press, Cambridge.
McIntyre, J. (1996) *Tools for ethical thinking and caring*. Community Quarterly, Melbourne.

[15] For example, the Australian Youth Policy Action Co-alition (1997) have stressed these rights.

key prevention factors.[16] The poor educational outcomes for Aboriginal people (see Bowden 1994) who have limited school readiness skills and are without the home resources to undertake homework[17] and the high crime rate and high incarceration rate in Alice Springs indicate the need for measures to promote community integration and recreational activities.[18]

Citizenship for those with a disability means that those who do not work or are physically or mentally unable to work have the same rights as those who are able-bodied and employed. They have the right to generic services and to a *valued role* in mainstream society.[19] Some citizens with physical or mental disability face multiple levels of disadvantage[20]

[16] White, R. and Alder, C. (1994) *The police and young people in Australia*. Cambridge University Press.

[17] Durnan, D. and Boughton, B. (1998) *Detour Project Consultancy: 1997 Evaluation and 1998 Planning. A Report to Tangentyere Council*. Tangentyere Council, Alice Springs.
Traves, N. (1998) *The Detour Project: An Intergenerational Place of learning for Aboriginal Youth: Background Paper*. Centralian College, Alice Springs.
Wright, K. (1998) ASYASS, The alternative family model of working with young people. ASYASS, Alice Springs.

[18] Australian Youth Policy Action Coalition (A.Y.P.A.C.) Newsletter at: http://alice. topend.com.au Gearn, S. (1997) Locking up the children *Balance Northern Territory* Women Lawyers' Association
Short, M. (1997) Mandatory Sentencing Balance Northern Territory Women Lawyers' Association
Alice Springs News 1998 Vol. 5, No 46 Dec 16th.
Schetzet, L. (1998) A year of bad policy Alternative *Law Journal* Vol. 23, No 3 June.
Cain, M. (1996), *Recidivism of Juvenile Offenders in NSW*, Department of Juvenile Justice in Darwin Community Legal Service, page 6.
Australian Bureau of Statistics (1998) *Year Book Australia*, National Figures on Crime and Punishment.

[19] Despite the criticisms of the work of Wolfensberger on valuing citizens with a disability because it is based on norms and benchmarks, rather than principles of respect and a shared sense of humanity, Wolfensberger's work has been widely used by Disability services in Australia (see Wolfensberger, W. 1991). A brief introduction to social role valorisation as a high order concept for structuring human services. Training Institute for Human Services Planning, leadership and Change Agency. Syracuse University, Syracuse, N.Y.
Wolfensberger, W. (1989) Human Service Policies; the rhetoric versus the reality. In L. Barton. (ed.) Disability and dependency. London, Falmer Press.

[20] Mental and physical morbidity per se are not the focus of this study. Numerous secondary research reports have been consulted and interviews with service providers and researcher confirm that physical and mental ill health are a considerable source of concern (O'Kane, Menzies School of Health, Jane White and Rhonda Renwick, Centrelink, Alice Springs).

because their life chances are also limited by discrimination by virtue of their age, gender, ethnicity or Aboriginality.

Women's citizenship rights means bearing in mind the life chance shapers of age, level of education and skills, culture and language.[21] For example, the rights of some women citizens seem to be in jeopardy as a result of many factors such as literacy and numeracy levels, language skills often linked with particular cultural and class backgrounds. The changes in gender role need to be re-negotiated in a changing world. Whilst some women have made strides towards independence many women (by virtue of a range of socio-cultural and economic factors) have been left behind. Opportunities for women to achieve leadership and self-determination need to be considered in development initiatives.

Similarly, men have particular needs in a changing world where the masculine identity (and potentially masculinity per se) as dominant provider and decision-maker is challenged by the increased independence of women and an increased acceptance of a wider range of roles for both men and women.

The health and well-being of both genders needs to be addressed in relational terms, that is, in relation to one another. The provision of learning opportunities for all citizens is essential for building a knowledge nation. As governments limit their areas of responsibility and privatise services, citizens are often seen as consumers who need to pay for services in a range of areas. This has a particular impact on residents who live on limited incomes or the social wage. The meaning of citizenship is changing as the welfare policy shifts from welfare defined as a universal right to welfare as a right for those who qualify in specific circumstances (e.g. policies of 'mutual obligation') rather than universal rights. The economic, market-driven response (known as economic rationalism, because of the belief in the economic logic of markets to determine costs and decision-making) has driven these changes.

Globally Indigenous peoples emphasise that they are the caretakers of the land for their children and that their health is linked to the health of the land. Indigenous concepts of citizenship rights and responsibilities include links *across social and environmental goals* (McIntyre-Mills 2000). Self-determination[22] is defined very differently by different stakeholders

[21] Weeks, W. in Wilson, Wilson, J., Thomson, J. and McMahon, A. (1996). *The Australian Welfare State: key documents and themes*. Macmillan, Melbourne.

[22] This 'exemplifies the notion of separateness through culture as a form of both resistance and persistence to use Keefe's (1988) terminology. The history of Aboriginality needs to be interpreted in terms of invasion, slaughter, and inability

ranging from government and business interpretations of user pays (for services and rent) to Indigenous concepts of land rights. Also for some Indigenous people in Alice Springs it means first gaining further control over their lives and feeling welcome and respected in public places. But at a deeper level it is also about making a difference through the civil arena and through constructing a future in the interests of future generations, rather than focusing on the past.

Analysis and decision-making based on theoretically and information literate processes, plus an understanding based on openness[23] can generate new, global knowledge. Literacy and numeracy are the basis of information literacy, which is in turn a cornerstone of knowledge, but by no means the only one.

Citizens can engage in lifelong learning through a range of media (print and non-print) and by means of access to recreational and learning resources and spaces, only by ensuring that the learning environment and resources reflect the needs of a socio-economically diverse group of citizens. Modern science developed along the lines of disciplinary specialisations. Working within narrow predefined areas or paradigms, in order to develop detailed knowledge extended the frontiers of knowledge in many areas. The attempt to apply modern scientific thinking in a closed and elitist manner in development contexts has failed (Hettne 1995). Research, policy and development based on the notion of controlling variables in order to engineer changes has been re-assessed in terms of thinking that recognises that the environment comprises multiple causal links across multiple causes and effects that are weblike rather than unilinear. In order to bring about changes a systemic and multisemic (addressing multiple sets of meaning) approach is needed that is not limited to bureaucratic organisations, but instead uses a broader range of arenas. Systemic interventions require networking in matrix teams that span local and wider national and international contexts.

Diversity management applied to public places of learning could ensure that educators are facilitators and open to the contributions of all the participants. Places of learning need to respond flexibly to a diverse environment, because it has open communication with many interest groups.

The social and economic environment in which citizens operate is changing fast and there is a need to build the capacity of people to respond to this environment creatively in order to enhance their life chances.

*Continued*

to control labour, exclusion from citizenship, segregation and exclusion from property rights' (Hollinsworth 1996).

[23] Webb, S.P. (1998), Knowledge management: Linchpin of change. The Association for information management. ASLIB, London.

Provision of space for diversity helps to prevent the homogenisation of culture. 'Development' and peaceful dialogue makes more sense if it can be framed in a meaningful way for all stakeholders. This takes time and a willingness to listen and to learn from one another. Goals need to be framed and reframed in terms of perceived and expressed need. The way data are collected and who collects the data will influence the quality of the data and the extent to which the definition of what constitutes knowledge is meaningful. Development is about empowering people to become involved actively as citizens and as workers, not merely passively as consumers.

Applying inclusive, integrated policy and planning for sustainable, integrated development by, with and for people within their environment can enhance life chances. Places of learning could play a key role as vehicles for development and peaceful, healthy settings providing welcoming, free access to a range of resources. But more importantly they could be places where people can represent themselves through a range of media. Thus democratising knowledge. This is about allowing people to represent many cultures, not only the dominant culture. In this way deep learning can be attempted, drawing on what Bateson (1972) called Level 1, 2 and 3 learning. Level 1 learning is about gathering information within one paradigm or frame of reference, level 2 learning is about being able to think critically about a paradigm and level 3 learning is about being able to develop new paradigms in response to the diversity of knowledge paradigms.

Such an integrated approach is the basis for bringing about change. The challenge is to find ways to encourage existing organisations (that tend to work separately) to work together. Globally, a range of geographic *living environments (ranging from undeveloped, rural environments, small villages, towns, cities and to vast metropolises)* are measuring their levels of health and development through the use of a wide range of socio, cultural, political, economic and environmental indicators.

To sum up ways to improve social-health outcomes involve:

- Researching the impacts of increased privatisation and ensuring that the negative impacts are addressed by keeping in place and developing the supports for marginalised people.
- Divisions and categories are not useful for problem-solving. Systemic thinking begins with ontology and epistemology and is spelled out in development policy and practice.

Problem-solving in human service organisations using open systems and non-hierarchical approaches to collaborative problem solving are the way forward.[24] This requires that organisations for learning are open and flexible not only places of learning. This is vital to redress the new conservatism associated with globalisation, managerialism, residualism in welfare and development and the control of social research. The valuing of people's ideas and the environment as resources need to be supported, because the opportunity costs of neglecting them are disastrous in social, political economic and environmental terms. Good management and good governance is about ensuring that the values of democracy imbued in the Australian constitution are implemented at the local level. Much could be learned about the ways of seeing and perceiving the world through engaging in a dialogue. Listening and learning from one through respectful communication, not only contributes to rational understanding, reconciliation and decision-making but also enriches transcultural understanding so as to improve the quality of our decision-making. Development responses need to avoid being at either end of the continuum of approaches that are merely economic rationalist or socialist in orientation, as neither of these models is as yet sufficiently systemic in nature.

All participants expressed the following vision for the future: to make Alice Springs a place where people feel they want to live and work and a place where people feel safe and happy. Participants stressed that the way forward from learned helplessness to optimism is to give young people and their families a stake in society. This could be achieved by building more opportunities for participation in decision-making and more linkages across education and employment pathways. Building social and environmental capital could enhance development capacity. This starts with building healthy settings in homes, neighbourhoods, schools, hospitals and places of work. Developing the social capital of Alice Springs is as important as developing the economic capital because the two are closely linked. Quality of life is dependent on creating opportunities through community and economic development initiatives for all.
Suggested strategies are:

(i) Empowering people to become involved actively as citizens with both rights and responsibilities, not merely passively as voters.
(ii) Being subjects of one's own future through participatory design that engages with the powerful (and the issues of power).

[24] Takala, M., Hawk, D., and Rammos, Y. (2001) On the opening of society: Towards a more open and flexible education system. *Systems research and Behavioural Science* 18, 291–306.

*TABLE 9.1 A*

*Vision for the Future: Policy to Achieve Social and Environmental Justice in Healthy Settings*

| Strategy | Process | Outcome |
|---|---|---|
| A focus on the identified SWOT across age, gender, level of education, language and specific access concerns. | Ongoing implementation in partnership with other organisations to address identified needs. | Greater social capital because of enhanced life chances. |
| Partnership and trust building using an integrated, sustainable approach to development. | Ongoing dialogue across sectors to address citizen's viewpoints in planning and practice. Strengthen links across government, non-government and private sector to address social, economic and environmental concerns. | Healthy, safe settings.[25] |
| Achieving accessible services. | Implement an access action plan (in terms of information, attitude, communication and infrastructure) in response to the Disability Discrimination Act of 1992. | Accessible, generic infrastructure in line with the Disability Discrimination Act (1992). Facilities to be considered not merely as infrastructure but as a means to promote the socio-economic development and well-being of citizens. |
| A safe environment. | Contribute to building an integrated crime prevention program with a focus on young people and their families. It is recommended that the specific projects should focus on crime prevention and alcohol management through working across the sectors of education, health, employment, recreation and community policing. | A reduction in the crime rate. |

[25] For example: A report on the responsibilities and resources for local government prepared by the Advisory Council for Inter-government Relations cited by House of Representative Standing Committee on Aboriginal Affairs (1989: 4–5) 1. Services to individual properties, 2. Services to individual households. 3. Neighbourhood infrastructure services.

*TABLE 9.1 A (Continued)*

| Strategy | Process | Outcome |
|---|---|---|
| | Encourage responsible government, community and individual approach to the use and availability of alcohol. Contribute to creating a culture of learning beyond the school walls. Currently the low retention rates in schools could be addressed through a development programme. Contribute to reconciliation by building youth leadership in all sectors of the population through culturally appropriate programs that build self-esteem and encourage reconciliation. | |
| Enhancing public health. | Use community events to promote social health and well-being messages. | Improve in terms of social health indicators. |
| | Partnership building with applicable organisations to address specific public health concerns. | |
| | Involve a wide range of interest groups in decision-making. | |
| Advancing social and economic well-being. | Enhance employment opportunities along a continuum from CDEP to programs to build job readiness skills, to cadetships and mentoring programmes for employment within the mainstream. | Build a sense of rights and responsibility through having a stake in society as defined in terms that are relevant to particular interest groups. |
| | Contribute to policy on job creation in conjunction with other levels of government, NGOs and business for those who are currently unemployed. | |
| Divisions across interest groups. | Public education and building participation, listening to one another and building trust. Representation within community organisations. | Enhanced opportunities for reconciliation and self-determination. |

(iii) Addressing and redressing the issues of access and equity by means of a dialogue based on things that matter, things about which people have strong emotions. This could be a generative dialogue that re-energises and envisages hope (Jope 2001, Banathy 2000).

(iv) Building systemic collaborations to develop sustainable designs that sweep in a range of social, economic and environmental considerations.Technical/instrumental, strategic and communicative knowledge (as per Habermas 1974, 1984) are linked. The technical problems can only be strategically addressed when communication is successful in the public sector or working at the grassroots level of promoting civil governance.

Essentially this case study attempts to address social planning to address so-called social problems through recursiveness and retroductive logic. These tools are explained in chapters 1–4 and their implications for addressing democracy and citizenship as they relate to gender, age, culture, class, race are discussed in chapters 5–8. At the end of the day it is important to return to the discussion of left–right hemisphere thinking and the relevance of remembering that all human beings have the ability to think in terms of categories and in terms of webs or connections. Although some cultures tend to emphasis one rather than the other or at least represent themselves as thinking in terms of binary oppositions or in terms of webs there are examples of oppositional thinking and weblike thinking. Thinking in terms of young versus old, male versus female are just two obvious examples that are considered problematic of many Australian citizens, irrespective of cultural background. The 'either or' thinking approach is one that needs to be avoided if meanings are to be co-created.

Perhaps the only lesson that emerges is that co-creation of meaning is not only worthwhile in ethical terms, but in ontological terms. In epistemological terms (how we know what we know) it can shape our approach to research. It is also vital for the way we approach knowledge and knowledge management across areas of health, development and peace studies (as per WHO and UN policies inspired by the Brundtland Report 1987 'Our Common Future' and the Ottawa Health Charter 1986). But finally in the last instance, co-creation is about civil democracy at the grass roots at the local level and how ordinary citizens treat one another and how they feel after leaving their homes and taking a trip to the local shops. Can they afford to get there? How do they get there? Can they access public transport? Can they afford to buy groceries and essentials so that they feel part of their own community? Are they treated with respect? Do they struggle to be served? Can they count their change? Are they asked to move along when they sit down on the pavement, because they can't afford to buy a cold drink and to sit down at a restaurant? Do they feel hopeful about the future?

## 9.5 *The Contributions of CSP*

(1) Holistic thinking and critique by seeing the social, political, environmental and economic implications for citizens.
(2) Processes for intervention and addressing diversity such as:

(i) Diversity management approaches to improve governance,
(ii) Healthy settings approaches through participatory action research, so that there can be shared learning,
(iii) Social movements that enable people to work in networks that span disciplines and sectors,
(iv) Participatory planning and design or as Banathy[26] calls it 'intelligent design for evolution',
(v) Introduction of transdisciplinary analogies from the natural sciences that are useful as long as they are seen as analogies and social systems are not treated as natural systems.

The critical systems approach to theory and praxis is still developing, but includes:

(1) The materialist and structuralist heritage of the Frankfurt School,
(2) The idealist heritage of Hegel's dialectic, Kant's idealism and ethics, Ulrich's critical heuristics, Churchman's broad ranging systems that embrace physics and psychology as they pertain to ethics and systemic design.

[26] Banathy, B (2001) Self guided evolution of society: The challenge: The self guided evolution of ISS *International Systems Sciences Conference*, Asilomar, USA. Searching together: The application of dialogue conversation in intentional collective communication. *International Systems Sciences Conference*, Asilomar, USA.
The AGORA Project: Self guided social and societal evolution: The new Agora's of the 21st century. *International Systems Sciences Conference*, Asilomar, USA.
We enter the Twenty-first century with schooling designed for the nineteenth *Systems research and Behavioural Science* 18: 287–290.

(3) Habermas who embraces Popper and the rationalist approach, in so far as testing propositions (statements and ideas) through dialogue,

(4) Foucault (who stresses the need for surfacing knowledge that has been suppressed through using critical dialogue and narrative discourses). For him power and knowledge are linked and by using this approach discarded knowledge of the least heard and most marginalised can be given attention.

(5) Ecohumanism[27] and critical systemic thinking is still evolving and making more explicit reference to gender, age, life chances and the

[27] 'Ecological Humanism ... builds in a specifically environmental focus and places faith not merely in the potential of people to bring about change but underline the way in which people and nature are systemically linked. As far as their understanding of social stratification is concerned EH thinkers believe that people can challenge social structures which support the interests of 'the haves'. No matter how great the risks, individuals and groups of people have lobbied for changes through the centuries. E.H. thinkers choose neither socialism nor capitalism as 'the answer'. They are always interested in confronting the 'bad news' (see Romm 1987) of an approach. It is understood that truth lies in the preparedness to listen to the viewpoint of others, whilst keeping a grasp on the key assumption of individual potential and creativity, and the common values of human rights and dignity. Social justice solutions are always an interim best fit for a specific time and place and in order to meet the common interests of groups of people within their ecosystems. Individuals and groups should be encouraged to be debunkers of taken-for-granted solutions. E.H. avoids becoming rooted at the micro level of people's subjective experience and neglecting the relevance of social structures in shaping experiences. Social structures do pose challenges and provide obstacles but if people believe in their creative ability (as demonstrated through the ages) they will have hope and optimism for the future. They will believe that they can continue to adjust the balance of power in order to ensure that democracy is enhanced, rather than eroded. Our goal as global citizens is to facilitate social action based on a sense of shared responsibility. ...

Tensions between feminism and humanism exist, but E.H reflects on the continuities and overlaps amongst these positions and attempts to draw out the "liberative potential" from feminism which is concerned with equal rights and is sensitive to the range of categories in addition to gender which can lead to discrimination. Patriarchy and capitalism are considered to be equally important shapers of the life chances of women. ... The paradoxical nature of many social issues is not brushed aside by EH. They are explored. Paradigm dialogue for systemic thinking could be regarded as a bridge across strands of humanism and strands of feminism. Fonow and Cook (1991) and Stanley and Wise (1983) meaning of life within the environment. It needs to address the questions of self, other, the environment and the machine. Where are our designs taking us? These are the big questions. Poverty is not merely equivalent to a lack of income or a lack of participation in society. It is also about being out of touch with our environment. The only immutable or absolute

*Continued*

discuss the key themes in feminist research, as follows. They see feminist research as advocacy research that gives a voice to women. They emphasise the value of combining qualitative and quantitative methods. They echo the themes addressed in the works of Freire, Polanyi and other critical theorists who have used ethnography in order to empower, rather than merely to record the meanings of informants. They see feminist research as advocacy research that gives a voice to women. They emphasise the value of combining qualitative and quantitative methods but do not remind the reader that these themes have been addressed in the works of Freire, Polanyi and other critical theorists who have used ethnography in order to empower, rather than merely to record the meanings of informants'.[8]

E.H. is unashamedly instrumentalist in so far as it provides a framework for theoretical thinking which will frame action which responds to the needs of people who are marginalised through understanding stakeholders' perceptions. It is substantivist in so far as all analysis and implementation occurs in a particular context and normative in so far as working with all of the stakeholders to find a workable non-violent solution. Whilst feminists have driven non-objective approaches they are not solely responsible for claiming new paradigms. But nor is Peter Reason (1989) who wrote Human Inquiry in Action in which he outlined a so-called 'new paradigm'. Guba and Lincoln (1989) who wrote 'Fourth generation research' which outlines the principles of a constructivist approach and Denzin and Lincoln (1994) who detail a host of qualitative research methodologies demonstrate the wide range of sources to which their thinking is linked. ...Feminist thinking is however unique in its explicit highlighting and acknowledgement of the emotional content and feelings of the research participants. There is no pretence at neutrality, instead emotions are discussed openly and an attempt is made to reflect on the implications of emotional content for their research, instead of pretending that emotions do not exist (Stanley and Wise 1993, Fonow and Cook 1991). There are of course both positive and negative aspects to this contribution. Emotion is recognised as being part of the research process and is not ' 'bracketed out'. This contributes to a greater sense of the connections between the researcher and the research subject(s). If we understand humanism as recognition of rights (irrespective of gender, race, socio-cultural categories) in other words if human diversity is celebrated then E.H. can be regarded as an essential thinking tool for bridging theories based on sectarian understanding. Johnson (1994) has argued along parallel lines for a radical rethinking of humanism and feminism. In this work common denominator is that by virtue of our shared biology and shared environment we have more to gain by co-operating with one another than by competing with one another. Competition with one another and with the environment expressed as industrial pollution, atomic eruptions, wars and economic rationalism have striven to expand territories, extract profit, to oppress and to shift the damage to the 'other' whilst keeping the gains for the winners. But conflict like pollution tends to have a 'boomerang effect' Beck (1992) and does not support a sustainable future.

## *9.6 Post script: Yeperenye Dreaming in Conceptual, Geographical and Cyberspace*

If we go the route of trying to find strands that link positions that strive for finding common denominators (by virtue of our shared humanity and shared plane) we can bind together the rifts between those that (i) think there is no ultimate truth, (ii) those who believe that 'the closest we can get to truth is through dialogue'[28] and (iii) those who believe in that great rational (and potentially oppressive truth). If all three positions accept that there is indeed a common foundation based on our humanity and our shared planet, then perhaps it is possible to find that ultimate healing synergy or dependent origination idea[29] (as per Buddhist philosophy and many indigenous ways of thinking).

We all share the same biology (at the moment) and one planet (at the moment), trans speciation, cyborgism and planetary travel are not impossible, but nor is hubris. The implications of future developments are

*Continued*

however, citizenship has been linked with a critical humanism based on a respect for diversity and a respect for the essential common denominators that link all humankind, namely one biosphere. As global citizens we need to forge webs of transcultural meanings if we are to achieve any form of social justice locally and internationally. The two common denominators: shared bodily needs and one environment need to be expressed in transcultural forms which recognise diversity and freedom to the extent to which it allows for the future survival of all McIntyre-Mills (2000: 113).

[28] See McIntyre-Mills (2000). Dialogical approaches have been stressed by critical thinkers in the social sciences and humanities, those with a concern about marginalised groups have stressed the need for more than one voice.

[29] See the work of Koizumi, T (2000) The Buddha's though and the systems science. Paper delivered at the International Systems Sciences, Asilomar.

unknown, except that we are human and that we need to respect the web of life if it is to be sustained in its current form. The mind, the machine and the body are part of a system and thus all cultures need to be mindful of the web of life.

> So long as the past stands unredeemed, the past and future remain disconnected from the present. Where we have been, our memories, are only fully related to who we are, our self-identity, when they contribute to where we are going. Redemptive action thus brings the three moments of time together, within the individual, in a special way. While certainly not offering a permanent solution ... these temporary redemptions ("temporary" also in the older sense of belonging to the temporal realm) at least offer a way to live within time that neither denies its presence nor finds it a crushing burden (Dienstag 1997: 184–185 in Fritzman 2001: 442).

Truth is respectful dialogue
Freedom is respectful dialogue
Our future is respectful dialogue
Social and environmental justice is respectful dialogue
Pragmatism and idealism are one
Preservation of the sacred web across the divides
of self, other and the environment
is the challenge of this millenium
Let compassion, humility and wisdom be our mandala.
Who am I to have a say?
Only as a non member of many categories
Only as a boundary worker
Not as a writer of maps!

### Endings and beginnings: Caterpillar dreaming, butterfly being
### A mandala of knowledge and healing

On a rainy morning I walked on a rocky hillock on the Golf Course that is built at the sacred site of the Mparntwe or caterpillar-dreaming story. The road in the distance is called Broken Promise Drive. The smell of the gums and the damp air provided a relief from the burning heat of the previous day. I sat down to enjoy the view. Next to me I saw a trail of caterpillars. I followed the trail back to the gossamer capsules merged with the bark of the Ghost Gum. The caterpillars marched on multiple legs, in single file, joined head to tail towards destiny. They would eat and become once again a chrysalis where they would dream and become a butterfly. Thus they close their mandala of being and provided this lesson:

Fly out Yeperenye
Brighten the circle of being with your colours
by reminding us that the mandala contains both the
caterpillar and the butterfly.
Together they heal the sadness
And encircle the land.

# Bibliography


ABS. 1993. *Disability, Ageing and Carers Survey.*

ABS. 1996. *Northern Territory's Young People.* Report No. 4123.7.

ABS. 1996. *Regional Statistics.* Report No. 1362.7.

ABS. 1997a. *Causes of Death.* Report No. 3303.0.

ABS. 1997b. *Mental Health and Wellbeing Profile of Adults,* Australia. Report No. 4326.0.

ABS. 1997c. *The Health and Welfare of Australia's Aboriginal and Torres Strait Islander People's.* Report No. 4704.0.

ABS. 1998. *Year Book.* Australia: National Figures on Crime and Punishment.

Ackoff, R.L. and Pourdehnad, J. 2001. On misdirected systems. *Systems Research and Behavioural Science* 18(3).

Albrow, M. 1996. *The Global Age.* Palo Alto, CA: Stanford University Press.

Alinsky, S.D. 1972. *Rules for Radicals: A Practical Primer for Realistic Radicals.* New York: Vintage books.

Amnesty International. 1999. *Submission to the Inquiry into Human Rights Mandatory Sentencing of Juvenile Offenders Bill.* 29 October 1999 of the Australian Senate Legal and Constitutional References Committee.

Attorney General's Department. 1999. Pathways to prevention: developmental and early intervention approaches to crime in Australia. *National Crime Prevention Publications.* Attorney General's Department: Canberra.

Australian Libraries: www.ALIA.org.au.

Australian Local Government Association. 1999. *Promoting Access and Equity in Local Government: Services for All.* March.

Australian Youth Policy Action Coalition (A.Y.P.A.C.) Newsletter at: http://alice.topend.com.au.

Banathy, B. 1991. *Systems Design of Education: A Journey to Create the Future.* New Jersey Englewood Cliffs: Educational Technology Publications.

Banathy, B. 1996. *Designing Social Systems in a Changing World.* New York: Plenum.

Banathy, B. 2000. *Guided Evolution of Society: A Systems View.* London: Kluwer/Plenum.

Banathy, B. 2001a. Self-guided evolution of society: the challenge: the self guided evolution of ISSS 45th *International Conference for the Systems Sciences.* July 8–13.

Banathy, B. 2001b. The Agora project: Self guided social and societal evolution: the new agoras of the 21st *Century 45th International Conference International Society for the Systems Sciences.* July 8–13 2001.

Banathy, B. 2001c. We enter the 21st century with schooling designed in the nineteenth. *Systems Research and Behavioural Science* 18(4): 287–290.

Bateson, G. 1972. *Steps to an Ecology of Mind: A Revolutionary Approach to Man's Understanding of Himself.* New York, NY: Ballantine.

Baum, F., Cooke, R., Crowe, K., Traynor, M., and Clarke, B. 1990. *Healthy Cities Noarlunga Pilot Project Evaluation* Southern Community Health Research Unit.

Baum, F., Bush, R., Modra, C., and Murray, C. 2000. Epidemiology of participation: an Australian community study. *Journal of Epidemiology and Community Health* 54(6): 414–426.

Bausch, K. 2001. *The Emerging Consensus in Social Systems Theory*. Kluwer/Plenum.

Beck, U. 1992. *Risk Society Towards a New Modernity*. London: Sage.

Beer, A. and Maude, A. 2002. *Community Development and the Delivery of Housing Assistance in Non-metropolitan Australia: A Literature Review and Pilot Study: A Final Report*. AHURI Australian Housing and Urban Research Centre.

Beer, S. 1974. *Designing Freedom*. Wiley: London.

Bell, D. 1987. *Daughters of the Dreaming*. Sydney: Allen and Unwin.

Berger, P. 1976. *Pyramids of Sacrifice: Political Ethics and Social Change*. Harmonsworth: Penguin.

Berger, P.L. 1977. *Facing up to Modernity*. Harmondsworth: Penguin.

Bertalanffy, Ludwig Von. 1975. *Perspectives on General Systems Theory: Scientific-Philosophical Studies*. Taschdjian, E. (ed.). New York: Braziller.

Boughton, B. 1998. *Pathways to Indigenous Development: Review of Research*. Australian National Training Authority Alternative VET.

Bourdieu, P. and Wacquant, L.J.D. 1992. *An Invitation to Reflexive Sociology*. Cambridge: Polity Press.

Bowden, M. 1994. *Will they be Aboriginals as We are Today? A Case Study of the Education of Arrernte Youth in Alice Springs*. Masters thesis submitted in partial fulfilment of the requirements of the Degree of Master of Education at the Northern Territory University.

Bradley, R. 2000. Agency and theory of quantum vacuum interaction. *World Futures*, 227–275.

Bradley, R. 2001. Bits and logons: Information processing and communication in social systems. Abstracts: International Systems Sciences Conference, Asilomar.

Braithwaite, J. 1998. *Crime, Shame and Reintegration*. Melbourne: Cambridge Press.

Buber, M. 1965. *Man and Man*. New York, NY Macmillan.

Cain, M. 1996. *Recidivism of Juvenile Offenders in NSW*. Darwin: Department of Juvenile Justice.

Castells, M. 1996. *The Rise of Network Society*. Oxford: Blackwell.

Castells, M. 1997. *The Power of Identity*. Oxford: Blackwell.

Camilleri, J. and Falk, J. 1992 *The End of Sovereignty: The Politics of a Shrinking and Fragmenting World*. Aldershot: Edward Elgar.

Caughlan, F. 1991. *Aboriginal Town Camps and Tangentyere Council. The Battle for Self-determination in Alice Springs*. BA. Diploma in Social Studies. Latrobe. Aboriginal Studies School of Humanities.

Central Land Cuncil (1997) Annual Report 1996–1997 Alice Springs.

Central Australian Women's Legal Service. 1999. *Submission to the Inquiry into Human Rights (Mandatory Sentencing of Juvenile Offenders Bill)*. 29 October 1999 of the Australian Senate Legal and Constitutional References Committee.

Chambers, R. 1983. *Rural Development: Putting the Last First*. New York, NY: Wiley.

Chesterman, J. and Galligan, B. 1997. *Citizens Without Rights: Aborigines and Australian Citizenship*. Cambridge University Press.

City of Albany. 2000. *Strategic Review of Library Services*.

Cochrane, J. 1999. *Technology's Shining Light for Orphans' Bleak Life*. Australian 17/1/99.

Coulehan, K. 1997. *Alice Springs: A Cultural Cross roads and a Service Centre and Transit Centre for Aboriginal People in Central Australia*. Paper prepared for the Office of Aboriginal Development, Darwin.

Churchman, C. West. 1971. *The Design of Inquiring Systems. Basic Concepts of Systems and Organisation*. New York: Basic Books.

Churchman, C. West. 1979a. *The Systems Approach*. New York: Delta.
Churchman, C. West. 1979b. *The Systems Approach and Its Enemies*. New York: Basic Books.
Churchman, C. West. 1982. *Thought and Wisdom*. California Salinas Intersystems publications.
Checkland, P. and Scholes, J. 1990. *Soft Systems Methodology in Action*. Chichester: Wiley.
Cox, E. 1995. *A Truly Civil Society*. Sydney, NSW: ABC books. Boyer lectures.
Crimgeour, D. 1997. Community Control of Aboriginal Health Services in the Northern Territory *Menzies School of Health Research* Northern Territory Issue No 2/97.
Cvetkovich, A. and Kelner, D. 1997. *Articulating the Global and the Local: Globalization and Cultural Studies*. Westview. Oxford.
Davies, J. and Kelly, M. 1993. *Healthy Cities: Research and Practice*. London: Routledge.
Davis, R. 1986. Development of library services in the Northern Territory. *Presentation to the Library*.
Denzin, N. and Lincoln, Y. 1994. *Handbook of Qualitative Research*. London: Sage.
Devitt, J. and McMasters, A. 1998. On the machine: Aboriginal Stories about kidney troubles. Alice Springs: IAD Press.
Dean, M. and Hindess, B. (eds). 1998. *Governing Australia: Studies in Contemporary Rationalities of Government*. Cambridge University Press.
Dewey, J. 1944. *Democracy and Education*. New York: Simon and Schuster.
Dewey, J. 1997. *How We Think*. Ontario: Dover.
Durnan, D. 1997. *Federation of Independent Education Providers Study*. Alice Springs: IAD press.
Durnan, D. and Boughton, B. 1998. *Detour Project Consultancy: 1997 Evaluation and 1998 Planning. A Report to Tangentyere Council*. Alice Springs: Tangentyere Council.
Edwards, M. and Gaventa, J. *Global Citizen Action*. Colorado: Boulder.
Estrella, M. (ed.). 2000. *Learning from Change: Issues and Experiences in Participatory Monitoring and Evaluation*. London: Intermediate Technology Publications.
Fals-Borda, O. and Rathman, M.A. 1991. *Action and Knowledge: Breaking the Monopoly with Participatory Action Research*. London: Intermediate Technology.
Fetterman, D. 1989. *Ethnography Step by Step*. London: Sage.
Fuerstein, M. 1986. *Partners in Evaluation*. London: Macmillan.
Firkin, R. 1999. *Alice Springs Public Library Internet Project*. Unpublished Report.
Flood, R. and Carson, E. 1998. *Dealing with Complexity: An Introduction to the Theory and Application of Systems Science*. New York: Plenum.
Flood, R. and Romm, N. 1996. *Diversity Management: Triple Loop Learning*. Chichester: Wiley.
Flynn, M. 1997. One strike and you are out. *Alternative Law Journal*.
Foley, D. 2002. Indigenous standpoint theory. *Sharing the space conference*. Flinders University.
Fonow, M. and Cook, J. 1991. *Beyond Methodology. Feminist Scholarship as Lived Research*. Bloomington: Indiana University Press.
Foucault, M. 1967. *Madness and Civilization: A History of Insanity in the Age of Reason*. London: Routlege.
Foucault, M. 1979. *Discipline and Punishment: the Birth of the Prison*. New York: Vintage.
Foucault, M. and Gordon, C. (ed.) 1980. *Power/Knowledge: Selected Interviews and Other Writings 1972–1977*. Brighton: Harvester.
Freire, P. 1982. Creating alternative research methods: Learning to do it. In Hall, B., Gillette, A., and Tandon, R. (eds), *Creating Knowledge: A Monopoly*? New Delhi: Society for participatory research in Asia. pp. 29–37.
in *The Action Research Reader*. Victoria. Deakin University publication.
Friedman, J. 1992. *Empowerment: The Politics of Alternative Development*. Oxford: Blackwell.
Fritzman, J. 2000. Redemption, reconciliation: Either/or, Both/And? *Human Studies* 23: 439–445 London. Plenum.second edition.
Gao, F. and Yoshiteru Nakamori. 2001. Systems thinking on knowledge and its management *45th International Conference for the Systems Sciences*, Asilomar, USA.

Gaventa, J. 2001. *Towards Participatory Local Governance: Six Propositions for Discussion*. Paper presented to the Ford Foundation, LOGO Program with the Institute of Development Studies. June.

Gaventa, J. 2002. Civil governance.

Gaventa, J. and Valderrama, C. 1999. Participation, citizenship and local governance: *Background note for workshop on 'Strengthening participation in local governance'. Institute of Development Studies*. June.

Gearn, S. 1997. Locking up the children *Balance* Northern Territory Women Lawyers' Association newsletter Law Society of Northern Territory. Darwin.

Gu, J., Zhu, Z. 2000. Knowing wuli, Sensing shili, Caring for Renli: Methodology of the WSR Approach. *Systemic Practice and Action Research* 13(1): 21–58.

Guba, E. and Lincoln, Y. 1989. *Fourth Generation Evaluation*. London: Sage.

Giddens, A. 1991. *Modernity and Self-identity: Self and Society in the Late Modern Age*. California: Stanford.

Geertz, C. 1973. Thick description: Towards an interpretive theory of culture. In Geertz, C. (ed.). *The Interpretation of Cultures: Selected Essays*. New York. Basic Books.

Gilley, T. 1990. *Empowering Poor People: A Consumer View of the Fanily Centre Project*. Melbourne: Brotherhood of St Laurence.

Gouldner, A.W. 1971. *The Coming Crisis of Western Sociology*. London: Heinemann.

Gouldner, A.W. 1980. *The Two Marxisms: Contradictions and Anomalies in the Development of Theory*. London: Macmillan.

Graham, P. 1991. *Integrative Management: Creating Unity from Diversity*. USA: Blackwell.

Habermas, J. 1974. *Theory and Practice*. London: Heinemann.

Habermas, J. 1984. *The Theory of Communicative Action: Reason and the Rationalization of Society*. Beacon: Boston.

Harding, S. 1992. After the neutrality ideal: science, politics and strong objectivity. *Journal of Social Research* 59(3): 567–587.

Harding R, 1993. Opportunity costs: Alternative strategies for the prevention and control of juvenile crime In Harding, R. (ed.), *Repeat Juvenile Offenders: The Failure of Selective Incapacitation in Western Australia*. Research Report No. 10, University of Western Australia, Crime Research Centre, p. 141.

Hauritz, M. 2000. *Dollars Made on Broken Spirits*. Unpublished report.

Helman, C. 1983. *Culture, Health and Illness*. Bristol: Wright.

Hettne, B. 1995. *Development Theory and the Three Worlds: Towards an International Political Economy of Development*. England: Longman.

Hillier, J., Fisher, C., and Tonts, M. 2002. *Rural Housing, Regional Development and Policy Integration: An Evaluation of Alternative Policy responses to regional disadvantage*. IHURI Final Report. ISBN 1877005398.

Hollinsworth, D. 1992. Discourses on Aboriginality and the politics of identity in urban Australia. *Oceania* 63: 137–154. Replies to Hollinsworth by Beckett, J., Atwood, B. and Mudrooroo Nyoongah. Wiping the blood off Aboriginality: the politics of Aboriginal Embodiment in Contemporary Intellectual debate Lattas, A. Australia. *Oceania 63*: 155–167. Coagulating categories: a reply to responses by Hollinsworth 168–171.

Hollinsworth, D. 1996. Community development in indigenous Australia: self determination or indirect rule Community Development Journal 31: 114–125.

Howitt, R. 1998. Recognition, respect and reconciliation: steps towards decolonisation? *Australian Aboriginal Studies* Number 1 (29).

Idriess, I.L. 1933. *Flynn of the Inland*. Angus and Robertson. Sydney.

International Council for environmental Initiatives, 1998. *Our Community our Future: A Guide to Local Agenda 21*.

International Council for Local Environmental Initiatives. 1996. *The Local Agenda 21 Planning Guide: An Introduction to Sustainable Development*. International Council for Local Environmental Initiatives.

International Council for Local Environmental Initiatives. 1998. *Model Communities Programme*. International Council for Local Environmental Initiatives, Canada.

Jackson, M. 1991. *Systems Methodology for the Management Sciences*. London: Plenum

Jackson, M. 2000. *Systems Approaches to Management*. London: Plenum.

Jamrozik, A. and Nocella, L. 1998. *The Sociology of Social Problems: Theoretical Perspectives and Methods of Intervention*. Cambridge: Cambridge University Press.

Jamrozik, A. 2001. *Social Policy in the Post Welfare State: Australians on the Threshold of the 21st Century*. NSW: Longman.

Janzen, J. 1978. *Quest for Therapy in Lower Zaire*. Berkely, CA: University of California Press.

Jope, S. 2001 *Understanding Poverty Project: Developing Dialogue and Debate*. Paper presented at the National Social Policy Conference Competing Visions. July.

Keating, P. 2000. *Engagement: Australia faces the Asia–Pacific*. Sydney: Macmillan.

Keeffe, K. 1988. Aboriginality: Resistance and Persistence. *Australian Aboriginal Studies* 1: 67–81.

Keeffe, K. 1992. *From the Centre to the City: Aboriginal Education, Culture and Power*. Canberra: Aboriginal Studies Press.

Keene, I. 1988. *Being Black: Aboriginal Cultures in 'settled' Australia*. Canberra: Aboriginal Press.

Kettner, P., Daley, J., and Nichols, A. 1985. *Initiating Change in Organisations and Communities*. Monterey: Brooks/Cole.

King, R., Bently, M., Baum, F., and Murray, C. 1999. Community Groups, Health Development and Social Capital Poster presentation at 31st Annual PHAA Conference: *Our Place, our Health: Local Values And Global Directions*.

Kickbusch, I.K. 1996. *Healthy People 2000 Consortium Meeting: Building the Prevention Agenda for 2010: Lessons Learned*. New York.

The Key Note address at WHO Archives Technical seminars on the Internet at: www.pro.who.int

Koizumi, T. 2001. The Buddha's thought and systems science *45th International conference of International Society for the Systems Sciences*, USA, Asilomar.

Lama Surya Das. 2000. *Awakening the Buddhist Heart*. Random House.

Levy Strauss, C. 1987. *Anthropology and Myth. Lectures* 1951–1982. Oxford: Blackwell.

Laszlow, A. 2001. The epistemological foundations of evolutionary systems design. *Systems Research and behavioural sciences Systems Research* 18: 307–321.

Lee, D. 1974. Writing in colonial space. In Ashcroft, B., Griffeths, G., and Tiffin, H. (eds), *The Post Colonial Studies Reader*. New York, NY: Routledge.

International Council for Local Governmental Initiatives 1996. Local Agenda 21. *Planning Guide: An Introduction to Sustainable Development*. International Council for Local Governmental Initiatives, Toronto. Canada.

Lyons, P. 1990. *What Everybody knows about Alice: A Report on the Impact of Alcohol Abuse on the Town of Alice Springs*. Report commissioned by Tangentyere Council.

Malinowski, B. 1922. *Argonauts of the Western Pacific*. London: Oxford University Press.

Mauss, M. 1990. *The Gift*, translated by Halls, W.D. foreword by Mary Douglas. London, New York, NY: Routledge.

Mckeon 1994, Midgley, G. 1996. What is this thing called Critical System's thinking? In Critical Systems Thinking: current research aid practice. Flood, R. L. and Romm, NRA (eds) Plenum, New York.

McCLung Lee, A. 1988. *Sociology for People: Toward a Caring Profession*. New York: Syracuse University Press.

McIntyre, B. 1986. *Public Libraries in the N.T.-Darwin City from 1874 to the Present*. Presentation to the Library Association of Australia Biennial Conference. Darwin.
McIntyre, J. 1996. *Tools for Ethical Thinking and Caring*. Melbourne: Community Quarterly.
McIntyre, J. 1998. Consideration of categories and tools for holistic thinking. *Systemic Practice and Action Research* 11(2): 105–126.
McIntyre, J. 2000. *The Quality of Life in Alice Springs. An Analysis for the Purpose of Policy Development and Planning with Specific Reference to Local Government*. Alice Springs Town Council. ISBN 0646390716.
McIntyre, J. 2001. Systemic integrated thinking and practice for social and environmental justice: a case study. *Proceedings of 45th Annual Conference of the International Society of the Systems Sciences*. Asilomar, California.
McIntyre-Mills, J. 2000. *Global Citizenship and Social Movements: Creating Transcultural Webs of Meaning for the New Millennium*. Netherlands: Harwood.
McIntyre, J. 2002a. Critical systemic praxis for social and environmental justice: A case study of management, governance and policy. *Systemic Practice and Action Research* February. 15(1): 1–32
McIntyre, J. 2002b. Teaching critical systemic thinking and practice by means of conceptual tools 46th *International Systems Sciences Conference.* Shanghai.
McCann, M. 1992. Epistemic games, transfer and the six party hats: what the world's great thinkers are thinking: 5th International Conference on thinking: exploring human potential. James Cook University, July. Australian Journal of Gifted Education. 1: 2.
Memmot, P., Stacey. R., Chambers, C., and Keys, C. 2001. Report to Crime Prevention Branch of the Attorney General's Department. Violence in Indigenous Communities. Commonwealth Government.
Midgley, G. 1996. What is this thing called critical systems thinking? In critical systems thinking: Current research and practice. Flood, R.L. & Romm, NRA (eds) Plenum, New York.
Mills, J.J. 1985. The possession state intwaso, an anthropological re-appraisal. *Journal of South African Sociology*, 16(1): 9–13.
McLuhan, M. and Powers, B. 1989. *The Global Village: Transformations in World Life in the 21st Century* Oxford: Oxford University Press.
Mills, C.W. 1975. *The Sociological Imagination*. Harmondsworth: Penguin.
Ministry of Maori Development. 1997. *Strengthening Youth Wellbeing. New Zealand Youth Suicide Prevention Strategy.*
Moss Kanter, E. 1989. Becoming PALS: Pooling, Allying and Linking across companies. *The Academy of Management Executive* 3: 3183–3193.
Mowbray, P. 2000. *Healthy Cities Illawara: Ten Years on. A History of Healthy Cities from 1987–1997.*
National Inquiry into the separation of Torres Strait Islander children from their families (Australia) 1997. *Bringing them home* (Commissioner Wilson, R.) Human Rights and Equal Opportunity Commission.
National Stolen Generation Workshop 1996. *The stolen generations* Alice Springs. The Stolen generations Litigation Unit of the North Australian Aboriginal Legal Aid Service. Darwin, NT.
Neate, G. 1989. *Aboriginal Land Rights Law in the Northern Territory*. vol. 1.
Negroponte, N. 1995. *Being Digital*. NewYork, NY: Vintage.
New Zealand Ministry of Health. 1997. *In Our Hands: New Zealand Youth suicide Prevention Strategy*. Published by the Ministry of Maori Development and the Ministry for Health.
Northern Territories *Correctional Services Annual Report* 1997/1998: 21:
Northern Territory Department of Education 1999. *Learning Lessons: An Independent Review of Indigenous Education in the Northern Territory*. Darwin.

Norum, E. 2001. Appreciative design. *Systems Research and Behavioural Science* 18: 323–333.

O'Kane, A. and Tsey, K. 1999. *Shifting the Balance-services for People with Mental Illness in Central Australia. Menzies School of Health Research.* Northern Territory.

O'Reilly, B. and Townsend, J. 1999. *Young People and Substance use in 1998.* Darwin: Territory Health Services.

Office of Women's Policy. 1999. *Northern Women's policy Domestic Violence Strategy: The financial and economic costs of Domestic Violence.*

Ogawa, H. 2000. *Lessons Learned from Regional Experiences: Healthy Cities in the Western Pacific Region.* Regional Advisor in Environmental Health, WHO Western Pacific Regional Office, Key Note Address, Australian Pacific Healthy Cities Conference June 2000.

Pearson, N. 1999. Positive and negative welfare and Australia's Indigenous Communities Australian Institute of Family *Studies in Family Matters* No 54. Springs/Summer.

Pixley, J. 1993. *Citizenship and Unemployment: Investigating Post-Industrial Options.* Cambridge: Melbourne.

Polanyi, M. 1962. *Personal Knowledge.* London: Routledge and Kegan Paul.

Popper, K. 1968. *The Logic of Scientific Discovery.* London: Hutchinson.

Public Health Association. 1999. In Touch: *Newsletter of the Public Health Association Inc,* vol. 16, No 1 February.

Reason, P. 1988. *Human Inquiry in Action.* London: Sage.

Reason, P. 1991. Power and conflict in multidisciplinary collaboration. *Complementary Medical Research* 5(3).

Reason, P. 2002. Justice, Sustainability and Participation. Inaugural lecture http://www.bath.ac.uk/%7Emnspwr/Papers/InauguralLecture.pdf Published in Concepts and Transformations 7(1) 7–29.

Report No 10, University of Western Australia, Crime research Centre page 141.

Romm, N. 1996. Inquiry-and intervention in Systems planning: probing methodological rationalities. *World Futures.* 47: 25–36.

Romm, N. 1998. The process of validity checking through paradigm dialogues *14th World Congress of Sociology,* Montreal.

Romm, N. 2001. Our responsibilities as Systemic Thinkers *45th International conference of international Society for the Systems Sciences,* USA, Asilomar.

Romm, N. 2001. Accountability in social research: issues and debates. London: Kluwer/Plenum.

Rowse, T. 1993. *After Mabo: Interpreting Indigenous Traditions.* Melbourne University Press.

Rowse, T. 1998a. Nugget Coombs and the contradictions of self determination. In Wright, A. (ed.) for Central Land Council, *Taking Power Like this Old Man Here.* IAD Press.

Rowse, T. 1998b. *White Flour: White Power: From Rations to Citizenship.* Cambridge: University Press.

Runcie M. and Bailie R. 2002. *Evaluation of Environmental Health Survey* on Housing Menzies School of Health.

Sam, M. and Secretariat of National Aboriginal Islander Child Care. Through blackeyes: a handbook of family violence in Aboriginal and torres Strait Islander communities. Queensland, SNAICC.

Stewart, R.G. and Ward, I. 1996. Politics one second edition. South Melbourne. Macmillan.

Sam, M. and Secretariat of National Aboriginal Islander child care through black eyes: a handbook of family violence in Aboriginal and Tones Strait Islander Community. SNAICC, Queensland.

Scott-Hoy, K. 2001. *Eye of the Other Within: Artistic Autoethnographic Evocations of the Experience of Cross-Cultural Health Work in Vanuatu.* Doctor of Philosophy, Division of Education, Arts and Social Sciences, University of South Australia.

Schetzet, L. 1998. A year of bad policy Alternative. *Law Journal* 23(3) June.

Shames Harley. 1985. *Community Needs Study*. Unpublished report for Alice Springs Town Council.

Simms, J. 2001. *Systems Science Fundamental Principles*. In Ragsdell, G. and Wilby, J. (eds). Understanding Complexity. New York: Kluwer.

Smith, S. and Williams, S. 1992. Remaking the connections. *Health Worker Journal* 16(16).

Stanley, L. and Wise, S. 1993. *Breaking out Again. Feminist Ontology and Epistemology*. London: Routledge.

Stone, S. 1998. *Reclaiming the Streets: Zero Tolerance Policing and the Northern Territory* Ministerial Statement. (August).

Stewart, R. G. and Ward, I. 1996. Politics One Second edition. South Melbourne. Macmillan.

Street Ryan. 1999. Alice Springs *Economic Profile*. Street Ryan and Associates. Ref 23/02/B99.

Strehlow, T. 1958. *Dark and White Australians*. Melbourne: Riall Bros Pty Ltd.

Strehlow, T. 1978. *Central Australian Religion: Personal Monototemism in a Polytotemic Community*. Australian Association for the study of religions.

Strehlow Research Centre. 1997. Occasional Paper No. 1 Strehlow Centre Board, Alice Springs.

Strehlow Research Centre. 1999. Occasional Paper No. 2 Strehlow Centre Board, Alice Springs.

Tangentyere Council. 1997–8. *Annual Report*.

Tangentyere Council. 2000. *Tangentyere Council Protocols*. Developed by the Centre for Remote Health and Tangentyere Council.

Toffler, A. 1990. *Knowledge, Wealth, and Violence at the Edge of the 21st Century*. London: Bantam.

Thompson, G. 1827. *Travels and Adventures in Southern Africa*. Forbes, V. (ed.), Cape Town: The Van Riebeck Society. (republished in 1967).

Thurow, L. 1996. *The Future of Capitalism: How Today's Economic Forces will Shape Tomorrow's World*. Allen and Unwin.

Traves, N. 1998. The Detour Project: *An Intergenerational Place of Learning for Aboriginal Youth*: Background Paper 1998 Centralian College: Alice Springs.

Tsey, K., Scrimgeour, D., and McNaught, C. 1998. An evaluability assessment of Central Australian mental Health Services: Towards a more informed approach to mental health service planning and evaluation in Central Australia Menzies *Occasional paper, issue No 1/98*.

Ulrich, W. 1983a. *Critical Heuristics of Social Planning: A New Approach to Practical Philosophy*. New York, NY: Wiley.

Utke, A.R. 1986. June. The cosmic holism concept: An interdisciplinary tool in the quest for ultimate reality and meaning. *Ultimate Reality and Meaning* 9: 134–155.

Van Gennep, A. 1960. *Rites de passage*. London: Routledge and Kegan Paul.

Van Gigch, J.P. 2003. Comparing the epistemologies of scientific disciplines in two distinct domains: modern physics versus social sciences. Part 1: the Epsitemology and knowledge characteristics of the physical sciences. *Systems Research and Behavioural Science* 19(3): 199–210.

Webb, S.P. 1998. *Knowledge Management: Linchpin of Change The Association for Information Management*. London: ASLIB.

Weeks, W. 1996. *The Australian Welfare State: Key Documents and Themes*. In Wilson, J., Thomson, J. and McMahon, A. (eds). Melbourne: Macmillan.

Wenger, E. 1998. *Communities of Practice: Learning, Meaning and Identity*. Cambridge University Press.

White, R. and Alder, C. 1994. *The Police and Young People in Australia*. Cambridge University Press.

White, L. 2001.'Effective Governance' Through complexity thinking and management science. *Systems Research and Behavioural Science*, 18 (3) Wiley.

White, L. 2002. Connection matters: Exploring the implications of social capital and social networks for social policy. *Systems Research and Behavioural Science* 19 (2): 255–270.

Wilson, J. Thomson, J., and McMahon, A. 1996. *The Australian Welfare State* edited by Melbourne: Macmillan.

Winch, P. 1958. *The Idea of Social Science*. London: Routledge and Keegan Paul.

Wilson, J., Thomson, J., and McMahon. 1996. *The Australian Welfare State Key Documents and Themes*. Melbourne: Macmillan.

Winter, I. 2000. *Social Capital and Public Policy in Australia.?* Australian Institute of Family Studies.

Wolfensberger, W. 1989. Human service policies; the rhetoric versus the reality In L. Barton. (ed.), *Disability and Dependency*. London: Falmer Press.

World Health Organisation (WHO). 1988a. *Healthy Cities Project Paper* No. 1. Promoting health in the urban context. WHO Healthy Cities Project Office: Copenhagen.

World Health Organisation 1988b. *Healthy Cities Project Paper* No. 3. A guide to assessing healthy cities WHO Healthy Cities Project Office: Copenhagen.

World Health Organisation Archives Technical seminars on the Internet at: www.who.org/archives/hfa/techsem/971128.htm.

Wright, K. 1998. Aboriginal Youth Violence: A Way of Life published in part in CARPA Newsletter Vol. 28 Nov.

Wright, K. 1998. ASYASS: The alternative family model of working with young people. ASYASS: Alice Springs.
www.slnsw.gov.au/plb
www.slq.qld.gov.au/pub/standard
www.who.org/archives/hfa/techsem/971128.htm

Young, R. 2001. *Postcolonialism: An Introduction*. Blackwell.Academic.

Zhu, Z. 2000. Dealing with differentiated whole: the philosophy of the WSR Approach. *Systemic Practice and Action Research* 13(1): 11–20.

Zimmerman, M.E. 1994. Chaos theory, ecological sensibility, and cyborgism. In *Contesting Earth's Future: Radical Ecology and Post Modernity*. Berkley. CA:

Zola, I.K. 1975. In the name of health and illness. *Social Science and Medicine* 9: 83–87.

# Glossary

The following concepts underpin the analysis:

**Abductive** analysis leaps beyond the frameworks that are usually used to solve problems. It redesigns the approach to issues (Banathy 1996, 2000).

**Aboriginality** is situational and historical and based on the recognition of kinship, marriage ties and links with the land. The sense of reality and the experiences of urban, rural based and tertiary educated, or minimally educated men, women, the elderly and the young can be different; but together make up a definition of Aboriginality.

**Assumptions** are ideas that we do not challenge and take for granted. The way we see the world is a result of our embodied selves, our gender, age, upbringing, level of education, level of income and a host of cultural factors. In other words, social, cultural, political, economic and environmental considerations are relevant to shaping our assumptions.

**Axial themes** are central themes that emerge from an analysis of the layers of narratives of all the stakeholders. Some aspects of narratives contradict one another; others complement one another and some overlap. It is the overlapping aspects that create axial themes that are relevant in a particular time and place.

**Citizenship** pertains to civil rights and individual freedoms. According to Chesterman and Galligan (1997) who quote Aristotle's definition 'a citizen is one who shares both in ruling and in being ruled'.[1] Citizenship rights apply irrespective of culture, age, gender, income or disability.

Indigenous citizenship is an expression of identity, heritage and political aspirations for self-determination across heterogeneous life chances, but with a clustering of many the least

[1] According to Chesterman and Galligan (1997) who quote Aristotle's definition "a citizen is one who shares both in ruling and in being ruled". According to Thomson and McMahon (1996) who quote Marshall (1963, 78) these include 'liberty of the person, freedom of speech, thought and the right to own property and to conclude valid contracts and the right to justice'. Political citizenship is 'the right to participate in the exercise of political power, as a member of a body invested with political authority or as an elector of the members of such a body" (op cit.). They quote social citizenship as defined by Marshall (1963: 78) as 'the right to a modicum of economic welfare security regardless of the position on the labour market and the right to share to the full the social heritage and to live the life of a civilised being, according to the standard prevailing in society".

advantaged citizens. The experience of Aboriginality is situational and historical and based on recognition of kinship, marriage ties and links to the land. The concept of Indigenous self-determination includes citizenship rights and Indigenous rights. Globally the 'nation within a nation idea' is contested in order to achieve self-management in smaller subsections. Ties with the land are closely linked with a sense of health and well-being and this cultural value has been used as a political vehicle for resistance by Aboriginal people internationally.

**Commodities** refer to material items of value and commodification of culture refers to perceiving or treating culture in terms of its material value as a means to an end, rather than as an end in itself.

**Community** (or polis as per Churchman 1979) can be defined in terms of a group of people sharing common interests based on: geography or location, gender, age, shared values and beliefs, language, religion, occupation, income or place of origin. Interest groups are forged as a result of a sense of shared meanings, for instance: young people share certain things in common, as do the elderly, women, men and people with disabilities. The life chances of groups vary considerably, however, depending on a number of other factors. Interest groups in communities overlap and diverge. Social, cultural, political, environmental and economic interests impact on shared rights and responsibilities.

**Community of Practice is a means to learn and practice more effectively in order to achieve group goals.** Wenger (1998) coined the concept of COP and based it on participant observation in a claims processing department in the USA. The original concept (developed as an interpretation of what actually happens in working life, on the basis of listening to narratives and close observation) has been used in a range of other contexts. The COP concept lends itself not merely to interpreting the way people work, but if the concept is transferred to computing networks; it can become a digital means to support workers by adding on their own existing networks and store knowledge and to relate it to wider contexts. Wenger explains he chose the low status and paper- chain-oriented work to establish how people engage in interpreting work instructions (reifications) by finding ways to make the rules workable. He argued that without establishing a working community (based on cooperation and conflict) that shares ideas and strategies, the rule bound process of work would be almost impossible. The essence of COPs is that people find ways to work across boundaries and create plausible outcomes. Participants in a COP find and create ways to deal with unique human tragedies. Fuzziness and messiness is their reality. Categorical order is the reified reality of management. Their task is to bridge the two in such a way that people can achieve their aims within the context of social, cultural and political challenges.

**Complementarism** means working with theories and methodologies through reflection, rather than within the limits of one approach (Jackson 2000).

**Critical systemic thinking and practice or praxis (CSP)** involves a dialectical process of considering social, political, economic and environmental aspects of decisions with all the stakeholders. It draws on the work of Jackson (1991, 2000) and Flood and Romm (1996) and applies the complementary approach to holistic problem-solving to working with organisations in a community context. To sum up CSP is characterised by: complementarism, co-creation, emancipation, critical reflection, 'systemic sweeping in' (see Ulrich 1983) and commitment to the enlightenment approach to rationalism and humanism. It is mindful of the contributions of idealism and materialism. The skills that are needed are: (i) participatory design and decision-making using tools for policy development, such as triple loop learning (a diversity management tool), (ii) an ability to think critically based on theoretical and methodological literacy of available statistical data and (iii) need to be able to apply qualitative and quantitative research methods to participatory action research approaches

(PAR) for establishing needs (normative needs, perceived needs and expressed needs current service usage). (iv) Communication skills, counselling, advocacy and negotiation, networking and lobbying skills.

**Culture** is a framework for living. It is all that we think, believe, do and create. All people use culture in a situational context. It is constructed or interpreted by people to ensure their survival and to make meaning out of their lives.

**Data** can be defined as Bits, (understood in terms of binary oppositions that can be computer read/interpreted as technical information) or as Logons (based on wave theory) (or both as per Bradley 2001), or as units of energy that are the basic unit of energy necessary to achieve life (Simms 2000). Integrating these definitions can lead us to define data as units of energy that that resonate as a continuum of life in all living systems.

**Development arenas** include groups, organisations, private and public sector bureaucracies, political parties, networks, matrix teams and social movements.

**Development for empowerment** is a process of responding to people's needs, which are shaped by their life chances. 'Development' is not necessarily going to make much sense unless it can be framed in a mutually acceptable way. This requires ongoing dialogue and political will amongst all the parties. Framing standards of living needs to be considered in terms of perceived and expressed need. The way data are collected and who collects the data will influence the quality of the data and the extent to which the definition of development is meaningful. Development is about empowering people to become involved actively as citizens and as workers, not merely passively as voters.

**Development models** refer to different approaches based on different sets of assumptions and values. Old models are based on top down approaches working within narrow specialisations (known as the modernisation approach). New models are based on thinking in terms of the links across disciplines and sectors. They advocate participatory approaches based on respect for the contributions of all stakeholders.

**Deductive analysis** makes meaning (that is implicit within the data) explicit. The logical analysis does not go beyond the available data.

**Dialectical** approaches to change address thesis (one argument), antithesis (an opposite argument) and synthesis (a co-created argument) that holds within it the dynamics for change.

**Discourses** are bodies of knowledge based on sets of arguments, which contain assumptions and values. Often we are unaware how these assumptions shape or construct our view of the world and our conclusions. Through engaging in dialogue we can explore our ideas and arguments and we can co-create shared discourses in some areas where agreement can be co-created.

**Diversity management** (Flood and Romm 1996) is open to the contributions of all the participants within an organisation and responds flexibly to a diverse environment. It is based on iterative questioning of task, process and rationale for decision-making. This approach adds to effectiveness. It also means that creativity is enhanced.

**Emic** means the insider's point of view. It is the way in which you see the world.

**Empowerment** is about working together with people, not about top-down development or making decisions on behalf of people. It is about achieving changes with, rather than for people. The process of involving people in discussions is more than merely occasional consultation that may or may not be addressed through policy. It involves working alongside people. Empowerment means helping people to achieve greater confidence and power in the following areas: resources, relationships, information and decision making (Gilley 1989).

Empowerment is about finding common denominators across interest groups within a geographical community by using networking skills to trace connections across groups and to identify the barriers and gaps in the networks. These are the areas where bridges need to be built. Development strives to address the issues of access and equity for all residents. It strives to address both opportunities for empowerment and outcomes measured by qualitative and quantitative indicators. Empowerment requires a promotive and preventative approach to social problems.

**Epistemology** addresses and questions how we know what we think we know and whether we are going about the process of designing our research and choosing appropriate methodologies and methods appropriately.

**Etic** means an outsider's point of view. It is the way in which other people see the world.

**Functionalism** is an approach based on the idea that the whole is greater than the sum of the parts. The emphasis is on the mechanics of the system and the way it functions. There was little or no recognition of feedback systems in the early work of functionalists.

**Global age** is a short hand term for the new way of life that relies increasingly on digital technology and knowledge management across the boundaries of disciplines and sectors. Knowledge (technical, strategic and communicative, as per Habermas 1984) needs to acknowledge this in day-to-day problem-solving. It is systemic in so far as the social and natural sciences begin to acknowledge that issues are linked and to apply systemic thinking to sustainable designs and solutions.

**Governance** is about ensuring that the values of democracy imbued in the Australian constitution are implemented at the local level. Governance is not merely about implementation of rules and procedures. It is about listening and responding to the diverse range of community needs by finding ways to make the best possible use of resources in partnership with all levels of government, the private sector and the third (or volunteer sector). Measuring governance requires measures of qualitative outcomes, based on perceptions and meanings, as much as efficiency outputs, counted in terms of numerical items. This can require a shift in organisational thinking.

**Health** is defined broadly following the World Health Organisation's Ottawa Health Charter of 1986 that linked health and development as integral. This approach means that development and planning approaches need to link health and development. There are direct links across life chances, life-style and the quality of the built and natural environment in which one lives. According to Davies and Kelly (1993) by promoting health and development through integrated policies well-being (salutogenesis) can be promoted through a number of linked program approaches, rather than about working in a reactive and separate manner through separate departments and separate projects. It also requires thinking about ways in which to enhance the quality of life of citizens. Life chances can be enhanced in multiple ways through inclusive, integrated policy and planning *by, with and for* people.

**Holistic thinking** connects across sectors and disciplines and avoids thinking in compartments or boxes.

**Holons or maps** of meaning are based on the personal constructs that people create through their stories. They explain the way in which they see reality. Holons in stories map or draw the connections that people make between issues and ideas. Understanding a point of view and developing shared ideas can be assisted through telling ,drawing and drawing out ideas.

**Identity** is based on the sense of self that is created through the meanings constructed through lived and learned experience. Subjective meaning is created through experience and intersubjective interaction in society is based on learning that is mutually agreed.

**Ideographic** approach to explanation is based on a detailed case study that unfolds meanings and attempts to explain the complexity of a local context by means of unfolding the multiple meanings of the narrators of stories. The richness of the stories is maintained.

**Inductive** analysis goes beyond the available data. It makes a leap in logic that is supported only by a sample. It can be said to be probably true, but not in any ultimate sense.

**Information** (see data) can be defined in terms of the way data is defined. Data can be defined in terms of Bits and Logons (as per Bradley 2001). Bits are the smallest unit of binary information (Bradley 2001 cites Claude Shannon's theory). The 'either or choice' or 'Boolean choice' is derived from computer language. This is the basis of information theory in Western social sciences. This is problematic because as I have argued elsewhere (McIntyre 2000) information is more than 'either or', it can also be 'both and' in some narratives. Meaning can be based on both categories (divisions) and on webs (connections).

**Interest groups** are formed situationally as a result of a sense of shared meanings and shared life chances.

**Interpolating** is tracing common points in narratives and connecting them to form a web of meaning that maps the shape of the landscape of ideas. It is also necessary to consider multisemic realities or multiple areas of reality when undertaking research in diverse cultural contexts. The maps of people are presented first and then attempts are made to find interpolations across maps that will create shared lines of reasoning and shared understanding (if not shared narratives, because they occupy different parts of one map in a shared landscape).

**Intersectoral** means working across sectors and across organizations using matrix teams as opposed to working and thinking within compartments. These approaches are referred to as an integrated or systemic approach that follows the World Health Organization's Ottawa Health Charter of 1986. It spelt out the links between health and development and is the basis for a new approach to development (see Davies and Kelly 1993).

**Intersubjective** refers to communication for building a shared sense of meaning. This is an essential process for enabling people to solve problems democratically.

**Knowledge narratives** for governance are defined in terms of co-created, contextual meanings that avoid either zealotry (one truth) or cynicism (no truth).

**Life chances** refer to opportunities in life experienced by people as a result of a host of demographic, socio-cultural, political and economic factors.

**Matrix teams** consist of participants comprising a range of different stakeholders. They are formed by working flexibly within and across departments, organisations or disciplines and the two-way communication in the teams is both vertical and horizontal.

**Modernist theories** are based on the assumption that there is one truth for which objective knowledge can strive. This can lead to communication, which is adversarial. Rightness is based on proving the other wrong. Justice is based on legal objectivism.

**Multiculturalism** is an Australian policy that has applied for more than two decades but it has been eroded by increased conservatism (in part due to immigration policy post September the 11th 2001). It recognised the diversity of Australians. Multiculturalism was preceded by an Assimilation policy in 1964 as a Section of the Department of Immigration. The Commonwealth accepted this formulation.

> For the same period Australia has had a universal immigration policy in the sense that race, ethnicity, culture, language and religion are not criteria for discrimination. Racial discrimination is illegal and institutions have been set

up to monitor this. The full citizenship status of Aborigines has been legally established since the 1967 referendum gave the Commonwealth power over Aboriginal affairs. Australia now has only two classes of resident, citizens and non-citizens ...[2]

**Multisemic** is a term that means multiple meanings (see McClung Lee 1988). People may mean quite different things when they use one term. Unfolding what people mean by the terms they use is part of the process of engaging in narrative dialogue that explores constructs and enables multiple meanings to be juxtaposed to one another.

**National identity, citizenship and the state** do not always overlap, Castells (1997). Nationality may be used to undermine or build states and some national identities may be recognised as full citizens and others not. The nation within a nation idea is being contested by a range of legislation, which is geared to encouraging self-management in smaller subsections. This is a key recommendation of the Reeves Report (1998). It is vital to note that in democratic nations such as USA, Canada and Australia, the rights to land continue to be debated. Ties with the land are closely linked with a sense of health and well-being and this cultural value has been used as a political vehicle for resistance by Aboriginal people internationally.

**Narratives** refer to the stories told by the participants. Layers of stories from participants give constructs of reality that complement and contradict one another. The multilayered narratives build up a sense that there are many views on a subject and many perspectives on reality. Some areas of overlap also occur and these are the axial themes. Power dynamics shape which narratives are heard and which are silenced.

**Need** is defined in terms of Bradshaw's definition.[3] *Felt need* is explored through consultation using interviews, focus groups and public forums to obtain qualitative data, which addresses the feelings of people. *Expressed need* is ascertained from statistics of service users by analysing annual reports. Comparative need is considered by comparing indicators with other parts of Australia. *Normative need* is addressed through analysing relevant policies locally and internationally.

**Ontology** addresses and questions the nature of reality.

**Paradigm** is a framework for understanding, based on specific sets of assumptions about the world.

**Paradigm dialogue** is the tool for creating shared webs of meaning which are essential for constructing a sense of global citizenship rooted in respect for self, the other and the environment.

**Participatory Action Research (PAR)** involves working with participants from the conceptualisation to the implementation and evaluation stage of development research. It involves (i) learning by doing and (ii) the ongoing feedback of ideas amongst the participants. The process of revising both ideas and practice on the basis of participation is ongoing and integral to PAR. It is a *process that involves all stakeholders as participants at all stages* of the research. It is not a straight-line approach to research and implementation; instead it is a *spiral approach.*[3]

**Positivism** is the process of testing the relationship between variables to establish whether there is a positive relationship between variables. It uses the hypothetico-deductive method.

[2] See Wilson, J. Thomson, J. and McMahon, A. (1996) *The Australian Welfare State:* Key documents and themes. Melbourne. Macmillan.

[3] See *Systemic Practice and Action Research Journal*. Plenum Press. New York and London.

This is a process based on testing out the possible relationship between variables. If it can stand up to testing then it is closer to the truth, or more probable than if it is untested.

**Postmodernism** is based on the assumption that there is no absolute truth, only subjective constructions of truth by stakeholders. At worst postmodernism can lead to abandoning a sense of common or international human rights, because it is assumed that only contextual understanding is possible.

**Post positivism** has revised the idea that truth can be established through testing in a narrow quantitative way. Meanings and perceptions need to be included using qualitative methods.

**Problem-solving, planning and participatory design**. By thinking creatively we can address problems in different ways. If we believe that there is more than one answer to a problem, we will allow ourselves to think beyond the limitations that we have set ourselves. 'It is vital that development be a process for securing improvements, rather than merely an outcome.'[4] Planning for development requires a process of including representatives of all interest groups. In this way an overview of different perspectives held by different stakeholders can be developed in order to address diversity and to enhance creative decision-making and problem-solving, based on iterative dialogue with a range of interest groups in the public, private and volunteer sectors. This sort of approach is likely to be more productive in terms of governance than an approach, that is top-down and less inclusive.

**Recursiveness** (as per Giddens 1991) is the process whereby we construct social reality and the way society shapes our thinking. Being mindful of this process can assist in participatory design and the promotion of social and environmental justice.

**Reflexivity** is the weighing up of ideas and considering their implications for all stakeholders.

**Retroductive** logic is used to explain the way things are in terms of the underlying social structures. It traces the connections across the institutions in society and demonstrates how society shapes life chances and the way things are. It also involves understanding what the terms mean and why in terms of the different stakeholders assumptions and values. It traces the connections across the institutions in society and demonstrates how society shapes the way we think, how we see ourselves, our life chances and the way things are.

**Self-determination**[5] is defined very differently by different stakeholders ranging from government and business interpretations of user pays (for services and rent) to Indigenous concepts of land rights translated into Australian Law.[6] Also for some Indigenous people in

[4] According to ATSIC 1995 cited in Fuller 1996. *Aspects of Indigenous Economic Development*. Paper presented to Regional Development Program Forum.

[5] Self-determination "exemplifies the notion of separateness through culture as a form of both resistance and persistence to use Keefe's (1988) terminology. The history of Aboriginality needs to be interpreted in terms of invasion, slaughter, and inability to control labour, exclusion from citizenship, segregation and exclusion from property rights". (Hollinsworth 1992). is defined very differently by different stakeholders ranging from government and business interpretations of user pays (for services and rent) to Indigenous concepts of land rights translated into Australian Law.

[6] Indigenous social, cultural, political and economic concerns are expressed through: Native Title Claims, The Land Act 1992, Amendments to the Pastoral Land Act 1992, Aboriginal Land Rights Act 1976 (NT) Sacred Sites Act 1989 and Aboriginal Heritage Act 1984 which is being amended.

Alice Springs it means first gaining further control over their lives in their own town camp area by successfully controlling the consumption and availability of alcohol on their camp. For some Indigenous people rights within their own community are as important as recognition that they are first and foremost Indigenous Australians. The self-determination of Aboriginal people means understanding that Aboriginality is multifaceted. According to Rowse, the concept of self-determination has two components: citizenship rights and Indigenous rights (1998: 210). The challenge is to address both in development initiatives (Rowse 1998). He argues that the notion of a detribalised, deracialised citizenship that is based on liberal, capitalist notions of rights and responsibilities needs to be balanced with the Indigenous land rights movements internationally.[7] It is clear from discussions with different interest groups amongst Aboriginal peoples, that for some rights within a community council area are as important as recognition that they are first and foremost Indigenous Australians. Self-determination[8] is defined very differently by different stakeholders ranging from top-down interpretations of user pays (for services and rent) to land rights. Striving for self-determination and real citizenship can be the struggle of user pays versus landrights. For some Indigenous people it means first gaining further control over their lives in their own town camp area. The starting point for a discussion on self-determination of Aboriginal people is understanding that Aboriginality is a vehicle for expressing identity, heritage and political aspirations. The fact that as a concept Aboriginality is situationally and historically sensitive does not lessen its value.

**Social capital** has the potential to empower groups of people who already have a basis of power (White 2002: 256), by virtue of their cultural capital (as per Bourdieu and Wacquant,

[7] Also for some Indigenous people in Alice Springs it means first gaining further control over their lives in their own town camp area by successfully controlling the consumption and availability of alcohol on their camp. For some Indigenous people rights within their own community are as important as recognition that they are first and foremost Indigenous Australians. The self-determination of Aboriginal people means understanding that Aboriginality is multifaceted. According to Rowse, the concept of self-determination has two components: citizenship rights and Indigenous rights (1998: 210). The challenge is to address both in development initiatives (op cit.). He argues that the notion of a detribalised, deracialized citizenship that is based on liberal, capitalist notions of rights and responsibilities needs to be balanced with the indigenous land rights movements internationally[5]. The Meek shall inherit the earth. Weekend Australian Review April 3–4 1999.

[8] "Struggles for social justice and cultural autonomy by Indigenous Australians have constituted some of the most far-reaching challenges to the Australian State. In the last twenty years Aborigines have gained official recognition as a people and support for self-management and self-determination policies. These apparent successes have resulted in an incorporation of indigenous communities and their politics into mainstream institutions in ways that can actually increase state supervision and threaten cultural independence. Partly this contradiction arises from the need to create peak bodies able to represent Aboriginal issues at the highest levels of government which run counter to the localised and land-based social networks which have enabled indigenous values to be maintained under welfare colonialism...." This quotation exemplifies the notion of separateness through culture as a form of both resistance and persistence to use Keefe's (1988) terminology. The history of Aboriginality needs to be interpreted in terms of invasion, slaughter, and inability to control labour, exclusion from citizenship, segregation and exclusion from property rights" (Hollinsworth 1992).

1992), it is thus more than merely a basis for trust in the community.[9] It has been summed up as follows: '*... features of social organisation such as networks and norms that facilitate co-ordination and co-operation of mutual benefit for the community. It builds the capacity to trust, and have a sense of security, social cohesion, stability and belonging*'. King, R., Bently, M., Baum, F. and Murray, C. "*Community Groups, Health Development and Social Capital* 1999 Poster presentation at 31st Annual PHAA Conference: "Our place, our health: local values and global directions". According to White (2002: 268) social capital is a concept that needs to be considered critically in terms of power, because conceptually it can mean different things to different interest groups. Thus merely studying social networks as if they were objective indicators of something uniform and meaningful for all the participants is mistaken from this critical and systemic point of view. Building networks of trust is indeed a worthwhile goal for enhancing civil governance, but it is by no means unproblematic from the point of view of the participants.[10]

**Social entropy** is the state of total equilibrium associated with dissipated energy within closed systems if the analogy of entropy is applied to the social world it refers to closure to ideas and the lack of creativity (Flood and Carson 1998).

**Social movements** are a means for bringing about social and environmental change based on wide ranging communication within and beyond formal organisations in the public, private and non-government sector.

**Structuralism.** An approach to understanding the way parts fit together and why they fit together in a particular way. Retroductive logic is used to explain the way things are in terms of the underlying social structures.

**Structural functionalism** is an approach that combines structuralism and functionalism and was used by social anthropologists such as Malinowski and Radcliffe Brown and sociologists such as Parsons.

**Sweeping in** (as per Churchman 1979, 1982) is considering the public, social, political, economic and environmental factors and the private, personal factors in problem solving.

**Systemic** means working with the knowledge narratives of all the stakeholders, including those of this writer, in order to establish shared or co-constructed stories and solutions for shared problems that make sense in a particular time and place.

**Technocratic thinking** and technology in itself is valuable but when thinking is limited (particularly in problem solving) to the hardware or the infrastructure without considering its impact on people and the way they perceive issues (software) or the environment, it is problematic. This is associated with thinking within boxes or disciplines, instead of thinking about the wider context (and the way each part of a system is inextricably linked with other parts).

**Tools for thinking** help to exercise the way we think so that we can make mental leaps outside a paradigm (Banathy 1996, 2000). They can help us to be more creative in our problem-solving, because discourses of thinking can be explored.

**Totalising calculus** attempts to apply systems thinking to the whole system of organic and inorganic matter without acknowledging that new physics is premised on fluidity and

[9] King, R., Bently, M., Baum, F. and Murray, C. "*Community Groups, Health Development and Social Capital* 1999 Poster presentation at 31st Annual PHAA Conference: "Our place, our health: local values and global directions".

[10] See the work of Whyte, W.F. 1995. Street Corner Society 2nd Edition. University of Chicago Press, Chicago.

change (Van Gigch 2002). Thus the social sciences need to acknowledge the potential for change in all social systems as basis for avoiding social entropy.

**Transcultural** is a concept that is different from cross-cultural, because it assumes that through paradigm dialogue (which will help us to think reflexively) we will build webs of shared meaning.

**Truth** is defined as shared meaning based on co-creation through reflexive weighing up of ideas within a particular context, but mindful of the wider context. The end result is the creation of and recognition of shared webs of meaning. Critical thinkers believe that the closest we can get to truth is through dialogue. Positivists believe that the process of falsification helps to establish whether an idea/fact can stand up to testing, or to another point of view.

**Unfolding** (as per Churchman 1979, 1982) is the process of thinking through the layers of meaning by asking: who, what, how, why, in whose opinion, in whose interests?

**Variable** is a changing factor in a research design.

**Vignettes** are sketches drawn from the context of participant observation and the informant's narratives. Vignettes are used in order to depict reality (as understood by the researcher) and thus are a mediated reality, based on systemic analysis of the key themes. The themes are presented as case studies that encapsulate the axial themes.

# Index